Probit analysis

Probit analysis

D. J. FINNEY

Professor of Statistics in the University of Edinburgh
Director of the Agricultural Research Council Unit of Statistics

Third edition

CAMBRIDGE
AT THE UNIVERSITY PRESS
1971

Published by the Syndics of the Cambridge University Press
Bentley House, 200 Euston Road, London N.W.1
American Branch: 32 East 57th Street, New York, N.Y.10022

Third edition © Cambridge University Press 1971

Library of Congress Catalogue Card Number: 78-134618

ISBN: 0 521 08041 X

First edition 1947
Second edition 1952
Reprinted 1962
Third edition 1971

Printed in Great Britain
at the University Printing House, Cambridge
(Brooke Crutchley, University Printer)

Contents

[v]

List of examples

[viii]

Preface to first edition

From the theory of probability, originally investigated in order
to explain nothing more important than the results of games
of chance, has developed the science of applied statistics. Over
one hundred years ago Laplace wrote that

...la théorie des probabilités n'est au fond, que le bon sens réduit
au calcul: elle fait apprécier avec exactitude, ce que les esprits justes
sentent par une sorte d'instinct, sans qu'ils puissent souvent s'en
rendre compte,

and these words might equally well be written of statistics
to-day. In many fields of scientific research, and especially in
the biological sciences, numerical studies are complicated by the
inherent variability of the material under investigation, and
conclusions must be based on averages derived from series of
observations. The estimation of these averages and the assessment of their reliability are statistical operations, in the
performance of which the experimenter inevitably employs
a statistical technique even though he himself may not always
recognize this fact. The operations may be simple or complex,
depending upon the circumstances; if, however, they cease to
be 'le bon sens réduit au calcul', they can no longer be expected
to contribute to the understanding of the problem under investigation.

The recent rapid advances in the application of rigorous
statistical methods to biological data began with the publication, in 1925, of R. A. Fisher's *Statistical Methods for Research
Workers*. Not only did Fisher develop exact methods for the
analysis of data from small samples to replace the older approximations from large-sample theory, but he also introduced new
and powerful techniques for making the most efficient use of
experimental results. Of equal importance to the growth of the
present-day philosophy of experimentation was Fisher's suggestion that the statistician should be consulted during the
planning of an experiment and not only when statistical analysis
of the results is required, as his advice on experimental design
may greatly increase the value of the results eventually obtained.

[xi]

Since Fisher's book first appeared, the principles of experimental design and the methods of statistical analysis have been extended so rapidly as to make it increasingly difficult for any but the professional statistician to be familiar with the variety of methods needed in biological problems.

Many books since Fisher's have been written with the aim of surveying a wide field of biological statistics, but these can give only an outline of some important topics. There is to-day a need for books in which the specialized statistical methods appropriate to certain branches of science will be discussed in sufficient detail to enable biologists to appreciate them and apply them to their own problems.

One subject requiring fuller discussion than can reasonably be expected in any general text-book of statistics is the method of probit analysis, for the development of which J. H. Gaddum and C. I. Bliss are largely responsible. This method is widely used for the analysis of data from toxicity tests for the assay of insecticides and fungicides, and also of data from other types of assay dependent upon a quantal response. In this book I have tried to give a systematic account of the theory and practice of probit analysis, including as much as possible of the most recent extensions and refinements, in such a form that it may be understood by biologists, chemists, and others who have some knowledge of elementary statistical procedure; at the same time, I have endeavoured to satisfy the mathematical statistician by showing the theoretical background of the method. The less mathematically minded reader will no doubt be content to omit, or at most to read cursorily, Appendix II and other sections concerned with the mathematical basis of the technique. Full understanding and appreciation of statistical methods can be gained only by experience in their use, but careful study of the numerical examples should enable many who were previously unfamiliar with probit analysis to apply it satisfactorily to their own data.

This book has been written as a result of several years of close collaboration with members of the Insecticides Department at Rothamsted Experimental Station, especially with Dr F. Tattersfield, Dr C. Potter, and, until he left Rothamsted, Dr J. T. Martin. I wish to express my gratitude to them for discussing with me a wide variety of their problems, for advising me on the experimental aspects of their results, and for the generosity with which they have permitted me to use their data both in this book and in earlier publications. I am also very grateful to my colleagues in the Statistical Department at Rothamsted for much helpful discussion, and particularly to Dr F. Yates for his detailed and constructive criticism in the preparation of my book. Others to whom my thanks are due include Miss G. M. Ellinger for assistance in German translation, Dr C. G. Butler for permission to use the numerical data of Ex. 33, Dr A. E. Dimond and Dr J. G. Horsfall for giving me very full information on the results discussed in §41 and for permission to use their data, Professor G. H. Thomson for assistance in tracing the history of the probit method, the Editors of the *Annals of Applied Biology* for permission to reproduce the first half of Table II, Professor R. A. Fisher, Dr F. Yates, and Messrs Oliver and Boyd, Ltd. for permission to reproduce Tables I, VI and VII from their book *Statistical Tables for Biological, Agricultural and Medical Research*, and my father, Robt. G. S. Finney, for very considerable help in the correction of proofs.

D. J. FINNEY

Rothamsted Experimental Station
August 1945

Preface to third edition

This book was originally planned to help biologists whose acquaintance with statistical method was slight and who were naturally hesitant to undertake calculations that seemed difficult, laborious, and highly idiosyncratic. After twenty-five years, familiarity with statistical ideas and principles is much more widely spread. Moreover, high-speed computers have completely changed the attitude to calculations that once were regarded as intolerably lengthy.

Improved understanding of statistical theory enables probit methods to be seen simply as a technique for a particular set

of estimation problems. The iterative calculation of a linear
regression is no more than a scaffolding that enables troublesome
equations to be solved. Old disputes about transformations relate to different models that can rarely be distinguished experimentally. Arguments relating to alternative routines of
calculation once seemed important. To-day, either they disappear because an approximation that saves a few seconds of
computer time is scarcely of interest, or they are shown more
clearly to represent fundamental differences of outlook on
whether or not the maximization of likelihood is the preferred
principle of estimation. One could indeed contend that there is
no longer a function for a book solely concerned with probit
methods. These and many other problems can be handled by a
general computer program that provides options for many
models and for different principles of estimation. After long
consideration, I concluded that the experimental situations for
which probits are commonly used still merit a special text,
which can seek to appeal to scientists in particular fields of
research and can also illustrate many facets of general maximum
likelihood estimation.

Although this edition contains much of the material in its
predecessors, I have completely rewritten and reorganized it.
Chapters 1 to 4 are of the same form as previously, though
brought up to date; the two Appendices that described the
classical computations in detail and outlined the mathematical
theory have been incorporated into Chapter 4. Chapter 5 is
entirely new, a discussion of the structure and use of a computer program. Thereafter, some numerical examples illustrate
the classical calculations but most also present the results of
analysis by computer. Later Chapters discuss estimations of
relative potency, the consequences of natural response rates,
the design of experiments, and a number of more complex
problems. The final Chapter is an introductory account of
models for the joint action of mixtures of poisons, mostly based
upon the publications of Hewlett and Plackett.

I have taken the opportunity of broadening the range of
examples, so as to illustrate the applicability of probit methods
to fields other than the testing of insecticides. These include the
analysis of data in pharmacology, immunology, radiology, and
physiology. Modern computing power makes possible the simul-

taneous estimation of many parameters, so making unnecessary approximations that were proposed previously. I have chosen to restrict myself almost entirely to the lognormal tolerance distribution, the probit transformation, and maximum likelihood. I believe this model and principle of estimation to be the most generally useful, but I do not intend to be dogmatic: the methods described are easily adapted to other choices, and presentation in terms of one consistent system seemed preferable to any attempt to cover all possibilities.

To all who have, in conversation, letters, or reviews, suggested improvements in or criticized earlier editions I am most grateful. They are too numerous to mention individually. I must record particular gratitude to Professor J. F. Crow for the results of an experiment used in Example 39, to my secretarial staff for much very careful typing and checking, and to my wife for invaluable help in various stages of preparing and checking the printed text.

D. J. FINNEY

Edinburgh
May 1970

1

Introductory

1.1 The quantal problem

Many problems of quantitative inference in biological and technological research concern the relation between a stimulus and a response. A typical example is that in which the stimulus is a drug or other chemical preparation administered to an animal and the response is the effect produced in the animal; interest will lie in the dependence of the magnitude of the response on the dose of the drug, or in some special aspect of this relation. Analogous situations are those in which the rôle of stimulus is played by a constituent of a manufactured product and the response is a measure of quality or performance, or the stimulus is the time for which an animal is exposed to a learning situation and the response is a measure of skill subsequently shown. The statistician becomes involved when the response is not *exactly* determined by knowledge of the stimulus, and repetitions of experiments or observations for fixed values of the independent variables do not all give the same magnitude of response. Regression analysis, in its many ramifications, has been developed to guide interpretation of such results.

This book presents statistical techniques appropriate to one exceedingly important type of response, that known as all-or-nothing or *quantal*. Quantitative measurement of a response is almost always to be preferred when practicable. However, certain responses permit of no graduation and can be expressed only as 'occurring' or 'not-occurring'. The most obvious example is death; although workers with insects may have difficulty in deciding precisely when an insect is dead (Tattersfield, Gimingham and Morris, 1925), in many insecticidal studies the only practical interest lies in whether or not a test insect is dead, or perhaps in whether or not it has reached a degree of inactivity such as is thought certain to be followed by early death. In fungicidal investigations, failure of a spore to germinate is a quantal response of similar importance. In studies of drug potency, the response may be the cure of a particular morbid condition, no partial cure being under consideration. The test of whether or not manufactured articles of a certain class are satisfactory may be their failure or survival under operating conditions. Many commonly presented examples relate to experiments on insecticides and fungicides, because the methods have been developed systematically with special reference to this field of research, and much of the standard terminology reflects that origin. The statistical techniques, however, are applicable to other data, both biological and non-biological.

1.2 Biological assay

Quantal records have been widely used in biological assay. Although not restricted to this topic, a brief outline of how they arise there may help to illustrate their practical importance. Biological assay is a set of techniques relevant to comparisons between the strengths of alternative but similar biological stimuli. Its statistical aspects have been discussed elsewhere (Finney, 1964). In its widest sense, the term means the measurement of the potency of any stimulus, physical, chemical or biological, physiological or psychological, by means of the reactions that it produces in living matter. The biological method of measuring the stimulus is adopted either for lack of any alternative, or because an exact physical or chemical measurement of stimulus intensity may need translation into biological units before it can be put to practical use.

Biological assay most commonly refers to assessment of the potency of vitamins, hormones, toxicants, and drugs of all types by means of the responses produced when doses are given to experimental animals. Estimation of the potency of a natural product, such as a drug extracted from plant material, in producing a biological effect of a certain type, is often impossible or impracticable by chemical analysis. Even if the chemical constitution of the material has been determined, there may be little knowledge of the magnitude of the effect which the constituents will produce. The difficulty is not confined to natural products but occurs also with many manufactured compounds, such as insecticides, made to precise chemical specifications yet of unknown biological activity. The material must in fact be tested and standardized by methods appropriate to its future use.

Biological assays often use responses (e.g. weights of bodily organs) measured on a continuous scale. Sometimes quantal responses are more conveniently obtainable. One feature of all biological assays is the variability in the reaction of the test subjects and the consequent impossibility of reproducing at will the same result in successive trials however carefully the experimental conditions are controlled. (Assays based entirely on physical or chemical measurements will be affected by variability, but this is generally of far less practical importance). The contrast between physical and biological techniques may be seen from a consideration of two methods for the estimation of the ratio of two unknown weights. The physical method is to balance each in turn against a set of standardized weights, and to take as the required estimate the ratio of their magnitudes. There may be technical difficulties in carrying out the operations of weighing to very high accuracy, and both the quality of the balance and the competence of the operator are important factors, but for most practical purposes the reproducibility of the results is not called in question; one measurement on each weight will usually suffice to determine the ratio with an accuracy far beyond that obtainable in any biological assay.

The physical assay of the ratio is so simple that no alternative is needed. Purely for comparison, one might imagine a biological technique, using quantal responses, in which the weights are dropped from a fixed height on to the heads of live rats. Data for the assay are provided by the records of death or survival. That the first weight, at its first trial, killed a rat, while the second weight did not, would not show with certainty that the first was the heavier, still less would it give any clue to their ratio; the effect would be influenced by the weight dropped, by the age, sex, size, and physical condition of the rat, and by other biological and environmental factors (as well as, of course, by the shape and elasticity of the weights, which will here be assumed the same for both). If batches of rats, chosen at random from the stock available, were tested with each weight, the proportionate effect of variation in susceptibility from rat to rat would be reduced with increasing size of sample, and the weights could be compared in terms of the two mortality rates.

Variability might be still further controlled, though never entirely eliminated, by using a carefully bred strain of rats, and selecting batches homogeneous for sex, age, and other relevant factors. When every test is made from the same arbitrary height, this assay cannot discriminate between weights too light to cause any deaths or between weights so heavy as to kill every rat. This difficulty could be overcome by making tests from a series of different heights and obtaining a range of mortalities for each weight. The weights are compared in terms of equivalent heights, or heights estimated to give the same (say 50%) mortality. The height scale then provides a basis for the biological comparison of any number of weights, but, without experimental or theoretical knowledge of the law relating mortality to height and the physical measure of weight, the results of the biological assay cannot be transformed to purely physical terms.

Despite its absurdity, this example illustrates the necessity for consideration of variability in any biological assay technique. The quantal nature of the responses is a complication, but quantitative responses by no means provide an escape from the problem. Equal doses of insulin will not produce equal effects on the blood sugar of different rabbits, or even on the blood sugar of the same rabbit at different times. Consequently, though two insulin preparations could be compared in terms of the magnitudes of the changes in blood sugar produced in two rabbits, only repetition of the tests on several rabbits for each preparation can give an estimate of the relative potency sufficiently precise to be of any practical value. When there is a large natural variability of response amongst the test subjects, any satisfactory analysis of numerical data for the estimation of the effects of applied treatments demands exact statistical techniques.

1.3   Purpose of this book

The first edition of this book was written to help experimental scientists (especially biologists) in the practice of a set of statistical techniques that was becoming steadily more important. The knowledge of mathematics, statistics, and computational techniques had to be assumed very slight, and the book was written to have minimal dependence on other texts. A later generation is more sophisticated in its appreciation of the rôle of mathematics in biology, and the development of high speed computers has removed the terrors from computations that once seemed intolerably laborious.

Nevertheless, an account of the statistical techniques now conveniently grouped under the heading of probit analysis is both intrinsically important to some scientists and methodologically interesting to a wider circle. The book that follows is completely revised. It still endeavours to explain the underlying concepts in terms that can be understood by those with little mathematical knowledge, though perhaps the minimum can be set a little higher than in 1945. It still explains the basic methods of computation in forms suited to desk calculating machines (or even to long-hand arithmetic if any scientists remain so inadequately equipped as to depend on this). It does not pursue all the previous methods devised for desk calculators. Instead, the problems are discussed in more general terms appropriate to electronic computers. The reader is shown how a few standard programs can be made to replace all the earlier heavy labour and to give the results required more accurately and quickly than was previously possible. Moreover, greater generality enables a broader range of problems to be incorporated.

Even for those not directly concerned in the planning and interpretation of experiments using quantal responses, the subject perhaps deserves study. It illustrates well the principles and practice of statistical estimation, including both the choice of a parametric model of phenomena and the aspects of the estimates that are almost independent of the model. It shows the essential unity of techniques devised for rather different problems. The statistical methods have developed in relation to various fields of science, and both the history of these and the parts played at times by different approximations throw light on the rôle of the statistician in experimental science. The biologist seeking better understanding of how statistical thought might help him and the statistical theorist needing experience of how to apply his theory to reality can here learn something of the interactions between their disciplines.

This book thus has two aims. Primarily it is planned as a manual of instruction in the statistical aspects of an important class of experiments, including both the planning and the analysis of results. Also, however, it provides illustration of statistical method, and of how standard techniques may be used in combination with special less familiar ones, that may interest a wider circle.

1.4   Statistical methods for the biologist

In many fields of research, experimental and observational results can be used to best advantage only by subjecting them to precise and critical statistical examination. When a programme of biological research involves the collection of numerical data, the problem of interpreting these is almost inevitably one of statistics. Some biologists still mistakenly imagine that they have power to choose whether figures shall be 'statistically analysed' or not. In reality, any condensation and summarizing of numerical data is a form of statistical analysis; the choice is only of whether the analysis shall be theoretically sound and able to extract all the relevant information, or inadequate and possibly unsound. The analysis appropriate to any body of data is determined by the inherent properties of those data, not by the whim of the statistician. Good experimental work should never be followed by a statistical treatment of the results so unsatisfactory that the conclusions are incomplete, unreliable, or even actively misleading! The function of the statistician is to supply that critical and objective judgement of numerical material which is a product of his specialized training and experience. An important aspect of his work is co-operation in the planning of an experimental programme so that, taking into account all relevant information already available, it is designed to give results of maximum utility and precision in relation to expenditure of time, labour, and materials. This cannot be achieved if the statistician is not consulted until after completion of the experimentation.

The methods of analysis used by the statistician are not esoteric mysteries, but are simply instruments for extracting the most important features from numerical data. The computational procedures appropriate to many types of data have been so far standardized that they can be applied by a biologist who has some understanding of their purposes, even though he may know little of their theoretical foundations. Blind application of formulae is to be avoided, however, for not infrequently the formulae will be used quite inappropriately; some familiarity with the basic methods of analysis of variance, linear regression, and contingency table analysis, as well as knowledge of the simpler experimental designs, is very desirable. Fortunately nowadays many good text-books are available. Among those suitable as preparation for the more specialized purposes of this book are four of contrasting types:

Fisher, R. A. (1969). *Statistical Methods for Research Workers* (14th edition). Oliver and Boyd Ltd., Edinburgh.

Freeman, H. (1963). *Introduction to Statistical Inference.* Addison–Wesley Publishing Company, Inc., Reading, Massachusetts, U.S.A.

Mood, A. M. and Graybill, F. A. (1963). *Introduction to the Theory of Statistics* (2nd edition). McGraw-Hill Book Company, Inc., New York.

Snedecor, G. W. and Cochran, W. G. (1968). *Statistical Methods* (6th edition). Iowa State University Press, Ames, Iowa, U.S.A.

The reader not satisfied by any of these may be able to obtain other suggestions from a statistical colleague.

2

Quantal responses and the
dose-response curve

2.1 The frequency distribution of tolerance

In every dose-response situation, two components must be considered: the *stimulus* (for example, a vitamin, a drug, a mental test, or a physical force) and the *subject* (for example, an animal, a plant, a human volunteer, or a metal sheet). The stimulus is applied to the subject at a stated *dose*, an intensity specified in units of concentration, weight, time, or other appropriate measure, and under environmental conditions as carefully controlled as is practicable. As a result, the subject manifests a *response* (growth, a colour change, a score, or signs of wear). Different stimuli can be compared in terms of the magnitudes of the responses they produce, or, more commonly and usefully, in terms of the intensities required to produce equal responses (§ 6.1).

If the characteristic response is quantal, occurrence or non-occurrence will depend upon the intensity of the stimulus. For any one subject, under controlled conditions, there will be a certain level of intensity below which the response does not occur and above which the response occurs; such a value has often been called a *threshold* or *limen*, but the term *tolerance* is now widely accepted. This tolerance value will vary from one subject to another in the population used. (When the characteristic response is quantitative, the stimulus intensity needed to produce a response of any given magnitude will show similar variation between individuals). If the concept of repetition of tests on an individual is meaningful, tolerance is likely also to vary from one occasion to another as a result of uncontrolled internal or external conditions.

Discussion of quantal response data therefore requires recognition of the frequency distribution of tolerances over the population studied. If the dose, or intensity of the stimulus, is measured

[8]

by z, the distribution of tolerances may be expressed by

$$dP = f(z)\,dz. \qquad (2.1)$$

This equation states the proportion, dP, of the whole population of subjects whose tolerances lie between z and $z + dz$ at the time of testing, where dz represents a small interval on the dose scale; the factor relating dP to the length of this interval is the *frequency function*, $f(z)$, uniquely determined for each possible value of z.

If a dose z_0 were given to the whole population, every individual whose tolerance was less than z_0 would respond. The proportion of these is P, where

$$P = \int_0^{z_0} f(z)\,dz; \qquad (2.2)$$

the measure of dose is here assumed to be a quantity that can conceivably range from zero to $+\infty$, response being certain for very high doses so that

$$\int_0^\infty f(z)\,dz = 1. \qquad (2.3)$$

The frequency distribution of tolerances, as measured on the natural scale, is usually markedly skew, but often a simple transformation of the scale of measurement will convert it to a distribution approximately of the familiar Gaussian or *normal* form.

A variate is said to be normally distributed when it takes all values from $-\infty$ to $+\infty$ with frequencies given by a definite mathematical law, namely, that the logarithm of the frequency at any distance d from the centre of the distribution is less than the logarithm of the frequency at the centre by a quantity proportional to d^2. The distribution is therefore symmetrical, with the greatest frequency at the centre; although the variation is unlimited, the frequency falls off to exceedingly small values at any considerable distance from the centre, since a large negative logarithm corresponds to a very small number (Fisher, 1969, §12).

In tests of insecticidal sprays, for example, the distribution of tolerance concentration of the toxic agent is seldom symmetrical, because a few insects with extremely high tolerances can provide an extended 'tail' (Fig. 2.1); normalization can often be effected by expressing the tolerances in terms of the logarithms of the concentrations instead of the absolute values (Fig. 2.2). Indeed,

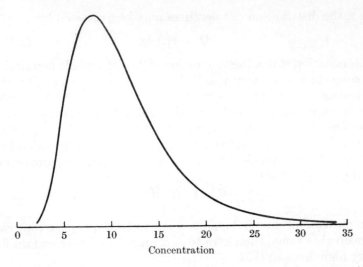

Fig. 2.1. Typical frequency distribution for the tolerance concentrations of a population. (The area between any two ordinates represents the proportion of subjects having tolerances between these two limits.)

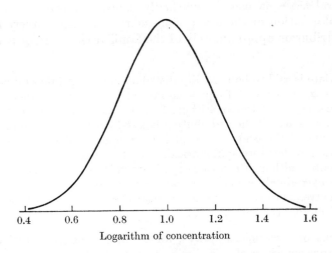

Fig. 2.2. Normal frequency distribution for the logarithms of the tolerance concentrations in Fig. 1.

this transformation is now standard practice for such trials (cf. Galton, 1879). Clark (1933), Hemmingsen (1933), Bliss (1935a), and others have sought an explanation of the normal distribution of log tolerances in the Weber–Fechner law and in adsorption phenomena, particularly as expressed by the Langmuir adsorption law, but these explanations are beyond the present scope. The validity and appropriateness of the logarithmic transformation in the analysis of experimental data are not dependent on the truth of hypotheses relating to adsorption. The normal distribution plays a central part in statistical science for many reasons, not least because frequency distributions of means and other derived quantities approximate to it even when individual observations behave differently. Some writers have suggested that a fractional power of tolerance (e.g. $z^{\frac{1}{2}}$ or $z^{\frac{1}{3}}$) might be normally distributed, but this must involve internal contradictions for low doses. The consequences of regarding the log tolerance frequency distribution as normal will be examined more closely later; for the present, the justification is the widespread applicability of the normal distribution as an adequate approximation to truth.

2.2 The dose metameter

The transformed scale of dose on which tolerances are normally distributed is known as a *metametric* scale, and the measure of dose is the *dose metameter* (Bacharach, Coates and Middleton, 1942). For this metameter, the symbol x will be used throughout the book. Thus the general definition is

$$x = \log z. \tag{2.4}$$

Whether common logarithms (base 10) or natural logarithms (base e) are used is immaterial, because results calculated with either are easily converted to the other base:

$$\log_{10} z = 0.43429 \log_e z. \tag{2.5}$$

For most practical purposes, common logarithms are more convenient and will be used unless otherwise noted, although theory is more tidily handled in terms of natural logarithms. Where there is no fear of confusion, x itself may be referred to as the dose.

In any situation for which tolerance can be satisfactorily

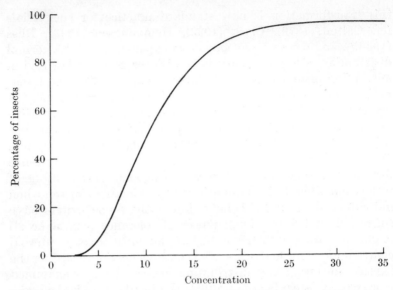

Fig. 2.3. Sigmoid curve derived from Fig. 1, showing percentage of subjects with tolerances less than a specified value.

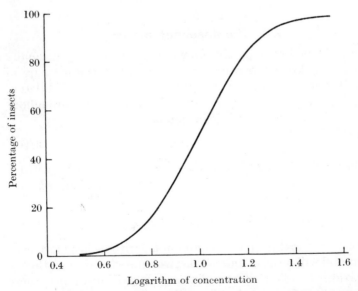

Fig. 2.4. Normal sigmoid curve derived from Fig. 2, showing percentage of subjects with log tolerances less than a specified value.

defined, every subject whose tolerance is less than a stated value of x will respond to that dose. Evidently, a graph of percentage responding against dose will give a steadily rising curve. The rate of increase in response per unit increase in dose is frequently very low in the region of zero or 100 % response, but higher in the intermediate region, so that the curve is sigmoidal (Fig. 2.3). When the stimulus is measured in metametric units, the curve takes the characteristic *normal sigmoid* form (Fig. 2.4). This curve approaches zero or 100 % response at infinitely low and infinitely high values of x respectively. Of course, infinitely low x represents zero dose, but infinitely high x must be regarded simply as a limiting concept.

2.3 Parameters

The normal frequency distribution of log tolerances, corresponding to equation (2.1) but rewritten in terms of the dose metameter, is:

$$dP = \frac{1}{\sigma\sqrt{(2\pi)}} \exp\left\{ -\frac{(x-\mu)^2}{2\sigma^2} \right\} dx \qquad (2.6)$$

for all values of x in the range $-\infty$ to $+\infty$. This is one of many equivalent ways of expressing the normal distribution.

The quantities μ, σ, the *mean* and *standard deviation* of the distribution, are collectively known as *parameters*. The parameters of any family of distributions are quantities occurring in the algebraic specifications, numerical values of which identify individual distributions within the family. Thus equation (2.6) represents all normal distributions. Insertion of any numerical value for μ and any positive (non-zero) value for σ produces one particular normal distribution.

Much of statistical methodology is concerned with formulating statements about parameters of distributions. This is certainly true of quantal response problems, as will soon appear.

2.4 Direct measurement of tolerance

Sometimes the tolerance of each test subject in respect of a stimulus can be measured directly. For example, in the 'cat' method for the assay of digitalis, anaesthetized cats are given a continuous slow intravenous infusion of digitalis until death occurs (Burn, Finney and Goodwin, 1950; Hatcher and Brody,

1910). If there is appreciable time lag between the introduction of the drug and its taking effect, the tolerance or lethal dose will be overestimated. Moreover, the dose required to cause death under conditions of slow infusion need not be the same as the tolerance for more rapid application. Despite these weaknesses, the technique has been used successfully for some types of bioassay.

An alternative method is to give to each subject successive doses of different intensities, allowing after every dose a suitable time interval for a return to normal and making the differences sufficiently small for a satisfactory determination of the lowest dose which causes the characteristic response. With an irreversible response, such as death, the doses would have to be given in an increasing series. For the method to be satisfactory, there must be no cumulative effect of doses already given, either as lowering or as increasing the resistance of the subject, a condition which severely limits its applicability.

With direct measurement of tolerance, the appropriate methods of statistical analysis are the same as for other types of biological measurements. If the tolerance of each subject has been separately and independently determined, the set of values obtained may be subjected to the same analytical processes as measurements of length or weight; the estimation of means and standard errors, the comparison of distributions, and the making of tests of significance present no new features. Bliss and Hanson (1939), for example, have discussed the application of the analysis of variance and covariance to assays based on the cat method. Finney (1964) has given examples and detailed theoretical discussion.

Direct tolerance measurement is often impracticable on account of the time the methods require. Even more commonly it is ruled out entirely by the nature of the problem: a direct measurement technique for the poison tolerance of an insect, or of a fungus spore, is scarcely conceivable. An entirely different approach must then be adopted, and the potency of the stimulus must be assessed from the proportions of subjects that respond, in random samples of the population tested at different doses.

2.5 The binomial distribution

Suppose that an insect, selected at random from a population, is exposed to a dose of a poison. The dose may be measured by the concentration of toxic substance, the absolute quantity used, the length of time of exposure to a fixed set of conditions, or some combination of these or other factors. The probability, P, that the insect dies is the proportion of deaths that would occur if the whole population received the same dose, and is given in general by an integral such as that of equation (2.2). The probability of non-response, $(1 - P)$, is usually denoted by Q. If two insects are exposed to the same dose, and if their reactions are completely independent, the probability that both respond is P^2, and the probability that both fail to respond is Q^2; the probability that only the first responds is $P \times Q$, and the probability that only the second responds is $Q \times P$. Thus the total probabilities of 2, 1, and 0 responding are P^2, $2PQ$, and Q^2 respectively, the successive terms in the expansion of $(P + Q)^2$. In a similar manner it may be seen that if a batch of n insects is exposed to the same dose, and all react independently, the probabilities of n, $(n - 1)$, $(n - 2)$, ..., 2, 1, 0 responding are the $(n + 1)$ terms in the expansion of the binomial $(P + Q)^n$. The probability of exactly r responding is therefore

$$Pr(r|n) = \frac{n!}{r!\,(n-r)!}\, P^r Q^{n-r}. \tag{2.7}$$

This well-known result, the *binomial distribution* of probabilities, is discussed more fully in almost every elementary text-book of statistics. If different batches of n insects are exposed to the dose, the average number of deaths per batch will be about nP, as might be expected. Many tables of the probability in equation (2.7) have been published (e.g. Harvard Computation Laboratory, 1955), but reference to these is not necessary for the methods of this book.

The reactions of separate members of a batch to the stimulus of a particular dose are not always independent. A correlation of response may result from incomplete randomness of selection of the batch, or from unsatisfactory control of experimental conditions causing the number responding to be seriously affected by some factor other than the dose. For example, if each batch consists of insects from a single brood, insects in one batch are likely to be more alike in tolerance than those in different batches, and the variation between batches in the numbers responding will be greater than that for the binomial distribution. Again, susceptibility to an insecticide might be greatly influenced by temperature; if the temperature during tests varied substantially from one batch to another, the variance of the numbers responding to a particular dose would be inflated. The extreme situation is that, in every batch tested, either all members respond or all fail to respond, so that the evidence from a batch is no more reliable than that from an individual. Whatever the cause, such heterogeneity must make the weight to be attached to the data less than is appropriate to the binomial distribution.

The result of testing a series of doses, each on a separate batch of insects, is to obtain for each dose a proportion

$$p = r/n \tag{2.8}$$

of insects that show the characteristic response, and that therefore have tolerances less than a known dose. Each p is an estimate of the corresponding P, the proportion in the population from which the batch was a sample. From the values of p, a *statistic*† will be derived corresponding to each parameter of the population. In general, both P and p will increase steadily with increasing dose but, if the number of test subjects in a batch is small, sampling variation may interfere with the regularity of the trend in p. For example (Trevan, 1927), if two batches of five subjects are given doses sufficient to cause 25 % and 75 % of deaths respectively in the whole population, in 2 % of repetitions of the experiment the lower dose would appear to be the more effective, in nearly 6 % the numbers of deaths in the two groups would be equal, and in a very small proportion (0.05 %) either none or all ten subjects would die. The larger the batches the greater is the assurance of satisfactory discrimination between the effects of different doses. In practice, a limiting factor to the size of the experiment may be the total number of subjects to be used:

† A 'statistic' is (Fisher, 1969, § 11) ' a value calculated from an observed sample with a view to characterising the population from which it is drawn'. In effect, each statistic will be chosen to estimate a parameter.

several batches of moderate size may then be preferred to two
or three large ones, in order that a wide range of doses may be
tested and an idea of the dose-response relation obtained.

2.6   The median effective dose

Early attempts to characterize the effectiveness of a stimulus in
relation to a quantal response referred to the *minimal effective
dose*, or, for a more restricted class of stimuli, the *minimal lethal
dose*, terms which fail to take account of the variation in tolerance
within a population. In exposing the logical weakness of such
concepts, Trevan (1927) said:

The common use of this expression in the literature of the subject
would logically involve the assumptions that there is a dose, for
any given poison, which is only just sufficient to kill all or most of
the animals of a given species, and that doses very little smaller
would not kill any animals of that species. Any worker, however,
accustomed to estimations of toxicity, knows that these assumptions
do not represent the truth.

It might be thought that the minimal lethal dose of a poison
could instead be defined as the dose just sufficient to kill a
member of the species with the least possible tolerance, and also
a *maximal non-lethal dose* as the dose which will just fail to kill
the most resistant member. Undoubtedly some doses are so low
that no test subject will succumb to them and others so high as
to prove fatal to all, but considerable difficulties attend determination of the end-points of these ranges. Even when the
tolerance of an individual can be measured directly, to say from
measurements on a sample of ten or a hundred that the lowest
tolerance found indicated the minimal lethal dose would be unwise: a larger sample might contain a more extreme member. When
only quantal responses for selected doses can be recorded the
difficulty is increased, and the occurrence of exceptional individuals in the batches at different dose levels may seriously
bias the final estimates. The problem is, in fact, that of determining the dose at which the dose-response curve for the whole
population meets the zero or 100 % levels of kill, and even a very
large experiment could scarcely estimate these points with any
accuracy.

Trevan suggested an escape from the dilemma by giving attention to a different and more satisfactorily defined characteristic, the *median lethal dose*, or, as a more general term to include responses other than death, the *median effective dose*. This is the dose that will produce a response in half the population. The median effective dose is commonly referred to as the ED 50, the more restricted concept of median lethal dose as the LD 50. Analogous symbols are used for doses effective for other proportions of the population, ED 90 being the dose which causes 90 % to respond. As will become apparent in later chapters, by experiment with a fixed total number of subjects effective doses in the neighbourhood of ED 50 can usually be estimated more precisely than those for more extreme percentage levels, and this characteristic is therefore particularly favoured in expressing the effectiveness of the stimulus; its chief disadvantage is that, especially in toxicological work, much greater interest may attach to doses producing nearly 100 % responses than to those producing only 50 %, in spite of the difficulty of estimating the former.

For any distribution of tolerances, the ED 50 is the value of z_0 that, inserted in equation (2.2), gives

$$P = 0.5.$$

If the parameters implicit in the function $f(z)$ are known, the equation can be solved. Of course, knowledge of the corresponding value of the dose metameter is equivalent: for the distribution represented by equation (2.6), the log ED 50 is the value of x_0 for which

$$0.5 = \int_{-\infty}^{x_0} \frac{1}{\sigma\sqrt{(2\pi)}} \exp\left\{ -\frac{(x-\mu)^2}{2\sigma^2} \right\} dx. \qquad (2.9)$$

The solution is well-known to be μ, so that the ED 50 is 10^μ.

Note that the ED 50 may alternatively be regarded as the *median* of the tolerance distribution, that is to say the level of tolerance such that exactly half the subjects lie on either side of it. On any metametric scale for which the tolerance distribution becomes symmetrical, the median and the mean will coincide, as here: the log ED 50 is also the mean of the distribution on the x-scale of dose measurement. If direct measurement of tolerance were possible, the mean of the distribution in respect of a logarithmic or other suitable metameter would be adopted as a feature having primary interest, almost without comment; thus to use the corresponding quantity when quantal responses are the only type observable is a natural and consistent policy.

The statistical analysis of quantal response data, using a logarithmic metameter, must concern itself with estimation of μ. The ED 50 alone does not fully describe the effectiveness of the stimulus. Two poisons may require the same rates of application in order to be lethal to half the population, but, if the distribution of tolerances has a lesser 'spread' for one than for the other, any increase or decrease from this rate will produce a greater change in mortality for the first than for the second. This 'spread' is measured by the variance, σ^2: the smaller the value of σ^2, the greater is the effect on mortality of any change in dose. Stimuli which produce their effects by similar means (in particular, poisons whose physiological effects are similar), often have approximately equal variances of log tolerances for any population of subjects, even though they differ substantially in their median lethal doses. Such stimuli can be compared in terms of median lethal doses alone (§ 6.1).

3

Estimation of the median effective dose

3.1 The occurrence of quantal responses

Examples of situations in which a relation between dose and the frequency of a quantal response must be investigated have already been mentioned. Others will appear below and in § 3.8. Typical is the outcome of an experiment on different doses of an insecticide, applied under standardized conditions to samples of an insect species. Table 3.1 shows one such set of results, in which the percentage of insects dead or seriously affected in each dose group has been entered alongside the raw data. An estimate of the ED50, the dose that on average would produce 50 % response, would be a convenient and useful summary, and calculations relevant to it will be discussed in this and the next chapter. Essentially the same problem arises in tests of fungicides, rodenticides, and the like.

Table 3.1 A test of the toxicity of rotenone to *Macrosiphoniella sanborni* (Martin, 1942)

Dose of rotenone (mg/l)	No. of insects (n)	No. affected (r)	% kill (p)
10.2	50	44	88
7.7	49	42	86
5.1	46	24	52
3.8	48	16	33
2.6	50	6	12
0	49	0	0

The rotenone was applied in a medium of 0.5 % saponin, containing 5 % of alcohol. Insects were examined and classified one day after spraying.

The testing and standardizing of therapeutic drugs can use similar techniques (Finney, 1964). The percentages of mice showing convulsions after injection with different doses of insulin

[20]

can be related to dose as a basis for standardization of insulin preparations. The testing of vaccines and toxoids provides further examples. Mice protected with different doses of a typhoid vaccine may be challenged by exposure to infection, the relation between death rate and vaccine dose being a basis for the comparison of vaccines (Yugoslav Typhoid Commission, 1962). Diphtheria toxoids have been compared by using various immunizing doses in guinea-pigs and recording the occurrence or otherwise of a characteristic skin reaction after an appropriate challenge procedure (Greenberg, 1953).

An interesting possibility is the use of quantal response techniques in the study of some phenomenon in a maturing animal that cannot be dated exactly but can readily be recorded as having occurred or not occurred in any one animal. To obtain reliable records of the age of menarche in adolescent girls is difficult, organizationally and possibly psychologically. On the other hand, a sample of girls distributed over the appropriate age range can relatively easily be classified according to whether or not each has yet menstruated. If age is regarded as analogous to dose, and 'having passed menarche' as response, the standard dose-response situation is established. The techniques to be introduced in this Chapter and Chapter 4 can then be applied to estimation of the median age of menarche and the standard deviation of the distribution. Wilson and Sutherland (1949) and Burrell, Healy and Tanner (1961) have reported this type of investigation. Parsons, Hunter and Rayner (1967) have studied the effect of the ram on ovulation in the ewe by essentially the same technique, and have also used some interesting extensions of the idea (§ 8.8). Leslie, Perry and Watson (1945) were able to discuss the distribution of body-weight at which female rats in the wild reach maturity; the percentage of specimens caught which were found to have developed corpora lutea was taken as response, and weight as dose. In another interesting application, Mollison and Armitage (1953) took records of infants suffering from haemolytic disease of the newborn, and examined percentage survival in relation to concentration of haemoglobin in the cord. They were thus able to compare sexes and also to look at differences between those infants treated by exchange transfusion and the others.

In the construction of standardized mental tests, there is need to assess how the proportion of correct answers to a single item changes with the general ability of the child, the latter being indicated by the total score on the many items comprised in the test (Ferguson, 1942; Finney, 1944b). Here the total score corresponds to dose. The non-independence arising because the item under study also contributes to the total is unimportant for a test that includes 100 or more items.

3.2   The N.E.D. and probit transformation

Modern statistical techniques for these problems began with Gaddum (1933), although the basic idea was proposed much earlier (§ 3.6). Gaddum proposed to measure the probability of response on a transformed scale, the *normal equivalent deviate* (or N.E.D.). This response metameter is Y, defined by

$$P = \frac{1}{\sqrt{(2\pi)}} \int_{-\infty}^{Y} \exp\{-\tfrac{1}{2}u^2\}\, du. \tag{3.1}$$

Thus the N.E.D. of any value of P between 0 and 1 is defined as the abscissa corresponding to a probability P in a normal distribution with mean 0 and variance 1.

Equation (3.1) determines either of P and Y uniquely from the other. Fig. 3.1 shows this graphically. The transformation stretches the extremes of the scale, so that, as P ranges from 0 to 1, Y ranges from $-\infty$ to $+\infty$. From integration of equation (2.6), if P is the probability of response at a dose whose metameter is a particular value X,

$$P = \int_{-\infty}^{X} \frac{1}{\sigma\sqrt{(2\pi)}} \exp\left\{-\frac{(x-\mu)^2}{2\sigma^2}\right\} dx,$$

which, by writing $X = \mu + \sigma u,$

becomes $$P = \int_{-\infty}^{(X-\mu)/\sigma} \frac{1}{\sqrt{(2\pi)}} \exp\{-\tfrac{1}{2}u^2\}\, du. \tag{3.2}$$

Comparison with equation (3.1) shows that

$$Y = (X-\mu)/\sigma. \tag{3.3}$$

Thus the relation between the dose metameter and the N.E.D. of the probability of response at that dose is a straight line. In Fig. 3.2, the normal sigmoid curve of Fig. 2.4 is reproduced,

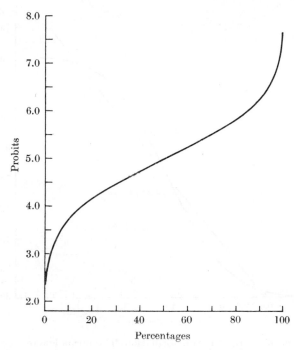

Fig. 3.1. Relation between percentages and probits.

together with the straight line obtained when its ordinates are re-plotted on a linear N.E.D. scale.

Bliss (1934b) suggested a slightly different response metameter. He defined the *probit* of P ('probit' = probability unit) as Y, where

$$P = \frac{1}{\sqrt{(2\pi)}} \int_{-\infty}^{Y-5} \exp\{-\tfrac{1}{2}u^2\}\,du. \qquad (3.4)$$

For any P, the probit is simply the N.E.D. increased by 5. All subsequent theory is essentially the same for the two metameters. The N.E.D., however, is negative if P is less than 50 %, whereas the probit is positive unless P is exceedingly small. At a time when most biologists lacked even simple calculating machines, and many had little skill in statistical arithmetic, avoidance of negative quantities was an appreciable practical advantage. To-day, vastly improved computing facilities and increased familiarity with statistical method would make the advantage

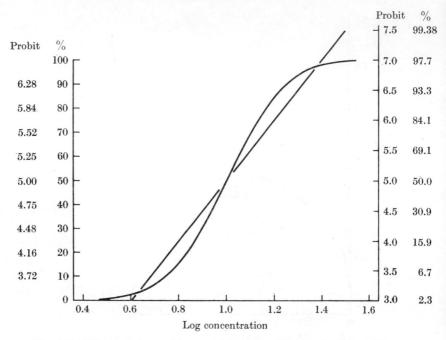

Fig. 3.2. Effect of the probit transformation. The normal sigmoid curve in Fig. 2.4 is transformed to a straight line when the ordinates are measured on a scale linear in probits instead of in percentages.

rest with the more simply defined N.E.D., but for the fact that 'probit analysis' is so well established as a name for a group of methods. All the tables required occupy about half as much space for the N.E.D. as for the probit, because of symmetry, and the 5 in the probit is untidy in theoretical discussions. To abandon probits now might cause confusion. In this book, as in previous editions, all tables and calculations will be in terms of probits; the reader who, very logically, prefers to use N.E.D. can readily adjust by subtracting 5 from every probit. All theoretical discussion, however, will assume Y to be related to P by equation (3.1).

The relation between the probit of the expected proportion of responses and the dose, corresponding to equation (3.3), is the linear equation

$$Y = 5 + \frac{1}{\sigma}(x - \mu).   \tag{3.5}$$

Experimental results can lead to an estimate of this equation, and the parameters of the tolerance distribution can then be estimated; in particular, the median effective dose is estimated by the value of x that gives $Y = 5$.

A table giving probits for specified values of P, prepared by Bliss (1935a), was reproduced by Fisher and Yates (1964, Table IX). A simplified table, sufficiently detailed for many purposes, is given as Table 3.2, and the full table is reproduced as Table I.

Table 3.2   Transformation of percentages to probits

%	0	1	2	3	4	5	6	7	8	9
0	—	2.67	2.95	3.12	3.25	3.36	3.45	3.52	3.59	3.66
10	3.72	3.77	3.82	3.87	3.92	3.96	4.01	4.05	4.08	4.12
20	4.16	4.19	4.23	4.26	4.29	4.33	4.36	4.39	4.42	4.45
30	4.48	4.50	4.53	4.56	4.59	4.61	4.64	4.67	4.69	4.72
40	4.75	4.77	4.80	4.82	4.85	4.87	4.90	4.92	4.95	4.97
50	5.00	5.03	5.05	5.08	5.10	5.13	5.15	5.18	5.20	5.23
60	5.25	5.28	5.31	5.33	5.36	5.39	5.41	5.44	5.47	5.50
70	5.52	5.55	5.58	5.61	5.64	5.67	5.71	5.74	5.77	5.81
80	5.84	5.88	5.92	5.95	5.99	6.04	6.08	6.13	6.18	6.23
90	6.28	6.34	6.41	6.48	6.55	6.64	6.75	6.88	7.05	7.33
—	0.0	0.1	0.2	0.3	0.4	0.5	0.6	0.7	0.8	0.9
99	7.33	7.37	7.41	7.46	7.51	7.58	7.65	7.75	7.88	8.09

The linear equations (3.3), (3.5) are commonly written

$$Y = \alpha + \beta x \tag{3.6}$$

where α, β become new parameters in place of μ, σ.

3.3   The probit regression line

When experimental data on the relation between dose and mortality have been obtained, either a graphical or an arithmetical process can be used to estimate the parameters. Both employ the probit transformation. The graphical approach is rapid and sufficiently good for many purposes, but for some more complex problems, or when an accurate assessment of the precision of estimates is wanted, the more detailed arithmetical analysis is necessary. In this Chapter only the graphical method will be presented, though the ideas introduced will be wanted again for the discussion of maximum likelihood estimation in Chapter 4.

The percentage response observed for each dose should first be calculated and converted to probits by means of Table 3.2.† The probits are then plotted against the dose metameter, and a straight line is drawn by eye to fit the points as satisfactorily as possible. In drawing the line and judging its agreement with the data, only the vertical deviations of the points must be considered: the line must be so placed that the differences between the probit values which are plotted and the probits given by the line at each x are as small as possible. Very extreme probits, say outside the range 2.5 to 7.5, carry little weight and may almost be disregarded unless many more subjects were used than in the batches giving intermediate probit values. The line is a graphical approximation to the *regression line* of response probit on x, equation (3.6).

This line may be used, as described in § 4.4, to initiate the arithmetical process of estimating a better fitting line. For many carefully conducted experiments, however, the empirical probits lie so close to a straight line that the provisional graphical line is good enough. Only experience of the field of research and of the statistical technique used can be a sound guide, but experimenters who use probit analysis should not feel obliged to undertake arithmetic when eye estimation would suffice. As will appear from § 4.4, the complete method for deriving the best estimate of the line is not difficult; it can be laborious if adopted as a routine measure without adequate computing aids.

If eye estimation alone is to be accepted, the log ED 50 is estimated from the line as m, the dose at which $Y = 5$. The slope of the line, b, is an estimate of $1/\sigma$, and is obtained as the increase in Y for a unit increase in x. These two estimates are then substituted for the parameters in equation (3.5) to give the estimated relation between dose and response. To test whether the line is an adequate representation of the data, a χ^2 test may be used, as in Example 1 below. A value of χ^2 within the limits of random variation indicates satisfactory agreement between theory (the

† Table I need not be used for these empirical probits, even when batches of test subjects are sufficiently large to justify the use of three or more places of decimals.

line) and observation (the data). A significantly large χ^2 may arise either because individual test subjects do not react independently, or because the straight line does not adequately describe the relation between dose and probit. If the former, the scatter of the points about the line will be wider than would occur had there been no correlation between the reactions of insects in the same batch; the precision of the line will be reduced (§ 4.6), though its position should be free from bias providing that adequate precautions have been taken in the conduct of the experiment (§ 9.1). If the latter, a systematic departure of the points from the line may indicate a curvilinear relationship; a new look at the particular problem, possibly with a change of metameters and possibly with a more drastic alteration of approach, is then essential.

A reader to whom the techniques of this book are new should note that they are appropriate only for data from subjects tested once each. An experiment in which six batches of 50 subjects are tested, one batch at each of six doses, is entirely different from one in which a single batch of 50 subjects (or of 300 subjects) is examined repeatedly at progressively increasing dose. The latter may be an excellent plan of experimentation, for example if 'dose' is measured by time of exposure to certain conditions of stress, but entirely different statistical techniques are appropriate. Probit and similar methods have been devised to meet the special requirement of the former situation, the almost inevitable plan of experimentation when dose is amount of an insecticide or of a therapeutic drug. The complete independence of the subjects tested at different doses, and of the binomial distributions associated with them, is implicit in the theory of all the methods presented in this book. In this context, Example 12 is illuminating: it analyses extensive cross-sectional data from 25 batches of girls in different age groups. A longitudinal study of a single cohort of girls followed from 9 to 18 years of age would require another (and simpler) form of analysis, in fact that familiar to statisticians for dealing with a grouped frequency distribution.

The probit is no more than a convenient mathematical device for solving certain equations. Though it may also be used to give a simple diagrammatic representation of the dose response relation, and though familiarity enables these diagrams to be interpreted directly, any suggestion that the statistical analysis is completed by the estimation of a probit regression line must be avoided. This was emphasized by Wadley (in Campbell and Moulton, 1943): 'The use of transformations carries with it a temptation to regard the transformed function as the real object of study. The original units should be mentioned in any final statement of results.'

Many of the numerical examples in this book have been chosen to illustrate special points of statistical technique, and, since the data have been removed from their original context, their discussion may not always be carried as far as a statement of conclusions in biological or chemical terms. In practice, results should finally be expressed by median effective doses, relative potencies, tolerance variances, or other suitable quantities, the units employed usually being dose (not log dose) and percentage response; at that stage, the word 'probit' need seldom be mentioned.

Ex. 1. *Fitting a probit regression line by eye to the results of an insecticidal test.* Martin (1942, Table 9) sprayed batches of about fifty *Macrosiphoniella sanborni*, the chrysanthemum aphis, with a series of concentrations of rotenone. His results appear in Table 3.1. The number affected is the total of insects apparently dead, moribund, or so badly affected as to be unable to walk more than a few steps; this classification was frequently used by Tattersfield and his co-workers at Rothamsted (Tattersfield, Gimingham and Morris, 1925). The total was taken as the 'kill', and normal or only slightly affected insects were considered to have survived.

Experiments like this should always provide for estimating the natural mortality among untreated insects. The last line of Table 3.1 records that, of forty-nine insects sprayed with the alchohol-saponin medium alone, all survived. The results for the five concentrations of rotenone will therefore for the present be assumed not appreciably influenced by the addition of a natural mortality of insects. As will be seen later (Chapter 7), adjustments to the statistical analysis are needed when natural responses augment the effects of the applied stimulus.

Table 3.3 summarizes the dose metameter, percentage kill, and empirical probit values for the experiment. The percentage kill, p, is defined by equation (2.8), and is an estimate of P, the average proportion for the whole population; of course, p is subject to sampling errors, in that repeated trials of the same dose will give different values of p distributed around P. In all formulae in the text, p and P denote proportions, the most convenient symbolism for theoretical interpretation; most users of these techniques prefer to do their arithmetic in terms of percentages, and in calculations the same symbols will be used even though $100p$ and $100P$ would be more correct. No confusion should be caused by this slight ambiguity of symbolism.

Table 3.3   Empirical probits, and approximate expected probits for data of Table 3.1

Log dose x	No. of insects (n)	% kill (p)	Empirical probit	Expected probit (Y)
1.01	50	88	6.18	6.30
0.89	49	86	6.08	5.83
0.71	46	52	5.05	5.10
0.58	48	33	4.56	4.58
0.41	50	12	3.82	3.90

Over the range of concentrations tested, the sigmoid nature of the relation between percentage kill and log concentration is not very apparent. The percentages are plotted against the dose in Fig. 3.3, together with the normal sigmoid curve fitted to them by the present analysis. Between 25% and 75% this curve is practically indistinguishable from a straight line; a line drawn to fit the five points would give $x = 0.68$ as approximately corresponding to 50% kill, in good agreement with the value obtained later for the log LD 50. A straight line for percentages must nevertheless be quite inadequate at more extreme values.

The probits of p (read from Table 3.2) have been entered in Table 3.3. When plotted against dose (Fig. 3.4) they lie nearly on a straight line, and such a line has been drawn by eye. From this line, probits corresponding to many different values of x have been read, converted back to percentages by using Table 3.2 inversely, and plotted against x in Fig. 3.3 to give a sigmoid curve; this facilitates prediction of either the mortality to be expected at any given dose, or the dose which will kill, on

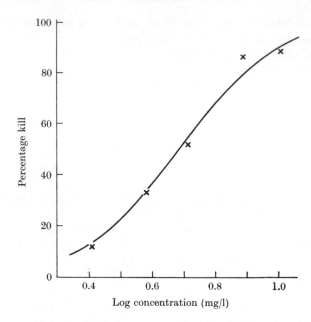

Fig. 3.3. Relation between percentage kill of *M. sanborni* and dose of rotenone (Ex. 1), showing normal sigmoid curve represented by equation (3.9).

average, a stated percentage of insects. In practice the sigmoid seldom need be constructed, as all predictions can be made directly from the probit diagram. For example, in Fig. 3.4 a probit value of 5.0 corresponds on the x-scale to a dose

$$m = 0.687; \tag{3.7}$$

this is the estimate of log LD 50, whence the LD 50 is estimated as a concentration of 4.86 mg/l. The value may be compared with 4.85 mg/l obtained by the arithmetical process of fitting a straight line (Example 6). Similarly, since the probit of 90 % is 6.28, the log LD 90 can be read from Fig. 3.4 as 1.003; hence the LD 90 is estimated as 10.1 mg/l.

Fig. 3.4 can also be used to give the slope of the line: an increase of 0.8 in x corresponds with an increase of 3.21 in probit. The estimated regression coefficient of probit on dose, or the rate of increase of probit per unit increase in x, is

$$b = 1/s = 4.01; \tag{3.8}$$

$s(=0.25)$ is an estimate of σ, the standard deviation of the log tolerance distribution (equation (2.6)). Hence the relation between probit and dose, equation (3.5), is estimated as:

$$Y = 5 + 4.01(x - 0.687),$$

or, in the form of equation (3.6),

$$Y = 2.25 + 4.01x. \tag{3.9}$$

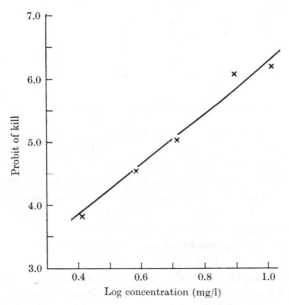

Fig. 3.4. Relation between probit of kill of *M. sanborni* and dose of rotenone (Ex. 1), showing probit regression line from equation (3.9).

Substitution of the actual values of x from the experiment gives the *expected probits*, the final column of Table 3.3. These are used in Table 3.4 to give the corresponding expected percentages, P; a probit of 6.30 in Table 3.2 falls between 90 % and 91 %, more exactly $(90 + 2/6)$ %. Multiplication of each proportion by n, the number of insects tested, gives the *expectation* or *expected number* of responding insects; this is the average number that would respond in a batch of n if equation (3.9) represented the true parametric relation. The numbers nP may then be compared with the observed r as a guide to the agreement of the experiment with theory.

Table 3.4   Comparison of observed and expected mortality in eye estimation for rotenone–*Macrosiphoniella sanborni* test

			No. of insects	No. affected Observed	No. affected Expected	Discrepancy	$\dfrac{(r-nP)^2}{nP(1-P)}$
x	Y	P	(n)	(r)	(nP)	$(r-nP)$	
1.01	6.30	90.3	50	44	45.2	-1.2	0.33
0.89	5.83	79.7	49	42	39.1	$+2.9$	1.06
0.71	5.10	54.0	46	24	24.8	-0.8	0.06
0.58	4.58	33.7	48	16	16.2	-0.2	0.00
0.41	3.90	13.6	50	6	6.8	-0.8	0.11

$$\chi^2_{[3]} = 1.56$$

Only the second concentration tested shows appreciable discrepancy, three more insects being affected than equation (3.9) predicts; as will be shown, this is within the limits of random variation. There is no indication of systematic departure from the line, such as might arise if the wrong normalizing transformation had been used and the true equation were not linear. A test of significance of the discrepancies may be obtained by squaring each, dividing the square by $(1-P)$, and again dividing by the tabulated value nP; slide rule accuracy is adequate. The sum of these quantities is, to a sufficiently close approximation if the line in Fig. 3.4 has been well drawn, a χ^2. Because the two parameters of equation (3.9) have been estimated from the data, the degrees of freedom are two less than the number of concentrations tested. In random sampling, the mean value of χ^2 is equal to the number of degrees of freedom; the value in Table 3.4, 1.56, is less than 3.0 and clearly sufficiently small to be attributed to random fluctuations about the relation specified in (3.9). Fisher and Yates's Table IV, of which a simplified form is reproduced as Table VII, shows that a value greater than 7.8 could occur by chance in 5 % of cases. Hence the probit regression line in Fig. 3.4 appears to be a very satisfactory representation of the results of the experiment.

3.4 Precision of estimates

The binomial distribution of probabilities (§ 2.5) determines the probabilities that n, $(n-1)$, ..., 2, 1, 0 subjects respond when all members of a batch of n react independently to a stimulus. As is well known from elementary statistical theory, the standard deviation of the observed proportion, p, about its mean value P, is $\sqrt{(PQ/n)}$. The square of a standard deviation is known as a *variance*; here the variance of p is PQ/n, a quantity inversely proportional to the size of the batch. The reciprocal of the variance, termed the *invariance* or *quantity of information*, is proportional to n and measures the weight attaching to the observations on the batch in respect of the information that p provides on P.

In § 4.2, the weight to be similarly attached to the probit of P is shown to be nw, where

$$w = Z^2/PQ; \qquad (3.10)$$

here Z (not to be confused with z in Chapter 2) is the ordinate to the standardized normal frequency function at the point corresponding to the N.E.D.:

$$Z = \frac{1}{\sqrt{(2\pi)}} \exp\{-\tfrac{1}{2}Y^2\}. \qquad (3.11)$$

Bliss (1935a) and Fisher and Yates (1964, Table XI) have tabulated w, the *weighting coefficient*, as a function of Y at intervals of 0.1. A shortened version of the table, Table 3.5, is adequate for many purposes; the column $C = 0$ of Table II contains more accurate values. Finney and Stevens (1948) have tabulated w at intervals of 0.01 in Y. Values of w are symmetrical about $Y = 5$: $Y = 3.8$ and $Y = 10 - 3.8 = 6.2$ both have $w = 0.370$.

The weighting coefficients may be used in estimating a standard error for the graphically estimated log ED 50. The value of Y corresponding to each dose used must be read from the regression line drawn by eye, and the weighting coefficient determined from Table 3.5 for each Y. Multiplication by n, the number of subjects at the dose, gives nw, which must be summed for all doses; the symbol S will be used to indicate such a summation. If the log ED 50 is not very different from the mean value of x in the experiment, its standard error is approximately $1/b\sqrt{(Snw)}$. This expression makes no allowance for sampling errors in the estimation of b, and it may be a serious underestimation if m, the estimated log ED 50, is far from $\bar{x}$, the weighted mean dose or $(Snwx)/(Snw)$. The variance of b is $1/Snw(x-\bar{x})^2$, and a more trustworthy value for the variance of m is therefore

$$V(m) = \frac{1}{b^2}\left\{\frac{1}{Snw} + \frac{(m-\bar{x})^2}{Snw(x-\bar{x})^2}\right\}. \qquad (3.12)$$

Replacement of m by the corresponding estimate of another ED value (e.g. ED 90) gives an appropriate variance for that estimate. In a rapid analysis, dependent primarily on the provisional line drawn by eye, the second term in equation (3.12) is scarcely necessary; if more extensive calculations are undertaken for fitting the line (§ 4.4), all quantities needed will be available and the full formula might as well be used. More rapidly calculated but still cruder approximations have been suggested (Litchfield and Fertig, 1941), but these to-day have little interest.

Table 3.5   The weighting coefficient, $w = Z^2/PQ$

Y	0.0	0.1	0.2	0.3	0.4	0.5	0.6	0.7	0.8	0.9
1	0.001	0.001	0.001	0.002	0.002	0.003	0.005	0.006	0.008	0.011
2	0.015	0.019	0.025	0.031	0.040	0.050	0.062	0.076	0.092	0.110
3	0.131	0.154	0.180	0.208	0.238	0.269	0.302	0.336	0.370	0.405
4	0.439	0.471	0.503	0.532	0.558	0.581	0.601	0.616	0.627	0.634
5	0.637	0.634	0.627	0.616	0.601	0.581	0.558	0.532	0.503	0.471
6	0.439	0.405	0.370	0.336	0.302	0.269	0.238	0.208	0.180	0.154
7	0.131	0.110	0.092	0.076	0.062	0.050	0.040	0.031	0.025	0.019
8	0.015	0.011	0.008	0.006	0.005	0.003	0.002	0.002	0.001	0.001

Precision has been discussed above as if the line drawn by eye were in fact the same as a properly calculated regression. To the extent that this is untrue, variance will be underestimated because of neglect of errors in placement of the line (Cornfield and Mantel, 1950); this is one of the hazards of a graphical procedure, and no attempt to take account of it will be made here.

If the n subjects in a batch tested at a dose do not respond to the stimulus independently of one another, the binomial probability distribution will not apply. This may show itself by a large value of χ^2, calculated as in Table 3.4. Such heterogeneous variation of observed response rates about expectations is another cause of underestimation of variance; the topic is considered more fully in § 4.6.

Ex. 2. *Standard error of a median lethal dose.* In Table 3.6, the values of Y predicted by equation (3.9) for each dose in Example 1 are tabulated to the nearest 0.1. From Table 3.5, the w corresponding to each Y is entered and multiplied by n to give nw; a slide rule gives sufficient accuracy. The sum of this column is

$$Snw = 119.6,$$

whence the approximate standard error of m is (χ^2 was well within the limits of random variation):

$$s_m = 1/4.01\sqrt{(119.6)}$$

$$= 0.0228. \tag{3.13}$$

Table 3.6 Calculation of standard error of log LD 50 in eye estimation for rotenone–*Macrosiphoniella sanborni* experiment

Log con-centration (x)	No. of insects (n)	Y	w	nw	nwx
1.01	50	6.3	0.336	16.8	16.968
0.89	49	5.8	0.503	24.6	21.894
0.71	46	5.1	0.634	29.2	20.732
0.58	48	4.6	0.601	28.8	16.704
0.41	50	3.9	0.405	20.2	8.282
				119.6	84.580

$$\bar{x} = 84.580/119.6 = 0.707$$

$$Snwx^2 = 64.427$$

$$(Snwx)^2/Snw = (84.580)^2/119.6 = 59.814$$

$$Snw(x-\bar{x})^2 = 64.427 - 59.814 = 4.613$$

For equation (3.12), $\bar{x}$ must be found; a new column, nwx, is formed in Table 3.6 as the product of x and nw, from which

$$\bar{x} = Snwx/Snw = 0.707. \tag{3.14}$$

Multiplication of each nwx by x (the separate products need not be recorded) and summation again gives $Snwx^2$. A well-known result then yields the sum of squares of deviations of x:

$$Snw(x - \bar{x})^2 = Snwx^2 - (Snwx)^2/Snw$$

$$= 4.613. \tag{3.15}$$

The arithmetic is summarized at the foot of Table 3.6. Since $\bar{x}$ differs little from m, the second term in (3.12) is small:

$$V(m) = \frac{1}{4.01^2}\left\{\frac{1}{119.6} + \frac{(0.687 - 0.707)^2}{4.613}\right\}$$

$$= (0.008\,36 + 0.000\,09)/4.01^2$$

$$= 0.000\,525. \tag{3.16}$$

The square root of this,

$$s_m = 0.0229, \tag{3.17}$$

is the standard error of m, which differs only negligibly from equation (3.13).

Now m is measured on a logarithmic scale (to base 10). One cannot derive a true standard error for the LD 50 on the original scale, but a rough value can be obtained from the equation

$$\text{s.e.}\,(10^m) = 10^m \times \log_e 10 \times s_m, \tag{3.18}$$

so that $\text{s.e.}\,(\text{LD}\,50) = 4.86 \times 2.30 \times 0.0229$

$$= 0.26.$$

Hence the LD 50 may be stated as $4.86 \pm 0.26\,\text{mg/l}$. However, in so far as the standard error is used to give an idea of the range within which the truth may lie, calculations should be on the metametric scale. A hypothesis that the true LD 50 of rotenone under the conditions of this experiment is $4.2\,\text{mg/l}$ could be examined by comparing the logarithm of this, 0.623, with m. The difference, 0.064, is nearly three times the standard error, 0.023, so that the discrepancy is clearly significant.

If the standard error of estimation of the log LD 90 is required the estimate is very different from $\bar{x}$ and equation (3.12) must be preferred to any cruder alternative:

$$V(\log \mathrm{LD}\,90) = \frac{1}{b^2} \left\{ \frac{1}{119.6} + \frac{(1.003 - 0.707)^2}{4.613} \right\}$$

$$= 0.001\,701,$$

so that the $\log \mathrm{LD}\,90$ is estimated as 1.003 ± 0.041.

None of these variance and standard error calculations should be regarded as other than a rough guidance. More trustworthy methods are introduced in Chapter 4.

3.5 Fiducial probability

When a parameter such as the median effective dose has been estimated, the experimenter may wish to infer within what limits its true value can reasonably be expected to lie. A statement about the probability of the true value lying between certain limits cannot be made in terms of the ordinary concept of probability, by which probabilities can be assigned only to statements about the occurrence of observations or of statistics calculated from the observations. In order to overcome this difficulty, the concept of *fiducial probability* has been introduced.

There is said to be a fiducial probability, F, that the true value of a parameter lies between calculated upper and lower limits if the lower limit is the lowest value and the upper limit the highest value which would not be contradicted by a significance test at the $\frac{1}{2}(1 - F)$ probability level; these are termed *fiducial limits* to the value of the parameter. Their meaning and calculation will be made clear by considering fiducial limits to the $\mathrm{LD}\,50$ for the data of Example 1.

The above is a much simplified statement of a complicated idea. Strictly, the fiducial argument requires that so-called *sufficient* estimators be available. A logically different argument leads to the definition of *confidence limits*, which are numerically the same as fiducial limits in many problems. The reader anxious to learn more about the theory (and the controversies) underlying these ideas should refer to general statistical texts. Here such theory would be out of place. In this field of applied statistics, the fiducial terminology is customary; Example 3 is only an illustration of the type of arithmetical procedure to be developed more exactly later.

Ex. 3. *Fiducial limits to the LD 50 in the rotenone*–Macrosiphoniella sanborni *experiment*. In Example 2 the log LD 50 for rotenone applied to *M. sanborni* was estimated as

$$m = 0.687 \pm 0.023.$$

To the order of approximation of this Chapter, standard errors may be considered in relation to the normal distribution. From the last line of Table VIII, there is a 5 % probability that the deviation from the true log LD 50 shall be at least 1.96 times the standard error, or a $2\frac{1}{2}$ % probability that the estimate shall be at least $1.96s_m$ (= 0.045) less than the true value and a $2\frac{1}{2}$ % probability that the estimate shall be at least 0.045 greater than the true value. In other words, if the true log LD 50 were 0.732 estimates as low as or lower than 0.687 would occur in only $2\frac{1}{2}$ % of trials such as that under consideration, and if the true log LD 50 were 0.642 estimates as high as or higher than 0.687 would occur in only $2\frac{1}{2}$ % of trials. There is therefore a 95 % fiducial probability that the true value lies between 0.732 and 0.642, which limits correspond to concentrations of 5.40 and 4.39 mg/l. Had limits been calculated directly from the LD 50 and its standard error as $4.86 \pm 1.96 \times 0.26$ mg/l, or 5.37 and 4.35 mg/l, they would have been equidistant from the estimated LD 50, whereas the limits derived on the logarithmic scale are not symmetrically placed on the concentration scale. The difference is trivial here, but with relatively larger standard errors the logarithmic calculation is preferable.

3.6   History of the probit method

Widespread use of the probit transformation in the statistical analysis of biological data began about 1935, but the underlying principle has been known for many years. The first ideas originated with psychophysical investigators, who were confronted with the problem of estimating the magnitude of a stimulus from statements by test subjects that it seemed to them greater than or less than the various members of a standard series. The proportion of answers 'less than' steadily increases as the scale of standard stimuli is ascended, and shows a sigmoid relation with the measure of these stimuli.

Fechner (1860) discussed the frequency with which a subject correctly identifies the heavier of two weights, and its dependence upon the difference between the weights. He suggested converting relative frequencies to deviates of a normal distribution with mean zero and precision† unity, by means of a table of the normal integral.‡ If steps are taken to eliminate biases due to the order, in time or space, in which the weights are picked up, when the weight difference is negligibly small the proportion of right answers should be one-half, and the normal deviate therefore zero. Fechner further suggested that a linear relation would be found between the weight difference and the normal deviate. Hence, if the proportion of right answers were known for one weight difference, the factor of proportionality with the normal deviate could be estimated; estimates could then be made of the proportions corresponding to any other weight differences or vice-versa. This appears to be the earliest reference to the reduction of a sigmoid response curve to a straight line by means of a transformation of the responses based on the normal integral. Fechner deserves credit for an ingenious statistical invention far in advance of his time.

Müller (1879) recognized that the transformation from proportions to the standardized normal deviates introduced a differential weighting.§ He proposed to determine the parameters of the distribution of threshold values by fitting a straight line to the transformed data, weighting each point by its 'Müller weight'. This weight is proportional to z^2, where z is the ordinate corresponding to the observed proportion, and not, as in equation (3.10), to the expected value from the fitted line. Müller's method became known as the *Constant Process* or the *Method of Right and Wrong Cases*.

Urban (1909, 1910) collected experimental data, of the type considered by Fechner, on the difference threshold for lifted

---

† The *precision*, $h = 1/\sigma\sqrt{2}$, was at one time commonly used instead of σ as a parameter of the normal distribution.
‡ Known to psychophysicists as ' Fechner's Fundamental Table'.
§ There is a danger of verbal confusion: weights are used as stimuli and, in the statistical analysis, the responses are assigned ' weights' proportional to their invariances.

weights. In his experiments, a standard weight of 100 g was lifted before each of a series of seven weights ranging from 84 g to 108 g and the subject gave a judgement of lighter than, equal to, or heavier than, for each trial. Urban described several methods for estimating the threshold of just perceptible weight differences, and for assessing the relation between the weight difference and the proportion of right answers. In presenting his $\Phi(\gamma)$ *Process*, he pointed out that the Müller weights failed to take account of the variability in the original proportions due to the binomial distribution of right and wrong answers. To allow for this, he introduced a factor $1/pq$, and weighted his normal deviates in proportion to z^2/pq.

Thomson (1914, 1919 a) drew attention to certain defects in Urban's treatment of the problem, especially in estimation of standard errors and in assessment of the fit of the line to the data. In a further remarkable paper (1919 b), Thomson presented clearly and concisely the essential features of maximum likelihood estimation as applied to this problem. The general statistical theory of maximum likelihood was not to be formulated until several years later, and computing techniques of the time were inadequate, but Thomson saw the possibility of estimation based upon maximizing the probability of the whole set of observations. Some years later (Brown and Thomson, 1940), he produced a scheme of computation very similar to the classical probit method, except that the weights were functions of p instead of P and the standard error calculations were more cumbersome. He made no provision for zero or 100 % values of p.

Independently of the psychophysicists, Greenwood and Yule (1914) introduced a transformation essentially equivalent to the probit; they sought to relate the percentage of inoculated subjects becoming infected in an epidemic with the severity of the epidemic as measured by the normal equivalent dose for uninoculated subjects. This ingenious idea was not allied with any formal estimation theory, and apparently has no connexion within any earlier or later work.

Also independently, Hazen (1914) and Whipple (1916) suggested graph paper with ordinates graduated according to a normal probability distribution, so that a proportion is plotted as its corresponding normal deviate. A normal sigmoid curve plotted on this paper is automatically transformed to a straight line. A modified form of paper has a logarithmic scale of abscissae, so that a logarithmic transformation of dose is automatically made at the same time. O'Kane, Westgate, Glover and Lowry (1930) used probability paper for plotting the results of insecticide tests.

In 1923 Shackell suggested that the normal integral might help in interpreting the results of toxicity tests. Three years later Wright (1926), also without knowledge of earlier work, proposed an inverse function of the normal probability integral as a means of simplifying the statistical treatment of certain frequency data. His paper, however, escaped the attention of biologists, who had to wait a further seven years for yet another rediscovery of the method.

In 1933 Gaddum published an important memorandum on the analysis of quantal assay data in biological investigations. He proposed to transform each percentage to its *normal equivalent deviation* (N.E.D.), defined by equation (3.1), a transformation essentially the same as that used by Fechner. Gaddum found the N.E.D. of the percentage kill of various animals to give a straight line when plotted against the log dose of the drug applied. He described the regression technique for fitting the line, in a form similar to that given earlier by Urban and Thomson, but his treatment of the standard errors of the parameters and associated quantities was much simpler.

Bliss (1934a) suggested the division of the interval between 0.01% and 99.99% into units of normal deviation which he called *probits*, the whole interval ranging from 0 to 10 probits and 50% being 5 probits. When he later saw Gaddum's publication, he modified his definition of the probit and redefined it as in equation (3.4), so that it became the N.E.D. increased by 5 (Bliss, 1934b). In two comprehensive papers (1935a, b), he presented the theory of probit analysis, with tables of probits and weighting coefficients, and discussed the estimation of relative potency. Fisher (1935) showed how maximum likelihood theory enables account to be taken of zero and 100% responses, and introduced the *working probit* (§§ 4.2, 4.3). Although Bliss had advocated computing weighting coefficients from values of P determined by a provisional line, the discussion on Irwin's paper (1937) indicated that some still argued for using the observed proportion, p. Bliss (1938) finally expounded the full maximum likelihood technique.

Two other developments deserve notice. In 1894, Henry transformed a cumulative normal curve to a straight line by use of
normal equivalent deviates, and suggested graph paper with
probability ruling. Haag (1926) gave examples but, like Henry,
discussed only data for which direct measurements were available and not the essentially quantal type. The work of Kapteyn
(1903) and Kapteyn and Van Uven (1916) owed something to
Fechner; their reaction curves are aids to the normalization of
distributions, but they also were not concerned with essentially
quantal data.

Meanwhile, psychologists apparently remained unaware that
biologists had adopted and refined their method. Ferguson (1942)
used the constant process for analysing data on the selection of
items for mental tests. Lawley (1943, 1944), in considering the
theory of this problem, came near to an independent derivation
of the maximum likelihood solution. Finney (1944 b) endeavoured
to unite the biological and psychophysical methods by illustrating the use of the probit method on Ferguson's data. In
reviewing the first edition of this book, G. H. Thomson (1947)
showed how the two streams eventually came together.

3.7   Various approximations

Like all techniques for statistical inference, the probit method
involves assumptions. In its full form, it preserves internal consistency; independently of it or as alternatives to it, however,
many less consistent methods of estimating the ED 50 have been
proposed, some becoming very popular. Some implicitly assume
linearity of regression of percentage response on dose metameter.
Most will commonly give much the same answer as probit analysis,
but can be seriously misleading if the range of doses tested is
markedly asymmetric about the log ED 50.

(A) *The method of Extreme Lethal Dosages* is intended for
the limited class of experiments in which subjects are tested
singly, one at each of a series of doses;† the interval, d,
between successive log doses must be constant, and enough
levels must be tested to cover the range from those prac-

---

† Application of the probit method to experiments in which only one
subject is tested at each dose is discussed in § 9.3.

tically certain to be ineffective to those at which response is practically certain. The log ED50 is estimated as the mean of two log doses, the highest for a non-response and the lowest for a response.

(b) *The Dragstedt–Behrens method*, proposed by Dragstedt and Lang (1928) and independently by Behrens (1929), originated in an attempt to increase the effective number of subjects. For each dose in the experiment is caclulated:

S_- = total number of subjects responding at doses less or equal to this,

S_+ = total number of subjects not responding at doses greater or equal to this.

Then $S_-/(S_- + S_+)$ is regarded as an improved estimate of the true response rate at the dose, and the ED50 estimated by graphical or numerical smoothing of this sequence of values.

The argument that, for example, 'This method enables one to obtain results in terms of a much larger number of animals than originally employed, and, therefore, results are indeed more accurate' (Barr and Nelson, 1949) is wholly fallacious. Winder (1947) saw this clearly. Nevertheless, the method may behave reasonably well if the data are equally spaced in x, have equal n at each x, and are moderately symmetrical.

(c) *The Reed–Muench method*, one of the most popular of all, is a slight variant of the Dragstedt–Behrens. Reed and Muench (1938) proposed interpolating in x to find a value at which S_- and S_+ could be expected to be equal. The intersection of graphs of S_- and S_+ against x gives the answer.

(d) *The Spearman–Kärber method* is another example of parallel development in psychology and biology. Spearman (1908) suggested a simple way of analysing data from the constant process; Kärber (1931) made essentially the same proposal for experiments in pharmacology. The method requires that doses extend over the whole range from zero to 100 % kill, but need not have equal spacing of doses or equal numbers of subjects per dose. The increase in response rate, $(p_{i+1} - p_i)$, between two successive doses x_i and x_{i+1} is supposed concentrated at the middle of the interval in x. If k doses are tested, and if $p_1 = 0$, $p_k = 1$ (the zero and 100 % responses), the mean value of x:

$$S(p_{i+1} - p_i) \cdot \tfrac{1}{2}(x_{i+1} + x_i) \qquad (3.19)$$

estimates the mean of the tolerance distribution. If the interval between successive x_i is a constant, d, this estimate is particularly easily calculated as

$$x_k + \tfrac{1}{2}d - dSp_i. \qquad (3.20)$$

(E) *The Moving Average method* (W. R. Thompson, 1947) is related to the Spearman–Kärber though different in origin. For a moving average of span 2, $(p_i + p_{i+1})/2$ is plotted against $(x_i + x_{i+1})/2$ for each i, and simple interpolation used to find the value on the x scale corresponding to 50 % response; similarly for span 3, $(p_i + p_{i+1} + p_{i+2})/3$ is plotted against $(x_i + x_{i+1} + x_{i+2})/3$. Thompson claimed that his method was 'free from assumption as to the precise type of fundamental curve involved', but symmetry of the tolerance distribution is important. Like the Dragstedt–Behrens method, moving averages might at first seem also to permit estimation of other percentage points such as ED 75, but even under the best of conditions serious biases can enter because the constructed dose-response relation is logically quite distinct from the truth except at 50 %.

The precision of these methods of estimating the ED 50 has been discussed elsewhere (Finney, 1950, 1964), at least for the class of experiments that has equal intervals between log doses, equal numbers of subjects per dose, and sufficient doses to ensure zero and 100 % responses at the extremes. This investigation relates to the normal tolerance distribution, but a parallel study of the logistic distribution has also been undertaken (Finney, 1953). The standard error of the estimated ED 50 is a complicated function, but provided that the spacing between doses is neither very wide nor very narrow it is roughly proportional to $\sqrt{(\sigma d/n)}$, in the present notation. The factors of proportionality are approximately:

Extreme doses	0.75,
Dragstedt–Behrens	0.81,
Reed–Muench	0.79,
Spearman–Kärber	0.75.†

Corresponding moving average factors are not easily presented concisely, because they are rather sensitive to the magnitude of d/σ, but they are always greater than for Spearman–Kärber. The probit method for the same experiment has a factor 0.74.

These standard errors cannot be used without knowledge of σ. For the first three methods, no estimate of standard error can be formed satisfactorily from a single experiment. For Spearman–Kärber, however, the standard error of the estimate (3.20) may be taken as

$$d \sqrt{\left\{ S \left(\frac{p_i q_i}{n_i - 1} \right) \right\}} \qquad (3.21)$$

(cf. Irwin and Cheeseman, 1939a, b). Similar but more complicated formulae can be devised for moving averages.

The only merit of the extreme dose, Dragstedt–Behrens, and Reed–Muench methods is their numerical simplicity. Spearman–Kärber calculations are at least as easy, the conditions for reasonable validity are no more stringent, the standard error is no greater, and the standard error can be calculated from the internal evidence of an experiment. There appears to be no good reason for continued use of any of the first three methods, even though Reed–Muench has such wide popularity.

The indication of only a trivial advantage for the probit method over Spearman-Kärber is entirely misleading; as Gaddum pointed out, this conclusion applies only when doses are so chosen as to permit the use of one of the approximations. If previous experience or preliminary trials give any clue to the value of the ED 50, to distribute subjects evenly over a wide range of doses, in order to ensure having zero or 100 % response at the extremes, is wasteful. A limitation to doses nearer the ED 50 leads to observations of much greater weight and a correspondingly more precise estimate of the ED 50 (§ 8.1). A theoretical objection

† More exactly $1/\pi^{\frac{1}{2}}$ (van der Waerden, 1940a, b).

to the approximate methods is that, unless the doses are symmetrically situated with respect to the true log LD 50, each gives a biased estimate. Irwin (1937; Finney, 1950) found the bias to be negligible, even when the interval between successive doses was as great as 2σ.

The Spearman–Kärber and moving average methods may sometimes help provisional estimation of the ED 50, especially in the analysis of pilot or exploratory trials. The argument that speed of calculation makes them preferable to probit methods no longer has much force. The graphical method (§§ 3.3, 3.4) is as quick, is not limited to special patterns of doses or numbers of subjects, and readily permits assessment of goodness of fit of the probability formula and precision of estimates. Moreover, the full probit method (§ 4.2) can be extended to a wide range of more complicated problems. Modern computing facilities can make the time spent on maximum likelihood or similar calculations very short (§ 5.1). Although some understanding of the iterative arithmetic described in subsequent chapters is valuable, no one to-day need contemplate laborious repetition of this type of arithmetic for routine analyses. The moving average method, however, may retain some merits for data that deviate systematically from equation (3.2) without being known to conform to a specific alternative.

Ex. 4. *Estimation of the median lethal dose.* In the experiment summarized in Table 3.1, the logarithmic intervals between successive doses did not differ greatly from the mean of 0.15, and the number of insects tested at each dose was about 50. If an assumption were made that the next higher dosage would have killed all the insects and the next lower would have killed none, four of the methods described could be used for estimating the log LD 50. (The assumption is purely for the purpose of providing an example, and no justification is implied). The percentages dead and alive should be used rather than the actual numbers, in order to overcome the complication of unequal values of n. The estimates, calculated from the final column of Table 3.1, are

Dragstedt–Behrens	0.682,
Reed–Muench	0.685,
Spearman–Kärber	0.678,
Moving average (span 2)	0.703.

Agreement with m in equation (3.7) is fortuitous, because the third of the five doses tested was very near the LD 50 and the other doses are almost symmetrically placed on either side.

3.8   Other transformations

Reviewers of earlier editions justly criticized the neglect of alternatives to equation (2.6) as a representation of the tolerance distribution. From one point of view, the normal distribution is no less (and no more) reasonable an assumption here than in the many other standard techniques of applied statistics based upon it. From another, even the concept of a tolerance distribution can be challenged.

In fact, the probit method rests directly upon equation (3.2) rather than (2.6): what matters is the dependence of P on dose and the unknown parameters, and the tolerance distribution is merely a substructure leading to this. Berkson (1944, 1951) has argued forcefully that in many circumstances the law relating probability of response to dose is more fundamental. For example, suppose that the stimulus is a source of radiation and response is occurrence of a mutation in a fly hatched from an irradiated egg. Here the notion of a tolerance is scarcely relevant, and mutation is determined by the chance of whether or not the appropriate locus is 'hit'. As Berkson has said, such a situation might be better described by the so-called logistic function; this involves replacement of equation (3.2) by

$$P = 1/\{1 + e^{-(\alpha + \beta x)}\}. \qquad (3.22)$$

Wilson and Worcester (1943a, b, c, d) and Worcester and Wilson (1943) discussed the theory of estimating the parameters in this equation from quantal data. Berkson (1944, 1946) showed that a *logit* transformation could be defined as analogous to the probit; this involves replacing equation (3.1) by

$$P = \frac{e^Y}{1 + e^Y}. \qquad (3.23)$$

(If $2Y$ is written in place of Y in this equation, the logit, Y, is usually not very different in numerical magnitude from the N.E.D. of P). All the theory and practice of probit analysis can be rewritten in terms of this alternative transformation. Indeed, one can easily find a frequency function to which (3.22) corresponds exactly as does (3.2) to (2.6), and this could be regarded as a tolerance distribution. There is no way of determining from this kind of data whether or not a tolerance distribution has a real existence; for the types of experiment described in § 3.1, it seems a reasonable basis of discussion, whereas for the mutation experiment it does not. The important point is that estimation procedures are not affected by whether that distribution or the formula for P is regarded as fundamental.

Although the probit and logistic models are the most attractive to the biologist, because each arises in many contexts, other equivalent deviate transformations have been suggested. For example, Urban (1910) made some use of

$$P = \frac{1}{2} + \frac{1}{\pi} \tan^{-1} Y. \tag{3.24}$$

Wilson and Worcester (1943c) suggested

$$P = \frac{1}{2}\left\{1 + \frac{Y}{\sqrt{(1+Y^2)}}\right\}. \tag{3.25}$$

Knudsen and Curtis (1947) proposed

$$P = \begin{cases} 0 & \text{for} \quad Y \leqslant 0, \\ \sin^2 Y & \text{for} \quad 0 < Y < \pi/2, \\ 1 & \text{for} \quad Y \geqslant \pi/2. \end{cases} \tag{3.26}$$

The sole merit of this is a certain amount of numerical simplicity in the classical form of calculation because the weighting coefficient is independent of P. Martin (1951) even proposed the equivalent of

$$P = \begin{cases} 0 & \text{for} \quad Y \leqslant -2.5, \\ 0.5 + 0.4Y + 0.08Y^2 & \text{for} \quad -2.5 < Y \leqslant 0, \\ 0.5 + 0.4Y - 0.08Y^2 & \text{for} \quad 0 < Y < 2.5, \\ 1 & \text{for} \quad Y \geqslant 2.5 \end{cases} \tag{3.27}$$

as a rough representation suitable for some approximations. This corresponds to approximating the tolerance distribution by an equilateral triangle.

For each of these, the probability relation corresponding to (3.2) is supposed to be given by inserting

$$Y = \alpha + \beta x$$

as in equation (3.6). For each, methods of statistical analysis can be developed exactly similar to any of the probit methods. How is a choice to be made? Almost certainly, an attempt to choose the transformation that best agrees with the results of a particular experiment is doomed to failure because they are practically indistinguishable over a wide range (Armitage and Allen, 1950; Berkson, 1950; Biggers, 1952; Chambers and Cox, 1967). If arithmetical simplicity is not a criterion, only the probit and logit seem worth serious consideration as no theoretical basis has been suggested for others. These two are very similar indeed in all respects except for very small or very large P, and extremely large experiments would be needed to show one as a better fit than the other. No one should believe that either formula for P represents perfect truth, and therefore perhaps nothing other than personal inclination can decide which is to be used. Throughout this book, probit methods are presented, but the essential formulae (and a fuller discussion of comparisons) for the logit and others have been given elsewhere (Finney, 1964).

Somewhat excessive claims for probits and for logits have been made by their respective advocates. This book concentrates on probits, but is not to be regarded as uncompromising argument for their universal applicability. If the manner in which data are generated makes equation (3.22) seem more appropriate than equation (3.2), corresponding logit formulae and methods can be used easily.

Cornfield and Mantel (1950) showed that Spearman–Kärber estimation is essentially the same as maximum likelihood under the logistic hypothesis. Of course, this still requires that the conditions for using Spearman–Kärber are fulfilled, and is therefore no argument for preferring equation (3.22) to equation (3.2): usually an experiment designed well for efficiency will not allow use of the Spearman–Kärber formulae.

4

Maximum likelihood estimation

4.1 Principles of estimation

Although graphical and other methods described in Chapter 3 are often adequate for estimation of the parameters from a simple set of quantal response data, these methods do not generalize easily to more complicated experimental situations. Moreover, a more objective analysis is needed for serious scientific work.

The problem of using a particular set of experimental results to give numerical values for the unknown parameters in the best possible manner is an instance of the general problem of statistical estimation. The first need is to decide what is meant by 'best possible', what criterion is to be satisfied by estimates in order that they may be recognized as best. The simple answer: 'nearest to the true values' does not help, because the true values are unknown. Perhaps the most widely adopted principle of estimation is that of maximizing the likelihood of the observed results, and this principle will be adopted here.

The likelihood is defined to be proportional to the joint probability of all the observations, that is to say proportional to the product of expressions like that in equation (2.7) for every dose in the experiment. The proportionality factor, an arbitrary constant with no practical importance, does not affect the argument. For any set of observations, the likelihood is dependent only upon the values of the parameters: to estimate the parameters as those values that maximize the likelihood has an intuitive appeal. What optimal properties does it possess?

This is not the place for a general account of the theory of statistical inference, and a few brief comments must suffice. Maximum likelihood estimators are *consistent*, in the sense that (for quantal response problems) they will approach the true values if the number of subjects per dose increases without limit. In general, they are not *unbiased*: repeated experiments with the same set of parameters will not produce estimates exactly correct on average. However, strictly unbiased methods of estimation are seldom available, and are less important than might at first appear. If α, β in equation (3.6) were estimated without bias, the estimator of the log ED 50 would still be biased. Except for very small numbers of observations, the bias is usually rather small, but more study of the question for this class of problem is desirable.

Maximum likelihood estimators are fully efficient in large samples. If the number of subjects per dose is large, general theory shows that no other method can give estimators of greater precision (smaller variance); this is true whatever the formulation of the tolerance distribution or of $P(x)$ in terms of the parameters (Fisher, 1922; Cramér, 1946; Kendall and Stuart, 1967). What constitutes 'large' enough is unknown. In some simple problems, the optimal property of maximum likelihood persists in very small samples; in the problem of relating a quantal response to dose, little is known about the distribution of the estimates in small samples. Probably the number of observations often suffices for the large-sample efficiency of maximum likelihood estimation to be nearly achieved, but at present there is no certainty of this. Just as little is known, however, about the efficiency of alternative methods of estimation in small samples, most of these are known to be not fully efficient in large samples, and many have other weaknesses (§§ 3.7, 5.9).

Berkson (1946) suggested estimation of the parameters in the tolerance distribution by minimizing χ^2 instead of maximizing the likelihood, and later (1949) developed this method more rigorously. It is known (Cramér, 1946; Kendall and Stuart, 1967) to be as efficient as maximum likelihood in large samples; if the correct algebraic formulation of the tolerance distribution is being used, estimates from the two methods and their variances will tend to equality as the size of sample is increased, though in general they will differ for small samples. Minimum χ^2 might therefore seem to have the same status as maximum likelihood for the problems of this book. For samples of moderate size, however, it appears to have at least one disadvantage not shared by maximum likelihood, namely the instability of χ^2 when some class numbers are small; the extreme instance is that of individual mortality records (§ 9.3), for which estimation by minimum χ^2 would seem an unpromising procedure. Rao (1961, 1962) has proved that maximum likelihood possesses advantages in precision over all alternatives. Nevertheless, there is no *a priori* reason why minimum χ^2 should not be superior to maximum likelihood in small samples, or why some third method should not be superior to either (Tukey, 1949), and investigation is clearly needed. For the normal log tolerance distribution, minimum χ^2 has no theoretical advantages, and therefore no tables to facilitate its use are provided in this book. For the logistic type of distribution (equation (3.22)), as Berkson has shown, minimum χ^2 fitting of the logit regression is computationally rather simpler than maximum likelihood, at least if his ingenious approximation be adopted.

A number of more practical objections to and comments on maximum likelihood are discussed in § 5.9. Before this, the computation of estimates from data needs consideration.

4.2   The iterative method

Many who use the probit transformation in the analysis of numerical data will be content to accept computational techniques without explanation of mathematical and statistical theory. This section outlines the derivation of the classical method of iterative calculation. A statistician will readily see how the equations that follow lead to the computations described in § 4.3, or will modify them to obtain the analogous techniques for the alternative distributions of tolerance discussed in § 3.8. The essential features of the maximum likelihood estimation of the parameters in the quantal response problem were contained in a note by Fisher (1935), though the first adequate formal presentation appears to have been that of Garwood (1941); Finney (1947a, 1949d, 1964) has given a more general result. As emphasized in § 4.1, the known optimal properties of maximum likelihood relate to large samples; although some alternative might be superior in samples of finite size, none has yet been demonstrated to have a general advantage.

Suppose that the probability of response to a particular dose z_0 is given by equation (2.2), implicitly in terms of certain unknown parameters. Whether or not the notion of a tolerance distribution is meaningful does not matter, but P must be differentiable with respect to each parameter and the tolerance terminology is convenient. If a batch of n subjects receives this dose, each reacting independently, the probability that r respond is the binomial function shown as equation (2.7). An experiment in which k doses were tested can be summarized by the quantities z_i, n_i, r_i, for $i = 1, 2, ..., k$. The joint probability of these results is proportional to e^L, where

$$L = Sr \log P + S(n - r) \log Q \qquad (4.1)$$

and S denotes summation over all doses; P, Q are to be evaluated for each z_i and associated with the corresponding n_i, r_i. The *likelihood* of the observations is defined (Fisher, 1922) as a quantity proportional to e^L but having a maximum value unity with respect to variations in the parameters.

The proportionality factor merely adds a constant to L, and therefore affects neither the determination nor the properties of the maximizing values of the parameters. The likelihood is a maximum when L is a maximum; hence, if θ is any parameter of the distribution of individual tolerances, the maximum likelihood estimate of θ must satisfy

$$
\begin{aligned}
0 = \frac{\partial L}{\partial \theta} &= S \frac{r}{P} \frac{\partial P}{\partial \theta} + S \frac{n-r}{Q} \frac{\partial Q}{\partial \theta} \\
&= S \frac{n(p - P)}{PQ} \frac{\partial P}{\partial \theta},
\end{aligned} \qquad (4.2)
$$

where p, defined by equation (2.8), is an empirical estimate of the value of P at a particular dose. If the tolerance distribution involves more than one unknown parameter, a set of equations of the form of (4.2) must be satisfied simultaneously.

Explicit solution of such equations is seldom possible, but iterative methods can give successive approximations converging to the solutions. Illustration of this for two parameters, θ and ϕ, will suffice. Suppose that θ_1, ϕ_1 are any approximations to the solutions of the equations (4.2), obtained perhaps by rough graphical or arithmetical estimation. By the Taylor–Maclaurin expansion, to the first order of small quantities, second approximations will be $\theta_1 + \delta\theta$, $\phi_1 + \delta\phi$, where $\delta\theta$, $\delta\phi$ are obtained from

$$
\left.
\begin{aligned}
\frac{\partial L}{\partial \theta_1} + \delta\theta \frac{\partial^2 L}{\partial \theta_1^2} + \delta\phi \frac{\partial^2 L}{\partial \theta_1 \partial \phi_1} &= 0, \\
\frac{\partial L}{\partial \phi_1} + \delta\theta \frac{\partial^2 L}{\partial \theta_1 \partial \phi_1} + \delta\phi \frac{\partial^2 L}{\partial \phi_1^2} &= 0;
\end{aligned}
\right\} \qquad (4.3)
$$

the addition of the suffix to θ, ϕ indicates that the first approximations are to be substituted after differentiation. The second-order differential coefficients

$$\frac{\partial^2 L}{\partial \phi_1^2}, \quad \frac{\partial^2 L}{\partial \theta_1 \partial \phi_1}, \quad \text{and} \quad \frac{\partial^2 L}{\partial \phi_1^2}$$

may be simplified by putting $p = P$ after differentiation, to give expected values instead of empirical. Evidently a second differentiation of an expression such as occurs in equation (4.2) followed by $p = P$ will yield zero except for those terms in which the factor $(p - P)$ has been differentiated. Equations (4.3) thus become

$$\left. \begin{array}{l} \delta\theta S \dfrac{n}{P_1 Q_1} \left(\dfrac{\partial P_1}{\partial \theta_1}\right)^2 + \delta\phi S \dfrac{n}{P_1 Q_1} \left(\dfrac{\partial P_1}{\partial \theta_1} \dfrac{\partial P_1}{\partial \phi_1}\right) = S \dfrac{n(p - P_1)}{P_1 Q_1} \dfrac{\partial P_1}{\partial \theta_1}, \\[3mm] \delta\theta S \dfrac{n}{P_1 Q_1} \left(\dfrac{\partial P_1}{\partial \theta_1} \dfrac{\partial P_1}{\partial \phi_1}\right) + \delta\phi S \dfrac{n}{P_1 Q_1} \left(\dfrac{\partial P_1}{\partial \phi_1}\right)^2 = S \dfrac{n(p - P_1)}{P_1 Q_1} \dfrac{\partial P}{\partial \phi_1}, \end{array} \right\} \quad (4.4)$$

two linear equations that are easily solved for $\delta\theta$, $\delta\phi$.

The process may now be repeated with

$$\left. \begin{array}{l} \theta_2 = \theta_1 + \delta\theta, \\ \phi_2 = \phi_1 + \delta\phi \end{array} \right\} \quad (4.5)$$

in place of θ_1, ϕ_1, and further cycles computed until the latest set of adjustments is negligible. General theory of statistical estimation states that, if the first approximations are of non-zero efficiency, the first cycle of computation will yield fully efficient estimates. This is strictly true only for large samples, and also is likely to be disturbed by any subjective element in determining θ_1, ϕ_1 by the graphical process usual in probit analysis. The wisest course, therefore, is to iterate the calculations until the solutions of (4.2) are approached; experience in the choice of first approximations will ensure that usually two or three cycles give numerical accuracy sufficient for practical purposes.

If only one parameter, θ, needs estimation, the variance of $\hat{\theta}$, the solution of equation (4.2), is asymptotically

$$V(\hat{\theta}) = -1 \left/ \frac{\partial^2 L}{\partial \theta^2} \right., \quad (4.6)$$

in which $\hat{\theta}$ is substituted for θ after differentiation. When more than one parameter must be estimated, the variances and co-variances of the estimates are the elements of the inverse matrix of second differential coefficients; thus, for two parameters

$$
V = \begin{pmatrix} -\dfrac{\partial^2 L}{\partial \theta^2} & -\dfrac{\partial^2 L}{\partial \theta \, \partial \phi} \\[2ex] -\dfrac{\partial^2 L}{\partial \theta \, \partial \phi} & -\dfrac{\partial^2 L}{\partial \phi^2} \end{pmatrix}^{-1}. \tag{4.7}
$$

This matrix, but with the previous set of approximations instead of $\hat{\theta}$, $\hat{\phi}$, is inverted as part of the last cycle of iteration since the differential coefficients appear as the coefficients of $\delta\theta$, $\delta\phi$ in equations (4.4). In practice, the variances and covariances may be taken as approximately equal to the elements of this inverse, so as to avoid the necessity of recalculation.

The equations above apply whatever the form of P, and easily extend to a greater number of parameters. Toxicity tests and other quantal response problems require their specialization to forms suitable for computation. The most important form is that for the estimation of the parameters of the tolerance distribution given by equation (2.6), or

$$
P = \frac{1}{\sigma\sqrt{(2\pi)}} \int_{-\infty}^{x_0} \exp\left\{ -\frac{(x-\mu)^2}{2\sigma^2} \right\} dx, \tag{4.8}
$$

where x measures dose on a logarithmic or other suitable meta-metric scale (x_0 being a particular value of x). As shown by equation (3.6), this is equivalent to a linear dependence of Y, the N.E.D. of P, on x:

$$
Y = \alpha + \beta x; \tag{4.9}
$$

the new parameters α, β, are related to μ, σ by

$$
\left. \begin{array}{l} \mu = -\alpha/\beta, \\ \sigma = 1/\beta, \end{array} \right\} \tag{4.10}
$$

and Y is as defined by equation (3.1). If Y is a probit, (3.4) replaces (3.1), and $(\alpha - 5)$ must replace α in (4.10).

Now from (3.1)

$$
\frac{\partial P}{\partial Y} = \frac{1}{\sqrt{(2\pi)}} \exp\left\{ -\tfrac{1}{2}Y^2 \right\} = Z, \tag{4.11}
$$

where Z is the ordinate to the normal curve at the point with abscissa Y. Therefore

$$\frac{\partial P}{\partial \alpha} = Z, \quad \frac{\partial P}{\partial \beta} = Zx, \qquad (4.12)$$

and equations (4.4) may be constructed in terms of the parameters α, β.† If

$$Y_1 = a_1 + b_1 x \qquad (4.13)$$

is a first approximation to the maximum likelihood estimate of equation (4.9), equations (4.4) become

$$
\left.
\begin{aligned}
\delta a Snw + \delta b Snwx &= Snw \left(\frac{p-P}{Z} \right), \\
\delta a Snwx + \delta b Snwx^2 &= Snwx \left(\frac{p-P}{Z} \right),
\end{aligned}
\right\} \qquad (4.14)
$$

where the weighting coefficient, $w = Z^2/PQ$, is as defined in equation (3.10). The equations for the adjustments δa, δb are in form the same as for calculation of a linear regression of $(p-P)/Z$ upon x with weight nw for each value of $(p-P)/Z$. Some methods of analysis have been based upon this (§ 5.9), but the classical procedure is to introduce the *working deviate* or *working probit* at each dose:

$$y = Y_1 + \frac{p-P}{Z}. \qquad (4.15)$$

The suffix on Y is a reminder that it and all quantities derived from it (P, Z, w) use the first approximations from (4.13). If $SnwY_1$ is added to each side of the first, and $SnwxY_1$ to each side of the second equation in (4.14), they become

$$(a_1 + \delta a)\, Snw + (b_1 + \delta b)\, Snwx = Snwy,$$

$$(a_1 + \delta a)\, Snwx + (b_1 + \delta b)\, Snwx^2 = Snwxy.$$

Writing a_2, b_2 for adjusted estimates, as in (4.5), and defining means

$$
\left.
\begin{aligned}
\bar{x} &= Snwx/Snw, \\
\bar{y} &= Snwy/Snw,
\end{aligned}
\right\} \qquad (4.16)
$$

† Why should not the original parameters μ, σ be used as a basis for equations (4.4)? As shown below, equation (4.9) leads to a close analogy with standard statistical theory of weighted linear regression, and thus encourages a belief that the distribution of estimates will be more nearly normal for α, β than for μ, σ. So far as is known, this question has never been thoroughly investigated.

standard algebra of weighted regression gives

$$b_2 = Snw(x-\bar{x})(y-\bar{y})/Snw(x-\bar{x})^2, \qquad (4.17)$$

and

$$a_2 = \bar{y} - b_2\bar{x}. \qquad (4.18)$$

In other words, the revised relation

$$Y_2 = a_2 + b_2 x$$

is calculated as the weighted linear regression of y on x, the weights being nw.

The iterative routine is now clear, for the whole cycle of arithmetic can be repeated with Y_2 in place of Y_1. Usually successive iterations converge rapidly, even if Y_1 is a poor approximation (for example, $a_1 = 0$ and $b_1 = 0$), but the process is much speeded by starting from moderately good graphical estimates. The limit must be the maximum likelihood estimates. If symbols now refer to the limiting values, equation (4.7) reduces to the standard regression formulae

$$V(\bar{y}) = 1/Snw, \qquad (4.19)$$

$$V(b) = 1/Snw(x-\bar{x})^2, \qquad (4.20)$$

with zero covariance. Strictly, these variances are applicable only for large values of each n, but general experience of compounding results from binomial distributions suggests that both normality and applicability of asymptotic variance formulae will be good even for small numbers.

All the statistical analyses of Chapters 4, 6 can be computed from these formulae and the obvious extensions based on sets of parallel regression lines. The analogy with regression works less simply for more complicated problems such as are discussed in Chapter 7.

4.3 Working probits

The first step in the standard calculations for the maximum likelihood estimates is to obtain a provisional equation to represent the relation between probit and dose. The easiest way of doing this may be to draw a line on a diagram exactly as described in § 3.3, except that a very rough guess will suffice. From this line should be read the expected probit corresponding to each dose (cf. Table 3.3), one decimal place usually being enough.

The expected probits are used in calculations of weights (§§ 3.4, 4.4) and also of the working probits introduced in § 4.2. These quantities are not corrections of the probits, as some writers have erroneously suggested; they arise solely as a device to aid the organization of the otherwise complicated solution of the maximum likelihood equations. In all graphical presentation of results, empirical probits should be shown, not working probits.

As usual, Y is an expected probit from the provisional line, P and Z are related to Y by equations (3.1) and (3.11), p is the empirical proportion of responses, equation (2.8), and Q, q are the complements of P, p. The working probit at this dose is then

$$y = Y + \frac{p-P}{Z} = Y - \frac{q-Q}{Z}, \qquad (4.21)$$

either formula being used according to convenience. Note that, although zero and 100 % responses do not have finite empirical probits, the working probits take finite values; if $p = 0$, y becomes the *minimum working probit*

$$y_0 = Y - \frac{P}{Z}, \qquad (4.22)$$

and, if $p = 1$, y becomes the *maximum working probit*

$$y_1 = Y + \frac{Q}{Z}. \qquad (4.23)$$

Table III shows y_0, y_1, and the *range*:

$$y_1 - y_0 = 1/Z \qquad (4.24)$$

at intervals of 0.1 in Y over the interval 1.1 to 8.9. Finney and Stevens (1948) tabulated these functions at intervals of 0.01 in Y. From Table III, any desired working probit can be rapidly calculated; alternatively, Table IV enables most of those commonly wanted to be read directly. The fact that y is usually closer to the empirical probit than is Y provides a useful check against gross errors, but in extreme instances this is not true.

Ex. 5. *Calculation of working probits.* The empirical probit corresponding to a 72.3 % response is 5.592. Suppose that a provisional line gives $Y = 6.2$. From Table III,

$$y_0 = 1.6429, \quad 1/Z = 5.1497.$$

Hence $y = y_0 + p/Z$

$$= 1.6429 + 0.723 \times 5.1497 = 5.366.$$

Alternatively

$$y = y_1 - q/Z$$

$$= 6.7926 - 0.277 \times 5.1497 = 5.366.$$

In practice, the better formula is usually that with the smaller multiple of $1/Z$ to be calculated. The same result may be obtained from Table IV. The column for $Y = 6.2$ shows working probits of 5.351 and 5.402 for 72 % and 73 % response respectively: hence, by an interpolation which may be carried out mentally,

$$y = 5.351 + \tfrac{3}{10}(5.402 - 5.351)$$

$$= 5.366.$$

Commonly y is required only to 2 places of decimals, or p is reliable only to the nearest 1 %, so that y can be read directly from Table IV with little or no interpolation. Occasionally the expected probits may be needed to 2 decimal places. Interpolation is then necessary for y_0 or y_1 and the range, after which y may be calculated as above. If in this example Y were 6.24, the maximum working probit and the range would be given as the sum of 6/10 of the values for $Y = 6.2$ and 4/10 of the values for $Y = 6.3$. Hence y may be written, in a compact form suitable for machine calculation, as

$$y = 0.6 \times 6.7926 + 0.4 \times 6.8649$$

$$- 0.277 \times (0.6 \times 5.1497 + 0.4 \times 5.8354)$$

$$= 5.319.$$

From Finney and Stevens (1948), the correct value is 5.323.

4.4   Regression calculations

After the weight and working probit corresponding to each dose have been obtained from the provisional line, a weighted linear regression of y on x will give an improved estimate of the line. If this differs markedly from the provisional line, it may itself be used as a new provisional line in a repetition of the process. The maximum likelihood estimate is the limit to which these estimates tend as the cycle of determining a new line with the aid of that last calculated is indefinitely repeated. The experienced worker will often draw his first provisional line so accurately that only one cycle of the calculations is needed to give a satisfactory fit, but if the empirical probits are very irregular two or more cycles may be needed.

The apparent complexity of the theory and the apparent laboriousness of the arithmetic have hindered adoption of probit methods for quantal response data. The theory, outlined in § 4.2, is admittedly difficult for some who need to use it. Chapters 2 and 3 have therefore presented a reasonably simple account of the underlying principles, and have shown how rapid graphical analysis can be completed. Commonly, full arithmetical analysis is essential. The labour can be minimized and accuracy secured, however, by systematic arrangement and an orderly pattern of work. Example 6 contains an example of the computing required for the simplest type of analysis, a single probit regression line; the arrangement is that used throughout this book, but the steps are described in detail with recommendations for adequate checking. With modifications appropriate to circumstances, the same plan may be used for any of the more complicated analyses presented later.

The reader is assumed to be working with a desk calculating machine, preferably electrical; for pen-and-paper calculations, minor changes will be needed, and the labour will be heavy. A machine is no complete safeguard against arithmetical error. One of the most frequent sources of error lies in copying from paper to machine or from machine to paper; inversions of the order of digits and similar mistakes are easily made. Computations should therefore be so planned as to minimize copying on paper figures that later must be restored to the machine. The newer electronic desk calculators, which now appear to be replacing the electromechanical, have improved 'memory' facilities that enable the amount of copying of figures to be less than is shown below. The surest means of preventing errors is to have all work checked by another worker on another machine, but one person using careful checks should be able to work satisfactorily. Any misreading of figures, wrong setting of the machine, or faulty working of the machine may be repeated in checking; if a part of the calculations must be repeated because it cannot be checked by an independent path, the order of setting the machine should be changed so as to avoid exact duplication of detail. For example, a column may be summed from the bottom instead of from the top, or a product found by interchanging multiplicand and multiplier.

Persons inexperienced in statistics frequently carry more digits in their computations than the accuracy of the original data warrants, and present their results to six places of decimals when three at most are justifiable. In machine calculations an increase in the number of digits does not increase the labour to the same extent as in pen-and-paper work, and undoubtedly to carry an additional decimal place may be easier than to decide how many can be justified. Nevertheless, moderation in the number of digits saves time and reduces copying errors. The general practice in the numerical examples of this book is to cut out all unnecessary digits from the early stages of an analysis (in doses, working probits, weights, etc.), to retain digits fairly fully at intermediate stages (sums of squares and products, elements of matrices, etc.), as an aid to checking, but to present results shorn of superfluous digits and free from spurious appearance of accuracy. Unless doses have been measured correctly to within 0.1 %, two or three significant digits in x suffice. Percentage responses based on batches of 100 individuals or less need not be expressed more precisely than to the nearest 1 %, thus enabling the tables of this book to be used without interpolation; there is little to be gained by calculating percentages to greater accuracy than the nearest 0.1 % unless the batches contain more than 1000 individuals. These are not rigid rules. If an experiment has n ranging from 70 to 135 at different doses, p should be taken consistently to the nearest 1 % at all doses.

Ex. 6. *Arithmetical details for analysis of rotenone–*Macrosiphoniella sanborni *experiment.* The data of Example 1 will now be used to illustrate the fitting of a probit regression line by maximum likelihood estimation. Table 4.1 summarizes the calculations, and their performance is to be described step by step in an arrangement suitable for use when one person is responsible for the whole analysis. Personal taste will suggest modifications; some will wish to check a result before many further calculations have been based upon it, others will prefer to delay a check so as to reduce the risk of unconscious repetition of mistakes. The aim should be to calculate correctly the first time, a check being a verification of correctness and not a means of discovering errors. A check that is independent of a previous calculation may be made immediately, but a check that requires repetition of arithmetic should be made only after a lapse of time.

The reader who adopts this systematic arrangement for simple data should have no difficulty in adapting it to the more complex analyses discussed in later Chapters. For example, when relative potencies are being estimated for several preparations tested in a single experiment (Chapter 6), the provisional lines (step (ix)) will be drawn parallel if the data seem likely to agree satisfactorily with the hypothesis that the probit regression lines have equal slopes. Thereafter the computations proceed as outlined for each preparation separately, until the values of S_{xx}, S_{xy}, and S_{yy} have been obtained for each.

Table 4.1   Maximum likelihood computations for rotenone–
Macrosiphoniella sanborni experiment

Dose (mg/l)	x	n	r	$p(\%)$	Empirical probit	Y	nw	y	nwx	nwy
10.2	1.01	50	44	88	6.18	6.3	16.8	6.16	16.968	103.488
7.7	0.89	49	42	86	6.08	5.8	24.6	6.05	21.894	148.830
5.1	0.71	46	24	52	5.05	5.1	29.2	5.05	20.732	147.460
3.8	0.58	48	16	33	4.56	4.6	28.8	4.56	16.704	131.328
2.6	0.41	50	6	12	3.82	3.9	20.2	3.83	8.282	77.366
0	—	49	0	0	—	—	—	—	—	—
							119.6		84.580	608.472

$1/Snw = 0.008\,361\,204$ $\bar{x} = 0.7072$ $\bar{y} = 5.0876$

$Snwx^2$	$Snwxy$	$Snwy^2$
64.427\,00	449.5685	3177.748
59.814\,18	430.3057	3095.637
4.612\,82	19.2628	82.111
		80.440
		$1.671 = \chi^2_{[3]}$

$$b = 4.1759$$

$$Y = 5.0876 + 4.1759(x - 0.7072)$$

$$= 2.134 + 4.176x$$

(i) In the first column of Table 4.1 enter, in suitable units (here milligrams of dry root per litre of spray fluid), the doses tested, in descending order from the highest to the controls or zero concentration.

(ii) In the column headed x enter for each dose its logarithm to base 10, correct to 2 decimal places. The doses may be multiplied or divided by a power of 10 throughout in order to make x take small positive values, but a compensating adjustment must be made in later stages of the analysis if the results are to be expressed in the original units of dose.

(iii) In the columns headed n and r enter for each dose the number of subjects tested and the number responding (Martin's badly affected, moribund, or dead).

(iv) Check that steps (i), (ii), (iii) are correct, beginning each check from the bottom of the column.

(v) Evidently responses rarely occur among subjects receiving zero dose. If, as often happens, a reasonable number of subjects kept as a control batch has shown no responses, subsequent neglect of any adjustment for a natural response rate is permissible; the alternative is considered in Chapter 7.

(vi) Calculate the percentage response, $p = 100r/n$, to the nearest whole number; if n exceeds 100 for many doses, give all percentages to 1 decimal place.

(vii) Check step (vi): multiply p by n to give $100r$.

(viii) Read the probit of each p, from Table 3.2 or Table I, and enter as the 'empirical probit' to 2 decimal places. If $p = 0$ or $p = 100$, enter $-\infty$ or $+\infty$ respectively.

(ix) Plot empirical probits against x. Draw a provisional straight line to fit the points, judging its position by eye. This has already been done as Fig. 3.3, to which reference should now be made. Infinite empirical probits may be indicated by vertical arrows at appropriate values of x (cf. Fig. 4.1), and the line drawn a little more steeply on their account. Only a rough drawing is needed, and the additional calculations that some authors have suggested for improving the provisional line are scarcely worth while.

(x) Enter in column Y the expected probit, read as the ordinate to the provisional line at each x used in the experiment; 1 decimal place suffices unless n is generally greater than 200.

(xi) Check steps (viii), (ix), (x).

(xii) From the column $C = 0$ of Table II, read the weighting coefficient for each Y, multiply by the corresponding n, and enter, to 1 decimal, in the column nw. A good working rule is to have the same number of decimal places for each dose in one analysis, determining this number as appropriate to the smallest value of n and using 2 decimals if this n is less than 20, 1 decimal if it is between 20 and 200, and the nearest whole number otherwise. In exceptional circumstances, such as a wide discrepancy between the provisional line and an observation at a very high or very low dose, more decimal places in nw may be desirable for the one dose.

(xiii) From Table IV, enter the working probit, y, corresponding to each Y and p; use 2 decimals if n is generally less than 200, otherwise 3. Thus from the second page of Table IV, the column headed 3.9 gives $y = 3.83$ on the row $p = 12$. If Y is less than 2.0, greater than 7.9, or in any other way Table IV cannot be used, determine y from Table III by equation (4.21).

(xiv) Check steps (xii), (xiii); note that y is usually nearer to the empirical probit than Y is. In this example, each y is so nearly the same as the empirical probit that use of y is scarcely necessary. Sometimes, the difference is large and important, and the standard practice must be to use y.

(xv) Place the first value of nw on the keyboard of the calculator, multiply by the corresponding x, and enter the product in the column nwx to its full number of decimal places; leave the keyboard unaltered but clear the product from the machine, multiply by y, and enter in the column nwy. Clear the machine and repeat for each dose.

(xvi) Place the last value of x on the right-hand side of the keyboard, unity at the left-hand side, and multiply by nw; without clearing the result, repeat with the next to the last value of x and so work to the top of the x column, thus accumulating the products at the right-hand side of the result register and the sum of nw at the left-hand. Enter the totals

$$Snw = 119.6,$$

$$Snwx = 84.580$$

in Table 4.1. Check by addition of the nw and nwx columns from the top. Before $Snwx$ is cleared from the machine, divide it by Snw to give $\bar{x}$ to 4 decimal places, and record $\bar{x}$.

(xvii) Repeat (xvi) with y in place of x, thus checking Snw and obtaining
$$Snwy = 608.472.$$

Divide $Snwy$ by Snw, and record $\bar{y}$ to 4 decimals.

(xviii) Find the reciprocal of Snw, and record it to at least 7 decimal places. Additional accuracy is needed, as this quantity is to be multiplied by large numbers.

(xix) Set $1/Snw$ on the keyboard and check that multiplication by Snw gives unity; clear the result and multiply by $Snwx$, so checking the value of $\bar{x}$; clear the result and similarly check $\bar{y}$.

(xx) Square $Snwx$ and divide by Snw, multiply $Snwx$ and $Snwy$ and divide by Snw, square $Snwy$ and divide by Snw; the answers 59.814 18, 430.3057, 3095.637 are entered in Table 4.1, 5, 4, 3 decimal places being right for most purposes. Check by squaring the second, dividing by the first, and obtaining the third as the quotient; the first being still on the keyboard, multiply by 119.6 and divide by 84.580 to give 84.580 as a quotient. These checks may fail in the last digit as a result of rounding off.

(xxi) Set the first entry for nwx on the keyboard, multiply by x, and repeat with successive lines of Table 4.1, accumulating the total:
$$Snwx^2 = 16.968 \times 1.01 + 21.894 \times 0.89 + \ldots\ldots\ldots\ldots$$
$$+ 8.282 \times 0.41$$
$$= 64.427\,00.$$

Enter this figure in the appropriate position. Multiply the last value in the nwx column, which is already on the keyboard, by y and repeat with successive entries up the table, to obtain
$$Snwxy = 449.5685;$$

enter this figure. Multiply nwy by x $down$ the table, in order to check $Snwxy$. Before clearing the machine, subtract from $Snwxy$ the second of the three quantities calculated in step (xx) without the machine, to give
$$S_{xy} = 19.2628;$$

enter below the table and check the difference by machine. Multiply nwy by y up the table to give $Snwy^2$, and enter.

(xxii) Check $Snwx^2$ by accumulating products of nwx and x up

3

the table. Before clearing, subtract the first of the quantities calculated in step (xx) without the machine, to give

$$S_{xx} = 4.612\,82;$$

enter in the table and check by machine. Similarly check $Snwy^2$ *down* the table, derive

$$S_{yy} = 82.111;$$

enter in the table, and check.

(xxiii) From equation (4.17),

$$b = S_{xy}/S_{xx} = 19.2628/4.612\,82$$

$$= 4.1759.$$

(xxiv) From equation (4.18),

$$Y = \bar{y} - b\bar{x} + bx$$

$$= 2.134 + 4.176x.$$

Rarely are more than 3 decimal places worth retaining, and often, as here indeed, consideration of the standard errors of the parameters suggests that only 2 are justified.

(xxv) Check steps (xxiii) and (xxiv).

This completes the estimation of the parameters α, β of equation (3.6). The next stage in the analysis is to examine whether the data agree satisfactorily with predictions from the parameters, and then to discuss the precision of estimates (Example 7).

4.5   Goodness of fit and precision

The use of a χ^2 test of the discrepancy between the observations and predictions from the parameters has been described in § 3.3, but a simpler method of completing the arithmetic is now possible. This is because

$$\chi^2 = S\frac{(r - nP)^2}{nPQ}$$

$$= S\frac{n(p - P)^2}{PQ}$$

$$= Snw\left(\frac{p - P}{Z}\right)^2$$

by use of equation (3.10). Therefore χ^2 is the weighted sum of squares of the difference between the empirical and weighted probits. By equation (4.21),

$$\chi^2 = Snw(y - Y)^2$$
$$= S_{yy} - S_{xy}^2/S_{xx}, \qquad (4.25)$$

with $(k-2)$ degrees of freedom (k is the number of doses).

Hidden in this derivation is the assumption that the expected probit on which the cycle of calculation is based and that obtained from the regression equation at the end are identical: both have been denoted by Y. Strictly, this requires that the iteration has been continued to the limit so that the two versions of Y have coincided, the calculated line being the same as the provisional in the same cycle. In practice, no great harm is done by referring to the quantity in equation (4.25) as χ^2 in a cycle of iteration in which Y has not changed greatly.

If the quantity $(S_{yy} - S_{xy}^2/S_{xx})$ is calculated at the end of each successive iterative cycle, it need not decrease steadily. Commonly it will diminish from cycle to cycle, but it may increase. Indeed, even if a true χ^2 of the form of

$$S\frac{(r - nP)^2}{nPQ}$$

were calculated each time it need not diminish. Maximizing the likelihood function is not the same as minimizing χ^2, although the final χ^2 value seldom differs much from the minimum possible for the data. Moreover, general statistical theory teaches that the χ^2 in equation (4.25) tends to follow the χ^2 frequency distribution (Table VII) in large samples, when obtained from the limit of maximum likelihood calculations.

Ex. 7. *Goodness of fit and precision calculations.* Table 4.2 has been calculated from the experiment in Example 6 exactly as was Table 3.4 for the graphical analysis in Example 1. The details need not be described again. The column Y should always be formed and compared with the last recorded column of expected probits, but for many analyses the remainder of the table is unnecessary. Often rough comparison of observed and expected numbers at the more extreme doses, taken in conjunction with the alternative calculation of χ^2, may suffice for the examination of heterogeneity (cf. Example 8). The story of the analysis can now be continued.

Table 4.2   Comparison of observed and expected mortality in maximum likelihood fitting for rotenone–*Macrosiphoniella sanborni* experiment

| | | | | No. affected | | | |
| | | | No. of insects (n) | Observed (r) | Expected (nP) | Discrepancy ($r-nP$) | $\dfrac{(r-nP)^2}{nP(1-P)}$ |
x	Y	P					
1.01	6.352	0.9118	50	44	45.59	-1.59	0.63
0.89	5.851	0.8026	49	42	39.33	2.67	0.92
0.71	5.099	0.5398	46	24	24.83	-0.83	0.06
0.58	4.556	0.3285	48	16	15.77	0.23	0.00
0.41	3.846	0.1242	50	6	6.21	-0.21	0.01
						$\chi^2_{[3]} =$	1.62

(xxvi) Compare the calculated values of Y in Table 4.2 with the initial set of expected probits, Y, in Table 4.1, either numerically or by drawing a new line on the diagram. If agreement is poor (e.g. at two or more doses Y has changed by as much as 0.2), construct a new version of Table 4.1 using the new values of Y rounded to 1 decimal and repeat all calculations. If agreement is good (e.g. no Y has changed by more than 0.1), continue. No absolute standards can be established; for immediate practical purposes, agreement need not be very close, but serious scientific research usually merits two or three cycles. In this example, agreement is good, no change in Y being greater than 0.06.

(xxvii) Using equation (4.25), calculate

$$(19.2628)^2/4.612\,82 = 80.440,$$

and subtract from S_{yy} to give

$$\chi^2_{[3]} = 1.671.$$

This has been included in Table 4.1. The difference from 1.62 in Table 4.2 is due in part to rounding errors and in part to the fact that iteration has not been continued far enough towards the limit.

(xxviii) Refer χ^2 to Table VII. If it is statistically significant, the ideas of §4.6 will need to be considered. Clearly 1.67 with 3 d.f. does not indicate heterogeneous deviations from expectation, and variance calculations can begin.

(xxix) From equations (4.19), (4.20), the variances are

$$V(\bar{y}) = 1/Snw = 0.00836,$$

$$V(b) = 1/Snw(x - \bar{x})^2 = 0.2168,$$

whence

$$\bar{y} = 5.088 \pm 0.091,$$

$$b = 4.176 \pm 0.466.$$

The slope has been altered from its provisional value of 4.01 by an amount equal to about one-third of its standard error. A further cycle of computations is desirable, though scarcely essential; the next value obtained for b is, in fact, 4.196, the alteration being only 4 % of the standard error.

(xxx) Find the log ED 50 as the value of x in the estimate of equation (4.9) that gives $Y = 5$ (or $Y = 0$ if N.E.D.s are used):

$$m = \bar{x} + \frac{5 - y}{b} \qquad (4.26)$$

$$= 0.7072 - 0.0210$$

$$= 0.686.$$

(xxxi) By equation (3.12), the approximate variance of m is

$$V(m) = \frac{1}{(4.176)^2}\left[0.008361 + \frac{(0.0210)^2}{4.6128}\right]$$

$$= \frac{0.008457}{(4.176)^2}$$

$$= 0.0004849,$$

and therefore

$$s_m = 0.0220.$$

(xxxii) The log ED 50 here found by maximum likelihood, 0.686 ± 0.022, agrees well with that obtained graphically, 0.687 ± 0.023, in Examples 1, 2. Taking antilogarithms and using equation (3.18), the ED 50 is estimated as 4.85 ± 0.25 mg/l.

This discussion of precision is not fully satisfactory, and a better summary of results is expressed by fiducial limits (Example 10).

4.6 Heterogeneity

The data in Example 6 were chosen as particularly regular and presenting no complications. The χ^2 test for heterogeneity of discrepancies between observed and expected numbers is valid only when the expected numbers are not 'small'. In many experiments, at the more extreme doses tested either P or Q is nearly zero, so that, with the usual numbers of subjects, either the expected number responding (nP) or the expected number not responding (nQ) is too small for χ^2 calculated in the usual manner to be referred to the distribution in Table VII.

In other applications of the χ^2 test, a value less than 5 for the expected number in any class has often been taken as a warning (Fisher, 1969). The number 5 has no special merit; in some circumstances lower expectations produce no ill effects, and in others the χ^2 distribution may be unreliable with higher expectations (Cochran, 1942). The chief risk lies in using χ^2 with data in which the expectations for most classes are moderately large but one or two are very small; when all or nearly all expectations are small, the disturbance of the distribution is unlikely to be so serious. Table 4.3, a rapid guide to danger levels, shows that a batch of 50 subjects will have an expectation of non-responders less than 5 when Y exceeds 6.28, less than 2 when Y exceeds 6.75. At the other extreme, the expected number responding (nP) will be less than 5 when Y is less than 3.72 ($= 10 - 6.28$), less than 2 when Y is less than 3.25.

No more definite advice can be given than to suspect any large χ^2 which includes large contributions from doses at which nP or nQ is small. When n is greater than 30, 'small' may be roughly equated with 'less than 5', but if most n are about 10 a lower standard of 'less than 2' may be reasonable.

The difficulty of small expectations may often be overcome by combining the results for extreme doses with the next highest or next lowest, so as to build up larger expectations, though some sensitivity is thereby lost. Equation (4.25) cannot then be used for calculating χ^2, and recourse must be had to the calculation of all expectations (Tables 3.4, 4.2). The degrees of freedom for χ^2 must be reduced by the number of groups lost through combination (Example 8).

Table 4.3   Table of maximum expected probit giving at least nQ expected not to respond in a batch of n

No. in batch (n)	Expected no. not responding (nQ)			
	10	5	2	1
5	—	—	5.25	5.84
10	—	5.00	5.84	6.28
20	5.00	5.67	6.28	6.64
30	5.43	5.97	6.50	6.83
40	5.67	6.15	6.64	6.96
50	5.84	6.28	6.75	7.05
100	6.28	6.64	7.05	7.33
200	6.64	6.96	7.33	7.58
1000	7.33	7.58	7.88	8.09

In experiments on homogeneous subjects, χ^2 will on average be equal to its number of degrees of freedom, here $(k-2)$. Previous discussion of weights, variances, and standard errors has implicitly assumed that observed frequencies of responses do not vary heterogeneously about their expectations, or that the points plotted in a diagram such as Fig. 3.4 do not deviate significantly from the regression line. A significantly large χ^2 indicates breakdown of this assumption. Two things may go wrong. First, individual subjects in a batch receiving one dose may not react wholly independently of one another. Members of a batch of insects may be genetically related to a closer extent than insects in a different batch, and their responses may be correlated. Mice used in a biological assay may interact with one another and, if those at the same dose are caged together, environmental or social factors may influence the responsiveness of all in a batch. Such heterogeneity may arise from the subjects themselves, from uncontrolled factors that are constant for all subjects at a dose but may vary from dose to dose (e.g. humidity, diet), or from interaction between these. It increases dispersion of p about P, or of points about the line, in a random manner, and in consequence inflates χ^2. If it is known to be the cause of heterogeneity, the quantity

$$h = \chi^2/(k-2) \qquad (4.27)$$

can be regarded as estimating the *heterogeneity factor* by which all weights have been overestimated. All variances can then be multiplied by h, and subsequently taken to be based upon $(k-2)$ degrees of freedom. In all uses of standard errors, the t distribution with $(k-2)$ degrees of freedom will then be used instead of the normal. This amounts to empirical assessment of standard errors, and admission of a wider range of values as within the limits of experimental error. A condensed version of the t distribution table, always to be entered with the number of degrees of freedom for χ^2, appears as Table VIII.

The greater fear is that a significant χ^2 arises because the underlying mathematical model is incorrect: essentially, equation (2.6) is untrue. This will tend to produce a systematic deviation of p from P; points in a diagram such as Fig. 3.4 would lie near a curve rather than be widely scattered about a line. If the consequence were simply that equation (2.6) could be replaced satisfactorily by some simple alternative, perhaps one of those mentioned in § 3.8, all would be well. Unfortunately, nothing so easy is often found. Rarely can one distinguish clearly which form of heterogeneity, random or systematic, is causing a large χ^2. Even if the evidence points to a systematic trend, little can be done except to suggest a much larger experimental programme.

The lesson is to treat a large χ^2 with extreme caution. The temptation simply to use the heterogeneity factor must be resisted unless the evidence is clear that no sign of systematic deviation from linear regression appears. Careful inspection of a diagram is essential. In the past, uncritical multiplication of variances by h has been too readily advised and accepted (not least in earlier editions of this book) as an escape from a difficult situation; it is never to be commended unreservedly.

Ex. 8. *Application of probit analysis to heterogeneous data.* As a second example of fitting the probit regression line, results obtained by Busvine (1938; Bliss, 1940) on the toxicity of ethylene oxide to the grain beetle, *Calandra granaria*, may be considered. Only the records of insects examined 1 hour after exposure to the poison will be used.

The data are shown in the first three columns of Table 4.4, x being the logarithm to base 10 of the concentration of ethylene oxide in mg/100 ml. The empirical probits were plotted against x (Fig. 4.1), and a provisional line

$$Y = 3.06 + 7.95x$$

was drawn by eye. Note how infinite probits at $x = 0.391$ and $x = 0.033$ are represented; existence of these observations above and below the line should be remembered when the eye estimate is judged. The maximum likelihood process led to the new approximation

$$Y = 2.948 + 8.600x. \qquad (4.28)$$

The large change in the regression coefficient suggested a need for a second cycle of computations. The details in Table 4.4 are completed with Y from equation (4.28) and working probits from Table IV; maximum and minimum working probits at $x = 0.391$ and $x = 0.033$ respectively are read either from the 100 % and 0 % lines of Table IV or from Table III. Computations as in Example 6 then give the equation

$$Y = 2.928 + 8.074x. \qquad (4.29)$$

As will be seen shortly, differences between (4.28) and (4.29) are negligible by comparison with the standard errors, and no further iteration is necessary.

Equation (4.25) gives a χ^2 of 32.42 with 8 degrees of freedom, which by Table VII is clearly significant. The validity of this test is suspect, because three expected probits exceeding 6.0 and one less than 4.0 give dangerously low values to nQ or nP (Table 4.3). The test may be modified as shown in Table 4.5 by grouping the expectations for the three highest doses and also those for the two lowest. Equation (4.25) no longer applies: the separate contributions to χ^2 must be calculated, as in Example 1, by squaring the difference between observed and expected numbers killed, multiplying by the total number, and dividing by the product of the expected kill and expected survivors. From the first group:

$$\frac{91 \times (0.9)^2}{81.1 \times 9.9} = 0.09.$$

The number of groups has been reduced by 3, and therefore

$$\chi^2_{[5]} = 19.74;$$

still highly significant, this confirms the evidence for heterogeneity of the departures from the regression line.

Table 4.4   Maximum likelihood computations for ethylene oxide–*Calandra granaria* experiment (second cycle)

x	n	r	$p(\%)$	Empirical probit	Y	nw	y	nwx	nwy
0.394	30	23	77	5.74	6.3	10.1	5.52	3.9794	55.752
0.391	30	30	100	∞	6.3	10.1	6.86	3.9491	69.286
0.362	31	29	94	6.55	6.1	12.5	6.45	4.5250	80.625
0.322	30	22	73	5.61	5.7	15.9	5.61	5.1198	89.199
0.314	26	23	88	6.18	5.6	14.5	6.06	4.5530	87.870
0.260	27	7	26	4.36	5.2	16.9	4.38	4.3940	74.022
0.225	31	12	39	4.72	4.9	19.7	4.72	4.4325	92.984
0.199	30	17	57	5.18	4.7	18.5	5.19	3.6815	96.015
0.167	31	10	32	4.53	4.4	17.3	4.54	2.8891	78.542
0.033	24	0	0	$-\infty$	3.2	4.3	2.74	0.1419	11.782

139.8   37.6653 736.077

$1/Snw = 0.007\,153\,076$ $\bar{x} = 0.2694$ $\bar{y} = 5.2652$

$Snwx^2$	$Snwxy$	$Snwy^2$
11.18778	207.3360	3986.265
10.14789	198.3159	3875.603
1.03989	9.0201	110.662
		78.241

$32.421 = \chi^2_{[8]}$

$b = 8.6741$ $Y = 2.928 + 8.674x$

Fig. 4.1 shows wide dispersion of the points about the regression line, but not the slightest hint of a systematic curvilinearity. One should be reluctant to draw important conclusions from a single experiment, and an experiment such as this ought to be considered in relation to others of a series. Nevertheless, at least as an illustration of the use of a heterogeneity factor, further calculations that assume random heterogeneous variation are of interest. Whether equation (4.27) should be applied to the original χ^2 or that of Table 4.5 is debatable, but perhaps

$$h = 32.42/8$$
$$= 4.050 \text{ with 8 d.f.}$$

is preferable.

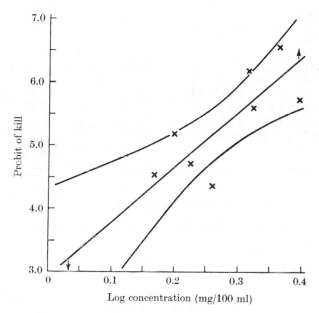

Fig. 4.1. Probit regression line and 95% fiducial band for toxicity of ethylene oxide to *C. granaria* (Exs. 8–10).

By equation (4.20), with incorporation of h,

$$V(b) = 4.050/1.040 = 3.894,$$

so that

$$b = 8.674 \pm 1.973.$$

From equations (4.26), (4.29),

$$m = 0.239,$$

and equation (3.12) leads to

$$V(m) = \frac{4.050}{(8.674)^2}\left[0.007\,153 + \frac{(0.030)^2}{1.040}\right]$$
$$= 0.000\,432,$$

so that the standard error of m is 0.021.

Table 4.5   Comparison of observed and expected mortalities
for ethylene oxide–*Calandra granaria* experiment

x	Y	P	n	r	nP	$r-nP$	Contribution to χ^2
0.394	6.346	0.911	30	23	27.3⎫		
0.391	6.320	0.907	30	30	27.2⎬	0.9	0.09
0.362	6.068	0.857	31	29	26.6⎭		
0.322	5.721	0.765	30	22	23.0	−1.0	0.19
0.314	5.652	0.743	26	23	19.3	3.7	2.75
0.260	5.183	0.573	27	7	15.5	−8.5	10.94
0.225	4.880	0.452	31	12	14.0	−2.0	0.52
0.199	4.654	0.365	30	17	11.0	6.0	5.17
0.167	4.377	0.267	31	10	8.3⎫		
0.033	3.214	0.037	24	0	0.9⎭	0.8	0.08

$$\chi^2_{[5]} = 19.74$$

4.7 Fiducial limits

The expected probit, Y, for a dose x may be written

$$Y = \bar{y} + b(x - \bar{x}); \qquad (4.30)$$

the variance of Y is

$$V(Y) = \frac{1}{Snw} + \frac{(x-\bar{x})^2}{S_{xx}}. \qquad (4.31)$$

Fiducial limits to Y, indicating a range within which the probit of the true response percentage is almost certain to lie, can then be calculated as $Y + t\sqrt{[V(Y)]}$ and $Y - t\sqrt{[V(Y)]}$, where t is the normal deviate for the level of probability to be used (1.96 for probability 0.95). If a heterogeneity factor is in use, the expression for $V(Y)$ in equation (4.31) must be multiplied by h and t must be altered to the t-deviate (Table VIII) with the appropriate degrees of freedom.

When fiducial limits of Y are plotted for each x, they produce two hyperbolic curves convex to the regression line and approaching the line most closely at $\bar{x}$. The more x differs from $\bar{x}$ in either direction, the greater is the contribution to $V(Y)$ from the second term of (4.31), which term represents errors of estimation of the regression coefficient b.

In many investigations, the absolute effect of a single preparation is of less interest than comparisons between the effects of two or more different preparations in the same experiment. Day to day changes in the susceptibility of subjects may alter considerably the response at any stated dose without seriously upsetting the relative effects of different preparations; consequently the fiducial band for one regression line may not be very helpful for inferring future behaviour (§ 6.1).

Ex. 9. *Fiducial bands for a probit regression line.* The plotting of fiducial limits as a band on either side of the probit regression line may be illustrated on the data of Example 8. Under the conditions of the experiment, the true response rate for the range of doses may be expected to lie within this band with a degree of assurance represented by the fiducial probability.

Formula (4.31) gives for these data

$$V(Y) = 4.05 \times \left\{ 0.007\,153 + \frac{(x-0.269)^2}{1.040} \right\};$$

variances and standard errors are shown in Table 4.6. Multiplication of the standard error, s_Y, by 2.31, the t-value for 8 d.f. and 95 % fiducial probability, gives the width of the fiducial band on either side of the regression line; the boundaries are shown in Fig. 4.1. Had the same calculations been made in Example 6, the boundaries would have differed only very slightly from two straight lines parallel to the regression line.

Table 4.6   Variance and standard error of expected probits in equation (4.29)

x	$V(Y)$	s_Y
0.00	0.3108	0.557
0.05	0.2158	0.465
0.10	0.1402	0.374
0.15	0.0841	0.290
0.20	0.0475	0.218
0.25	0.0304	0.174
0.269	0.0290	0.170
0.30	0.0327	0.181
0.35	0.0545	0.233
0.40	0.0958	0.310

The method of calculating fiducial limits to m described in Example 3, though often sufficiently good, is unsatisfactory as a general technique and will not be used hereafter. The limits required, for the ED 50 or any other percentage point, can be read from a diagram such as Fig. 4.1: they are the values of x at which a horizontal line at the appropriate probit value cuts the two curves bounding the fiducial band.

This method (Cochran, 1938) was subsequently shown by Fieller (1944; Finney, 1964) to be a special case of a general theorem. Fieller's theorem is stated here because applications of it occur frequently in the problems of this book, although only a simple form is wanted immediately. Suppose that α, β are two parameters and
$$\mu = \alpha/\beta; \qquad (4.32)$$
these need not be the α, β of equation (4.9), and are to be considered a special notation for this paragraph. Suppose further that a, b are estimates of α, β subject to normally distributed random errors and that v_{11}, v_{22}, v_{12} are joint estimates (all from the same data and with f degrees of freedom) of the variances of a, b and their covariance. The natural estimate of μ will be
$$m = a/b; \qquad (4.33)$$
a, b, m also need not be the quantities so symbolized elsewhere in the book. Consideration of the expectation and variance of $(\mu b - a)$ shows that the fiducial limits to μ are the roots of the quadratic equation
$$\mu^2(b^2 - t^2 v_{22}) - 2\mu(ab - t^2 v_{12}) + (a^2 - t^2 v_{11}) = 0, \quad (4.34)$$
where t is the appropriate deviate with f degrees of freedom for the chosen probability level. The limits may be written:
$$m + \frac{g}{1-g}\left(m - \frac{v_{12}}{v_{22}}\right)$$
$$\pm \frac{t}{b(1-g)} \sqrt{\left[v_{11} - 2mv_{12} + m^2 v_{22} - g\left(v_{11} - \frac{v_{12}^2}{v_{22}}\right)\right]}, \quad (4.35)$$
where
$$g = t^2 v_{22}/b^2. \qquad (4.36)$$

In probit problems, unless a heterogeneity factor is in use, t is a normal deviate (from the last line of Table VIII). Often, but not always, $v_{12} = 0$. When g is small, the limits given by (4.35) are almost identical with those from use of a variance formula analogous to (3.12). When g is not small, the true limits obtained by the present method can be much wider.

In practice, little harm will come from basing fiducial limits on the variance formula (as in § 3.5) when $g < 0.05$ or even $g < 0.1$. However, the saving of labour in systematic computation is slight, and from this point onward the limits will always be calculated by the appropriate version of (4.35). Application of the general theory to the log ED 50, equation (4.26), gives the limits as

$$m + \frac{g}{1-g}(m - \bar{x}) \pm \frac{t}{b(1-g)} \sqrt{\left[\frac{1-g}{Snw} + \frac{(m-\bar{x})^2}{S_{xx}}\right]}, \quad (4.37)$$

where now the standard notation is in use and

$$g = t^2/b^2 S_{xx}. \quad (4.38)$$

If a heterogeneity factor is appropriate, it multiplies the right-hand side of (4.38) and the whole of the expression within the square-root in (4.37), and of course t must have the right degrees of freedom. Formula (4.37) can be used for log ED 90, or any similar value, by taking m as the appropriate quantity.

With almost all good sets of data, g will be substantially smaller than 1.0, and seldom greater than 0.4. If $g = 1.0$, the limits explode, and the whole range of x is included. If $g > 1.0$, the 'range' of plausible values for the parameter μ consists of those values of x lying *outside* the calculated limits; this unusual situation, which does not have much practical importance, was discussed by Fieller (1954).

Ex. 10. *Fiducial limits for a median effective dose.* This theory will now be illustrated on the data analysed in Examples 6, 8.

(xxxiii) Example 6 showed no heterogeneity, and therefore for 95 % fiducial limits

$$g = (1.96)^2/[(4.176)^2 \times 4.613]$$

$$= 0.048.$$

(xxxiv) The log ED 50 in this example was 0.686; the fiducial limits are, by equation (4.37),

$$0.686 - 0.050 \times 0.021 \pm \frac{1.96}{4.176 \times 0.952} \sqrt{\left[\frac{0.952}{119.6} + \frac{(0.021)^2}{4.613}\right]}$$

or 0.729 and 0.641. (Compared with 0.729 and 0.643 obtained from the approximation $m \pm 1.96s_m$: for so small a value of g, the difference is trivial).

(xxxv) The conclusion is that, under the conditions of the experiment, the estimated median effective dose of rotenone to *Macrosiphoniella sanborni* was 4.85 mg/l; with fiducial probability 0.95, the true ED 50 may be asserted to lie between 5.36 mg/l and 4.38 mg/l. Agreement with the rough approximation in Example 3 is close, better than could usually be hoped!

In example 8, g is much larger; using h and the value of t with 8 d.f.,

$$g = 4.05 \times (2.31)^2/[(8.674)^2 \times 1.040]$$

$$= 0.276,$$

certainly not negligible. Here $m = 0.239$, and the 95 % limits are

$$0.239 - 0.381 \times 0.030 \pm \frac{2.31}{8.674 \times 0.724} \sqrt{\left[4.05 \times \left(\frac{0.724}{139.8} + \frac{0.0009}{1.040}\right)\right]}$$

or 0.228 ± 0.058. These limits, 0.286 and 0.170, are appreciably wider than the values of 0.276 and 0.180 obtained by the variance formula. They could be read directly from the boundaries of the fiducial band in Fig. 4.1. On the concentration scale, the estimate is 17.3 mg/l and the limits are 19.3 and 14.8 mg/l.

5

Computer and other techniques

5.1 The computer program

Most persons who are closely concerned with the statistical analysis of quantal response data already have, or soon will have, access to a modern high-speed computer. With relatively little expenditure of programming effort, such a computer can replace the admittedly lengthy calculations of the classical probit analysis by a method giving far greater accuracy in very little time. All that need be required of the user is a set of punched cards containing the data, provisional values for the parameters, and coded instructions for the choice between various options in the analysis.

The program can go far beyond the basic statistical analysis described in Chapter 4. Alternatives can be provided for the more complicated problems introduced in later Chapters. Considerable care is needed in the construction of the program, in testing the correctness at all points, and in planning for the output of useful information while avoiding overwhelming the user with irrelevant detail. A program was written specially for use in the preparation of this book, of necessity at the time in a little known language (Atlas Autocode) for an obsolescent computer (KDF 9); the program is not of intrinsic interest, but an account of its main features may help the reader to develop or obtain one best suited to his type of work. Once a program exists, analysis of any set of data will require nothing but 5 minutes for the preparation of appropriate punched-card instructions and a few seconds of computer time. The following sections describe the main structure of a program such as a good programmer could readily prepare in any standard computer language.

One point to be remembered is that the method of analysis described in Chapter 4 was devised for desk calculating machines and for operators familiar with the calculation of regression equations. The central objective, to maximize the likelihood function, need not be achieved by way of the regression analogy:

the problem can be approached solely from the point of view of what is best suited to the computer.

5.2   Input of data

Obviously the program must first read and store a full specification of the data. This should be in the form of a listing of each dose together with the number of subjects responding, r, and either the total number, n, or the number not responding, $(n-r)$. The dose should be in the original units, but the program should convert each value to logarithms (natural or base 10) unless information is supplied that the metameter is not to be logarithmic. For convenience, also, the number of dose groups should be read and used as a check. Problems discussed later will require insertion of data for several preparations at one time (Chapter 6) with a clear indication of how these are classified, data from 'control' subjects that received zero dose (Chapter 7), data relating to two or more distinct dose variates (§ 8.6) with clear indication of the number of these and how they are to be distinguished from response frequencies, and so on. Whether these facilities are included as options within one program or whether separate similar programs are written will depend upon judgement of the pattern of work to be expected.

With the data should be read whatever descriptive information is required. As a minimum, this will include adequate identification of the source so that the reader of the output runs no risk of confusion. Provision for reproduction of a piece of free text is convenient, in order that notes about the data can be tied to the analysis, but the importance of this depends upon the volume of work to be handled.

The coding of data for input and the steering instructions are the chief demands on human labour, and therefore should be minimized. A simple pattern is easily repeated and keeps opportunities for mistakes small. For a typical experiment, not more than 10 cards need be required, in addition to free text descriptions requiring normal typing skill and not demanding absolute accuracy.

5.3   Steering instructions

The program must also read various instructions on how to conduct the analysis and on the output that is wanted. Usually these will begin with a heading that states what analysis is being made, particularly important if two or more analyses of the same data are planned.

In addition to the data themselves, the major information that must be supplied to the program is:

(i)   Whether the analysis is to be based upon probits (N.E.D.s), on the logistic function, or on some alternative (§ 3.8);

(ii)   A set of initial values for the parameters, with which iteration can begin;

(iii)   For some maximizing routines, a set of 'step lengths', the amounts by which the parameters are to be varied in preliminary exploration (§ 5.4);

(iv)   Definition of dose metameters, logarithmic or other;

(v)   The maximum number of iterations to be performed before analysis stops;

(vi)   Coded indicators of parts of the program to be used in more complicated problems;

(vii)   Coded indicators of optional pieces of output for which the user of the program may call, and which may require special computations.

The programmer has freedom to organize these as he wishes, but he should give thought to all the categories. The importance of (ii), (iii), (v) will become clearer in § 5.4, and that of (vi) in § 5.7. Later chapters will illustrate (vii).

5.4   The maximization routine

Any computer installation used extensively for scientific research will require for many purposes a good method of maximizing (or minimizing) an arbitrary function of several parameters. The computations described in § 4.4 are one way of maximizing the likelihood function, taking advantage of its continuity and generally smooth character; they are particularly suited to a desk calculator and to an operator familiar with regression calculations. For an electronic computer, no special convenience lies in this arithmetical pattern. Indeed, if many iterations are intended, the necessity of frequent recalculation of y, w, S_{xx}, S_{xy} may make it a relatively inefficient method. One may instead consider the general problem of maximizing the log-likelihood without restriction of method.

The program employed in this book incorporated a routine slightly modified from that described by Nelder and Mead (1965). This has two phases, the simplex and the quadratic. In the simplex phase, the log-likelihood is evaluated many times for different combinations of the parameters and attention progressively shifted to those that are currently best. The process begins by evaluating the log-likelihood for the initial values of the parameters and for the alternative sets of values obtained by increasing one parameter in turn by its step length (§ 5.3, (ii) and (iii)) in each possible way. The initial values of the parameters are replaced by those corresponding to the largest of the three (for more complex problems, 1 + the number of parameters) log-likelihoods, and the step lengths are appropriately modified for a repetition of the process. Usually the step lengths are reduced, to extents dependent on the changes just made in the parameters, but there is provision for increase if a parameter appears to be far from its maximizing value.

This seemingly crude process is so programmed as to ensure that each change of parameters is an advance towards the maximum. Likelihood functions are usually 'well-behaved', not having more than one maximum and changing smoothly in the neighbourhood of the maximum. Therefore the simplex phase will usually approach the maximum rapidly; it will end either when an internal test of adequate convergence is satisfied or when a pre-selected number of iterations is exceeded. The routine then switches to the quadratic phase, with a final sequence of calculations closely related to those of Chapter 4. The scaffolding of weighting coefficients and working probits is discarded and equations (4.3) are solved for calculating adjustments to the parameters. In the course of the solution, the inverse matrix of the coefficients is calculated:

$$
\begin{pmatrix} -\dfrac{\partial^2 L}{\partial \alpha^2} & -\dfrac{\partial^2 L}{\partial \alpha\, \partial \beta} \\[2ex] -\dfrac{\partial^2 L}{\partial \alpha\, \partial \beta} & -\dfrac{\partial^2 L}{\partial \beta^2} \end{pmatrix}^{-1} = \begin{pmatrix} v_{aa} & v_{ab} \\ v_{ab} & v_{bb} \end{pmatrix}, \qquad (5.1)
$$

and its elements are the variances and covariance of the estimates a, b. From here on, the analysis may be continued and completed essentially as in Chapter 4. The program actually used enabled the sequence of simplex and quadratic phases to be repeated several times, with decreasing step lengths, three repetitions usually being satisfactory.

The routine as used provided for the differential coefficients required in (5.1) to be calculated from the appropriate mathematical formulae. However, it also permits these exact values to be replaced by approximations if for any reason the formulae are not easily obtainable. If $L(\alpha, \beta)$ is the log-likelihood evaluated for particular values of the parameter (not necessarily the 'true' values), then $\partial^2 L/\partial \alpha^2$ can be approximated by

$$[L(\alpha + c, \beta) - 2L(\alpha, \beta) + L(\alpha - c, \beta)]/c^2, \qquad (5.2)$$

and $\partial^2 L/\partial \alpha \partial \beta$ can be approximated by

$$[L(\alpha + c, \beta + d) - L(\alpha - c, \beta + d) - L(\alpha + c, \beta - d)$$
$$+ L(\alpha - c, \beta - d)]/4cd, \quad (5.3)$$

provided that c, d are very small. Many computer analyses reported in this book have used the appropriate formulae for differential coefficients, but (5.2), (5.3) and analogous approximations are wanted in more complex problems. The program must then be written in a way that ensures that c, d are small but not so small that the divisions by c^2, $4cd$, and so on produce great numerical inaccuracy.

5.5 The log-likelihood function

The routine described in § 5.4 needs frequently to evaluate the log-likelihood for a set of trial values of the parameters. This is achieved by including in the program a routine, which can be called as often as required, that evaluates

$$L = Sr \log P + S(n - r) \log Q, \qquad (5.4)$$

where

$$P = \int_{-\infty}^{\alpha + \beta x} \frac{1}{\sqrt{(2\pi)}} e^{-\frac{1}{2}u^2} du. \qquad (5.5)$$

The program written in fact provided the alternative of using a logistic model instead of the normal, by defining P as in equation (3.22). Merely by alteration of a single character in the steering instructions, logistic analysis could be substituted for the normal.

5.6 The parameterization

One may ask why, in equation (5.5) and therefore in the analysis as a whole, α and β are used as the parameters instead of the μ and σ of equation (3.3). Certainly the computer maximization could proceed in terms of the other parameters, and might seem to have the advantage of estimating the log ED 50 more directly. In addition, the trouble with the variance of the estimate would not arise; no longer would a ratio be used and no longer would Fieller's theorem be needed.

Undoubtedly one reason why this is not the general practice is that a long tradition attaches to estimation of α, β and, although computers have removed the immediate practical advantage, no clear reason for change has emerged. More important, however, is the consideration that the statistical sampling errors in the estimation of α, β are more likely to be approximately normal in distribution than are the corresponding errors in direct estimation of μ, σ. The theory of maximum likelihood teaches that in large experiments the errors of estimation tend to the normal form, but it is well-known that in small sets of data the error distribution may be far from normal and seriously skew. Probability statements made about the ED 50 or other properties of the tolerance distribution in practice have to be made in terms of the asymptotic normality of maximum likelihood estimators. The analogy with weighted linear regression strongly supports the idea that the estimates represented by a, b will be approximately normal in distribution, in which case the ratio required for the log ED 50 must be seen as calling for Fieller's theorem. The asymmetry inherent in the Fieller limits would then be present in the true distribution of a direct estimate of μ obtained by the alternative parameterization, and this would be concealed if the variance based upon the inverse matrix that replaced equation (5.1) were used as though it pertained to a normal distribution.

To summarize, the alternative pairs of parameters are exactly equivalent so far as the estimation itself is concerned. If iteration is continued until the likelihood function is very close to its maximum, the estimates by the μ, σ method can be made to agree as closely as desired with $-a/b$, $1/b$ respectively. Fiducial limits and other calculations that use normal distribution theory in association with variances and covariances estimated by an inverse matrix, however, seem likely to be more nearly correct if α, β are regarded as the primary parameters for estimation. The difference will be unimportant in data involving large numbers of subjects, and indeed even for small numbers the whole question merits more detailed study, but the α, β method will here be regarded as the standard.

5.7 Output

The ease with which a computer can present different aspects of a statistical analysis brings a temptation to program for excessive amounts of output. The purpose of any statistical analysis is to *summarize* the data as an aid to *interpretation*: this is not achieved if the computer output is vastly more complicated than the data. On the other hand, the cost of including in a program one or two additional steps or a little extra output is so small that considerations of economy will rarely dictate what shall be excluded. An ideal program will always produce at least the minimum of essential output, and will provide as options, controlled by the initial steering instructions, various other types of information.

For quantal response data, the essential features of the computer output are the estimates of the parameters and the variance–covariance matrix. Also very desirable is a table that compares observed and expected frequencies (cf. Tables 4.2, 4.5), with the associated value of χ^2. These two items should be produced every time the program is used.

Although the log ED 50 can be rapidly calculated from this output, automatic computation of it and its approximate variance or its fiducial limits should be available as an option. Usually it is wanted, and little extra computation is necessary, but occasionally greater interest will lie in the ED 90 or in some other function of the parameters. Possibly options should be provided for several of these. If the program is to be used for analysing several sets of data simultaneously, the problems of Chapter 6 will almost certainly arise and results pertaining to the ratio of two ED 50's will be wanted, as illustrated in § 6.2.

Tabulation of the log-likelihood is sometimes useful. The program should permit tabulation intervals for each parameter to be specified, and should produce at will *either* a table centred on the initial parameter values before maximization *or* a table centred on the final estimates after maximization. In some problems, assessment of reasonable initial values with which to begin iteration is difficult; a table on a widely spaced grid, say 20 values of α at intervals of 0.25 and 10 values of β at intervals of 0.5, will usually indicate roughly the region in which a search for a, b should be concentrated. This is more true of some of the methods of later chapters (e.g. Chapter 7), where a really bad start may give a non-converging iteration. Such tabulation can be undertaken without maximization. Usually the log-likelihood changes slowly as the parameters vary in the neighbourhood of the maximum likelihood estimates, and is well-approximated by a quadratic function of the parameters. A table centred on the estimates can be helpful confirmation that the function behaves in this manner, and consequently that interpretations based upon the variances and covariances will be trustworthy. Occasionally a problem is encountered for which the log-likelihood does not change in a smooth and symmetric manner around the maximum. This is a warning that simple interpretations may be misleading: it scarcely points to any specific alternative, but at least demands that the data and analysis be carefully re-examined. Occasionally, again, tabulation may disclose that the true maximum has not been achieved, possibly because the iteration has adopted excessively small step lengths before the maximum has been closely approached.

Yet another useful option in the computer output is information on the history of the iterative maximization. A listing of the value of the log-likelihood at regular intervals during iteration, together with corresponding values of the parameters and the step lengths, tells the user how successful he has been in his choice of initial conditions and number of iterations. It is particularly valuable in the early stages of using a new program, or in analysis of results that involve many parameters, but is certainly not wanted as standard output for every analysis.

The user of statistical technique for quantal responses in a particular field of science will rapidly gain experience of the behaviour of his data. In consequence, he will become able to judge whether a particular set of data is likely to give any difficulty in convergence, either because suitable initial values cannot easily be assessed or because the data are erratic and will not determine a maximum very well. Tabulation of log-likelihoods and output of the history of the iteration are then very helpful. Moreover, to run two or three analyses consecutively on the same data, using different initial values, step lengths, and numbers of iterations, will provide confirmation that all is well. As experience grows, most sets of good data will be recognized as easily analysed, a suitable number of iterations will be judged, and the program can be applied without dependence on some of its extra facilities.

5.8   Results by computer

An example of computer output is reproduced later (§ 6.6). For the present, some results of computer analysis on the Examples of Chapter 4 will be summarized.

Ex. 11. *Computer analyses of two sets of data.* The experimental results presented in Example 1, and analysed by the classical probit technique in Examples 6, 7, 10, have been put through the computer program described above. Indeed, they were analysed many times in tests of the program; inevitably numerical estimates obtained differed slightly because of differences in starting conditions or number of iterations, but the variations amount only to about 0.0001 in the two parameters and about 0.00001 in m. The regression equation

$$Y = 2.1126 + 4.2132x$$

may be compared with that in Example 6. Expected frequencies differ very little from those in Table 4.2, but χ^2 is a little larger:

$$\chi^2_{[3]} = 1.729.$$

The estimated log ED 50 and its approximate standard error are

$$m = 0.6853 \pm 0.0220.$$

The estimated ED 50 is 4.845 mg/l, with fiducial limits found by the method of Example 10, at 5.354, 4.364 mg/l. Agreement with Example 10 is excellent.

The data of Example 8 have been similarly analysed. The regression equation,

$$Y = 2.9383 + 8.6325x,$$

is very like equation (4.29). The heterogeneity test gives

$$\chi^2_{[8]} = 33.385,$$

whence is obtained the heterogeneity factor

$$h = 4.1731 \text{ with } 8 \text{ d.f.}$$

From the regression equation,

$$m = 0.2388.$$

The estimated ED 50 is 17.33 mg/l, and its 95 % fiducial limits are at 19.32, 14.70 mg/l.

5.9   Other computational techniques

When desk calculators were the only mechanical aid to probit calculations, many variants on the original Bliss–Fisher method of maximizing the likelihood were proposed. Some were mathematically equivalent but were asserted to have advantages in speed of convergence, some were frankly approximations to save time and reduce arithmetic, and some adopted different principles of estimation.

For example, Black (1950) advocated a different method of tabulating maximum and minimum working probits that leads to essentially the same analysis as that of § 4.4 but constructs the column nwy directly. Graphical representation and the linear regression analogy are sacrificed, with the compensation that the amount of arithmetic is reduced a little; the method is similar to scoring methods commonly used in genetical and other maximum likelihood problems (Finney, 1949c). Garwood (1941) compared use of expected values of the second differential coefficients in equations (4.3) with the use of empirical values calculated from the complete expressions for these differential coefficients. Although iteration must converge to the same estimates of parameters, variances are not identically estimated. In trials with several sets of data, empirical coefficients gave quicker convergence, in the sense that fewer cycles were needed in order to attain a specified closeness to the limiting values. This advantage of the empirical coefficients is counter-balanced by greater algebraic complexity and the consequent greater labour of calculating them, so that the time required for computing one cycle becomes greater than for the expected coefficients. Cornfield and Mantel (1950) suggested an improved computing scheme for the empirical coefficients and provided tables for it; one cycle of their method is more laborious than that of Chapter 4, but the advantage in speed of convergence could suffice to outweigh this disadvantage. A satisfactory computing routine for a desk calculator could be based on any of these methods, but detailed discussion of them seems pointless now that advances in computers have made possible entirely different and much quicker analyses.

In the class of approximations may be placed the purely graphical procedure of § 3.3, and various modifications of it. De Beer (1941, 1945) developed an ingenious sytem of scales and nomographs. With their aid, many of the expressions required as steps in the estimation of the ED 50 and its fiducial limits can be read directly from the diagram showing the provisional regression line. The method does not completely distinguish between empirical and working probits, nor does it obtain fiducial limits from the exact formula (4.35); nevertheless it merits serious consideration by anyone wanting an improvement on § 3.3 that could be good enough for rapid routine assessments. Berkson (1960) produced a nomographic method for the logistic model (§ 3.8) that gives maximum likelihood estimates except for inaccuracies of reading.

The Litchfield and Fertig method (§ 3.4) was modified by Miller and Tainter (1944), and again by Litchfield and Wilcoxon (1949). In its latest exposition, it could scarcely be said to look simple, but in routine use it could be rapid. All these methods employ a line drawn by eye to fit points plotted for each dose and response, preferably on logarithmic probability paper (§ 3.6). Systems of nomographs then simulate the maximum likelihood calculations for variances, fiducial limits, and contributions to a heterogeneity χ^2; even without the inevitable inaccuracies of reading from diagrams, the correspondence is not exact, because the fitting of the line cannot take proper account of the working probits and weights. For good data, however, with points close to a straight line and no responses near to zero or 100%, the results are often very close to those from maximum likelihood estimation.

Some approximations undoubtedly have merits for routine analyses, despite internal inconsistencies of assumption that will occasionally give trouble. The experienced statistician may safely use a rapid approximate method: he can recognize when some unusual feature of the data makes the method misleading or inapplicable. Unfortunately, the very simplicity of a method is an attraction to the experimenter who has little knowledge of statistical theory and who must therefore apply his chosen method uncritically. Moreover, under the assumption of a normal distribution of log tolerances and probably also under any other reasonable assumption about the form of this distribution, the approximate methods either do not make fully efficient use of the available data or require that the experiment be planned in a manner that does not make the best possible use of the materials. The labour of statistical analysis is usually only a small part of the total labour of an experiment, and, therefore, efficient analysis of a minimal number of observations is generally more economical than non-efficient analysis of a larger number. The argument sometimes advanced that elaborate statistical analysis of a few observations is not worth while seems to be based upon a misconception; scarcity of observations is likely to be a result of high cost of obtaining one observation, just the circumstances in which full utilization of each observation is economically most important. In this book, chief emphasis is placed on efficient methods of analysis, but the reader who has mastered them should have little difficulty in understanding the simpler methods that, whatever their shortcomings, can help rapid provisional evaluation of data. The simplicity of computer analysis as a routine now makes the approximations less interesting than they were when first proposed. Of course it must be conceded that, when data in a new field are scarce, the evidence on which to base a choice of model (such as the normal log tolerance distribution) will be lacking, and refined methods of analysis may convey a spurious impression of reliability. The situation is different for small bodies of data in a well-understood field of study.

Somewhat different considerations arise in connexion with alternative principles of estimation. Any technique that chooses values for α, β in such a way as to minimize discrepancies between p and P collectively for all doses can be regarded as a candidate. It is known that many different functions, when minimized with respect to α, β, yield estimates that are efficient in large samples, but knowledge of general properties in small samples is slight. The only function that has been much used in this way is χ^2 or a variant of it. As stated in § 4.1, the special theoretical merits of maximum likelihood estimation are those applicable in large samples. Although general experience of maximum likelihood estimators in many fields of statistics has been good, there is always the fear that, in a particular class of problem, an alternative technique may perform better in small samples. In the context of probit-type problems, various practical objections to maximum likelihood can be raised: the labour of computation, the possibility of biased estimation, the existence of possible experimental results for which no estimates can be found, the extent to which the iteration influences the numerical values stated for the parameters.

In a series of important papers, Berkson (1946, 1949, 1950, 1953, 1955a, b, c, 1957) advanced arguments in favour of minimizing χ^2. To minimize

$$S \frac{n(p-P)^2}{PQ}$$

requires iterative processes analogous to those for maximum likelihood. By altering the weight associated with each $(p-P)^2$, that is to say by using another multiplier instead of n/PQ, the optimal large-sample properties might be altered to only a negligible extent yet the more realistic needs of data of finite amount might be better met. Berkson explored several possibilities, but based his main argument on the simplest variant on χ^2:

$$S \frac{n(p-P)^2}{pq}.$$

He has been more concerned with the logit transform than with the probit, but many of his ideas apply with equal force to both.

A first order approximation to this quantity is

$$S \frac{nz^2}{pq} (y-Y)^2, \tag{5.6}$$

where, *in this section only*, y is the empirical probit of p, the observed proportion responding, and z is the ordinate corresponding; that is

$$p = \int_{-\infty}^{y} \frac{1}{\sqrt{(2\pi)}} \exp\{-\tfrac{1}{2}u^2\}\, du, \qquad (5.7)$$

and

$$z = \frac{1}{\sqrt{(2\pi)}} \exp\{-\tfrac{1}{2}y^2\}, \qquad (5.8)$$

Y still being defined as the expected probit. The expression in (5.6) Berkson terms the 'normit χ^2' (or, with the other transform, the 'logit χ^2'). Estimates of parameters obtained by minimizing the normit χ^2 may be proved to have the same limiting properties in large samples as do maximum likelihood estimates. They are certainly easier to compute, because no iteration is required, a and b being calculated for a simple weighted regression of y on x (Berkson, 1955*b*).

The question of bias is difficult. Despite the consistency of maximum likelihood estimation, situations are known in which bias can be quite important in small samples. Surprisingly little attention has been given to bias in the quantal response problem. Berkson (1957) studied a hypothetical experiment in which $n = 10$ at each of three equally spaced log doses. He obtained the estimates for each of the 1331 possible sets of results (indeterminate results are discussed below), and so was able to examine the frequency distribution of a, b by maximum likelihood and by minimum normit χ^2. He repeated the whole investigation with the central dose assumed to have true response rate $P = 0.5, 0.6,$ 0.7, or 0.8. He found both methods generally to show biases in their expectations of the two estimates of parameters, the bias tending to become larger when P for the central dose is far from 0.5. In relation to the small size of the experiment, the biases were not great, but they were quite definitely smaller for the normit χ^2 method than for the likelihood method. Also, the variances of a, b, representing the extent to which estimates vary between the possible sets of results from the same true parameters, were smaller for the normit χ^2 method. In an earlier paper (1955*a*), Berkson reached similar conclusions for the logit transform.

This might seem strong evidence in favour of the normit χ^2. Papers by Cramer (1962, 1964) disturb the simple conclusion.

He chose to estimate the parameters μ, σ of the tolerance distribution, so using the relation (3.3) rather than the α, β parameters of equation (4.9). His estimates from any data will correspond exactly with the other parameterization, according to equations (4.10). His findings, on hypothetical experiments similar to but not identical with Berkson's, are the direct contrary of Berkson's: the maximum likelihood estimates tended to agree with the true values more closely than the minimum normit χ^2. In particular, he verified that the behaviour of the estimate of σ (i.e. $1/b$) was the opposite of the behaviour of b. An experiment based on only 10 subjects at each of three doses is small relative to what would commonly be undertaken (though large for the types of enumeration that Berkson and Cramer had to make), and the differences between the two methods are likely to become less important for larger experiments. Whether the Berkson and Cramer findings are general or closely related to the particular numerical specifications is not known, but some of Berkson's work suggests a wide generality. Evidently one cannot conclude that either method of estimation has absolute advantages, and possibly the differences are negligible in practice; if a tolerance distribution is relevant, μ and σ are intrinsically more interesting parameters, α and β being primarily scaffolding for the estimation, and thus maximum likelihood seems favoured.

The special problems of observed responses $p = 0$ or $p = 1$ must be mentioned. Any such response will make an infinite contribution to the normit χ^2. Berkson proposed to escape this difficulty by taking these extreme responses to be $1/2n$ and $1 - 1/2n$ respectively, n being the number of subjects at the dose. In his analyses, he makes such adjustments wherever necessary and uses the normit χ^2 as though these had been observed. The procedure is equivalent to an assumption that an observation of 0 subjects responding or failing to respond may be interpreted as a 'frequency' of $\frac{1}{2}$. No theoretical justification is presented. Certainly this is not comparable with the use of working probits, which are purely a computational device for solving certain equations and do not distort the solutions obtained. Possibly this adjustment of extreme responses tends to improve the estimation of the parameters, but even for the particular numerical examples under discussion this has not been adequately tested. The Berkson and Cramer studies have compared the normit χ^2

method adjusted in this fashion with the standard likelihood method. Cramer's finding that expression of the model in terms of different parameters can appreciably affect the relative merits of alternative estimation methods discourages any idea that the arbitrary adjustment of extreme responses will profoundly affect the practical problem, but further theoretical investigation could be interesting.

A further difficulty lies in the possibility that data fail to determine estimates of the parameters. If every dose gives either $p = 0$ or $p = 1$, no estimation of α is possible, and estimation of β may tend to infinity. Other indeterminate data can arise, such as p being the same at all doses, or every dose except one having $p = 0$ or $p = 1$ (Table 1 of Cramer, 1964). Although such occurrences will be rare in experiments with well-chosen doses unless the number of subjects per dose is small, the probability for each configuration is non-zero. Berkson has criticized maximum likelihood because for some data it does not produce unique estimates or may even fail to give any estimates. However, his normit χ^2 method avoids this weakness (for most possible experimental results) only by its arbitrary adjustment of the extreme responses, and of course the same gain in tidiness could be affected by using the adjustment with the likelihood method. To claim an advantage for normit χ^2 because it escapes the indeterminacy difficulty seems scarcely fair when the escape is achieved by an adjustment that could equally well be used with likelihood.

Although the existence of possible experimental results that do not lead to estimates of parameters is 'untidy' and inconvenient, it is not something peculiar to quantal response data. Further development of the theory of statistical inference may lead to better understanding. Some of the criticisms that Berkson has raised are much to the point, but the problem cannot be resolved merely by an artefact that calls the troublesome response rates something different. On this issue, also, the claim that a method has been found that is superior to maximum likelihood when the numbers of subjects are small remains unproven.

Another objection that has been stated against maximum likelihood, or any other estimation process dependent upon iteration, is that the numerical conclusions eventually quoted depend to some extent on the starting point of iteration and the number of cycles calculated. Without a computer, this has point. Statistical writings, including earlier editions of this book, have encouraged the idea that one or two cycles of the iterative probit-regression method will commonly suffice to approximate to the maximum likelihood estimates. Fairly naturally, the numerical values obtained will in some degree reflect the provisional values with which the calculations begin (§ 4.4), and whether one or two or three iterations are used will further affect the outcome. Even so, practical conclusions are seldom affected appreciably by these factors. An example has been given elsewhere (Finney, 1951) of how very different, and even deliberately poor, provisional lines after one cycle of iteration give numerical estimates that do not differ by amounts that matter in practice.

With a computer, certainly the practical issue can be adequately covered. Sufficient iterations can be completed to give numerical accuracy to at least as many digits as the data justify. The philosophical problem remains. The likelihood function will be maximized by an unknown but unique combination of parameter-values. Any single sequence of calculations aimed at maximizing the likelihood will yield values whose arithmetical closeness to the truth cannot be assessed unless by reference to another sequence in which more digits are retained. Two runs of a program with different starting values or different numbers of iterations, may give log-likelihoods agreeing to 10 digits, yet the estimates of parameters may differ slightly in the fifth or sixth digits. The likelihood function usually changes very slowly as α, β are changed in the neighbourhood of the maximum. Nevertheless, this difficulty of arithmetical exactness is scarcely in itself a reason for changing to another method of estimation (such as minimum normit χ^2) that more readily determines estimates uniquely but is of questionable merit in other respects.

5.10   Comparison of probits and logits

This Chapter will be ended with a comparison between probit and logit models for one especially extensive series of records, consisting of 3900 subjects distributed over 25 'dose' levels. Both models fit the data well, the estimates are in close agreement, and no evidence appears to give preference to either. It is only fair to comment that one other shorter series of similar records

(Burrell, Healy and Tanner, 1961) showed markedly better agreement with the logit model.

Table 5.1   Age of menarche in 3918 Warsaw girls

| Mean age of group (years) | No. of girls | No. having menstruated | | |
		Recorded	Expected (probit)	Expected (logit)
9.21	376	0	0.10	0.76
10.21	200	0	1.08	2.06
10.58	93	0	1.25	1.74
10.83	120	2	2.81	3.34
11.08	90	2	3.53	3.72
11.33	88	5	5.51	5.36
11.58	105	10	10.05	9.33
11.83	111	17	15.56	14.19
12.08	100	16	19.70	18.06
12.33	93	29	24.72	23.15
12.58	100	39	34.51	33.26
12.83	108	51	46.64	46.27
13.08	99	47	51.69	52.46
13.33	106	67	64.78	66.67
13.58	105	81	72.95	75.42
13.83	117	88	90.00	92.79
14.08	98	79	81.56	83.51
14.33	97	90	85.65	86.97
14.58	120	113	110.61	111.45
14.83	102	95	96.89	97.05
15.08	122	117	118.26	118.00
15.33	111	107	109.01	108.55
15.58	94	92	93.06	92.61
15.83	114	112	113.39	112.87
17.58	1049	1049	1048.98	1048.40

Ex. 12. *Comparison of probits and logits in a study of Warsaw girls.* Milicer and Szczotka (1966) reported records of a sample of 3918 Warsaw girls in 1963, showing for each her age and whether or not she had reached menarche (cf. § 3.1). These were classified into 3-month age ranges (with wider intervals at the extremes), and probits and logits were used (independently) to estimate parameters of the frequency distribution of age at menarche. In this instance, no logarithmic 'dose' metameter was used, as

general evidence is that age itself gives a good linear relation with the response metameter. For statistical analysis, ages have been regarded as grouped at the centres of the 3-month ranges.

Table 5.1 contains the data and expected frequencies obtained in the two analyses. Both series appear to agree well with the observed frequencies: the χ^2 values, without any grouping at the extremes, were almost identical, 21.901 (probits) and 21.870 (logits) with 23 degrees of freedom. The regression equation for probits is

$$Y = -6.8189 + 0.9078x \qquad (5.9)$$

(the original authors, almost certainly after fewer iterations, obtained $Y = -6.513 + 0.879x$), and that for logits is

$$Y = -10.6131 + 0.8160x. \qquad (5.10)$$

The difference between equations (5.9), (5.10), after allowing for the 5.0 in probits, is not particularly relevant. In fact, the two estimates of median age are almost identical. Values of g are so small as to justify quotation of standard errors: for probits the median age is 13.019 ± 0.039 years, with 95 % limits at 13.09 and 12.94 years, and for logits 13.007 ± 0.039 years, with 95 % limits at 13.08 and 12.93 years. From the probit computations, the standard deviation of age at menarche is estimated as $1/b$, or 1.10 years.

Milicer (1968) subsequently published further data and analyses of this kind.

6

The comparison of effectiveness

6.1 Relative potency

In many experimental situations, the level of tolerance of test subjects may vary substantially from day to day, or with minor changes in environment. Consequently the estimate of the ED 50 on a particular occasion may be untrustworthy as a guide to the future. However, if a preparation to be studied is compared with a standard preparation, its potency may be expressible in terms that are more stable from one occasion to another. In the second of his early papers on the probit method and its applications (1935b), Bliss considered the measurement of differences between two or more comparable series of dosage mortality records. His suggestion of measuring differences in terms of relative potency has been widely adopted in the comparison of toxicity data, and is in fact an example of the standard procedure of biological assay (Finney, 1947a, 1964).

When the tolerances of the test subjects in respect of a stimulus are normally distributed on a logarithmic scale, the variances for closely related stimuli are often nearly equal. This equality of variances is manifested by parallelism of the probit regression lines, and the comparison of different series of data is then particularly simple. For any non-logarithmic dose metameter, parallel regression lines are without meaning and are less likely to be found.

If two series of quantal response data yield parallel probit regressions against the logarithm of the dose, the difference between values of x that produce the same response rates is constant. This implies a constant *relative potency* at all levels of response, where the relative potency of two stimuli is defined as the ratio of equally effective doses. The relative potency then provides a convenient description of the difference between the two series. The constant x-difference, usually denoted by M, is equal to the difference between the log ED 50's. This *relative dose metameter* of the second series, or the amount by which a log dose in the first series exceeds an equally effective log dose in the second, is

$$M_{12} = m_1 - m_2 = \bar{x}_1 - \bar{x}_2 - (\bar{y}_1 - \bar{y}_2)/b, \qquad (6.1)$$

where suffices indicate the two series and b is the common slope of the regression lines. Of course, M is an estimate of an unknown true value. The relative potency may be symbolized by ρ_{12}, in the sense that a dose z of preparation 2 has the same average effect as a dose $\rho_{12}z$ of preparation 1, and this parameter is estimated by

$$R_{12} = 10^{M_{12}}. \qquad (6.2)$$

The variance of M_{12} is approximated by

$$V(M_{12}) = \frac{1}{b^2}\{V(\bar{y}_1) + V(\bar{y}_2) + (\bar{x}_1 - \bar{x}_2 - M_{12})^2\, V(b)\}; \qquad (6.3)$$

the standard error, s_M, derived from this variance may be used for assigning approximate fiducial limits to M, exactly as in Example 3. However, the general method of §4.7 for obtaining fiducial limits is usually to be regarded as superseding variance formulae. Unless b has been estimated with such precision that $g = t^2 V(b)/b$ is small, use of equation (6.3) is unsafe and will give limits that are too narrow. Exact fiducial limits (Cochran, 1938), determined as in §4.7, take the form

$$M_{12} + \frac{g}{1-g}(M_{12} - \bar{x}_1 + \bar{x}_2) \pm \frac{t}{b(1-g)} \sqrt{[(1-g)\{V(\bar{y}_1) + V(\bar{y}_2)\}}$$
$$+ (M_{12} - \bar{x}_1 + \bar{x}_2)^2\, V(b)]; \qquad (6.4)$$

t is the normal deviate for the significance level. In these and other formulae, if a heterogeneity factor is being used, the factor h must be incorporated into all variances, and t must be a t-deviate (Table VIII) based upon the appropriate degrees of freedom. Equation (6.4) reduces to $M_{12} \pm ts_M$ when g is very small.

Numerical estimation of M is accomplished by extension of the methods of Chapters 3 and 4. The provisional regression lines are drawn parallel, though the hypothesis involved will later be tested (§6.2). Working probits are formed and sums of squares and products calculated for each series separately. A new approximation to the common slope, with which the computations may be repeated if necessary, is given by adding components from each series, and taking

$$b = \frac{{}_1S_{xy} + {}_2S_{xy}}{{}_1S_{xx} + {}_2S_{xx}} = \frac{\Sigma S_{xy}}{\Sigma S_{xx}}, \tag{6.5}$$

where ${}_1S$ and ${}_2S$ refer to the preparations and Σ indicates summation over the two. When a satisfactory estimate of b has been obtained, heterogeneity may be tested in the usual manner by comparing observed and expected numbers of subjects; Example 13 illustrates the simplest calculation of χ^2. If there is no heterogeneity, the variances to be inserted in equations (6.3), (6.4) are

$$V(\bar{y}_1) = 1/{}_1Snw, \tag{6.6}$$

$$V(\bar{y}_2) = 1/{}_2Snw, \tag{6.7}$$

$$V(b) = 1/\Sigma S_{xx}. \tag{6.8}$$

The logic of the parallelism test is important, yet it defies exact definition. If β were *known* to be the same for both preparations, no test would be needed. If there were no reason to expect equality of variances, absence of significant evidence of non-parallelism would be an illogical basis for concluding that the regressions should be parallel and that ρ is a meaningful parameter; indeed non-significance could always be secured by conducting a small or ill-planned experiment! The difficulty lies between these extremes. Often the nature of two preparations (chemical or pharmacological) gives *a priori* reason for thinking parallelism likely, and a significance test in a good experiment can then be regarded as reassurance that nothing untoward has occurred rather than proof of a special property. Preferably such judgements should rest on the evidence of a group of experiments, as a safeguard against optimistically subjective decisions.

Though the relative potency of two preparations can be estimated from data on only two doses of each, provided that the regression lines do not markedly depart from parallelism, a minimum of three doses in each series is desirable. In the first discussion of relative potency for probit analysis, Bliss (1935 b) gave expressions for M when only two doses have been used and also when a single dose of the test material is assayed against a series of doses of a standard. The latter is an unsatisfactory method of assaying relative potency, because it permits no test of the assumption of parallelism that is implicit in the estimation and interpretation of ρ. Experiments depending on only two doses of each preparation should not be used unless the regression relation is already known to be linear, since the data themselves can provide no information on the existence of a curvature (Finney, 1964).

6.2   Numerical calculation of relative potencies

The theory outlined in § 6.1 readily extends to experiments in which more than two preparations are included. The central feature is the obvious generalization of equation (6.5), after which any pair of preparations can be compared. Example 13 illustrates the whole procedure.

Ex. 13. *Relative potencies of four analgesics.* Grewal (1952) reported an experiment in which the analgesic potencies of amidone, phenadoxone, and pethidine were estimated relative to morphine. The technique was to record how many standard electric shocks could be applied to the tail of a mouse before the mouse squeaked; a specified dose of drug was then administered and a new shock-count recorded. If the number of shocks increased by 4 or more, the mouse was classed as responding. Large numbers of mice (60–120) were tested at several doses of each drug.

Table 6.1 contains the data for this experiment, with the basic calculations as in Table 4.1. Fig. 6.1 illustrates the empirical probits and provisional regression lines; this is comparable with Fig. 3.3, but the four lines have been constrained to be parallel to the accuracy of a rough drawing. These lines have the approximate equations:

$$\left.\begin{aligned}
Y_M &= 3.70 + 2.30x, \\
Y_A &= 3.88 + 2.30x, \\
Y_{Ph} &= 4.95 + 2.30x, \\
Y_P &= 2.60 + 2.30x.
\end{aligned}\right\} \qquad (6.9)$$

By arithmetic or from the graph, the log ED 50 can be obtained for each line, for example for morphine and amidone 0.57 and 0.49 respectively. Equation (6.1) then gives 0.08 for the logarithm of relative potency. Similarly the values for phenadoxone and pethidine relative to morphine are 0.55 and − 0.47. The antilogarithms of these quantities, provisional estimates of potencies relative to that of morphine, are 1.20, 3.55, 0.34 respectively.

Table 6.1   First stage of calculations for estimation of potencies of three analgesics relative to morphine

x	n	r	$p(\%)$	Empirical probit	Y	nw	y	nwx	nwy	Y (new)
				Morphine						
0.18	103	19	18	4.08	4.1	48.6	4.08	8.748	198.288	4.0
0.48	120	53	44	4.85	4.8	75.3	4.85	36.144	365.205	4.8
0.78	123	83	67	5.44	5.5	71.5	5.44	55.770	388.960	5.5
						195.4		100.662	952.453	
				Amidone						
0.18	60	14	23	4.26	4.3	31.9	4.26	5.742	135.894	4.3
0.48	110	54	49	4.97	5.0	70.0	4.98	33.600	348.600	5.0
0.78	100	81	81	5.88	5.7	53.2	5.87	41.496	312.284	5.8
						155.1		80.838	796.778	
				Phenadoxone						
− 0.12	90	31	34	4.59	4.7	55.4	4.59	− 6.648	254.286	4.7
0.18	80	54	68	5.47	5.4	48.0	5.47	8.640	262.560	5.4
0.48	90	80	89	6.23	6.1	36.4	6.22	17.472	226.408	6.2
						139.8		19.464	743.254	
				Pethidine						
0.70	60	13	22	4.23	4.2	30.2	4.23	21.140	127.746	4.2
0.88	85	27	32	4.53	4.6	51.0	4.53	44.880	231.030	4.6
1.00	60	32	53	5.08	4.9	38.1	5.08	38.100	193.548	4.9
1.18	90	55	61	5.28	5.3	55.4	5.28	65.372	292.512	5.3
1.30	60	44	73	5.61	5.6	33.5	5.61	43.550	187.935	5.6
						208.2		213.042	1032.771	

Table 6.2 contains the next stage of calculations, sums of squares and products being formed for each preparation separately exactly as in Example 6 and Table 4.1. The corre-

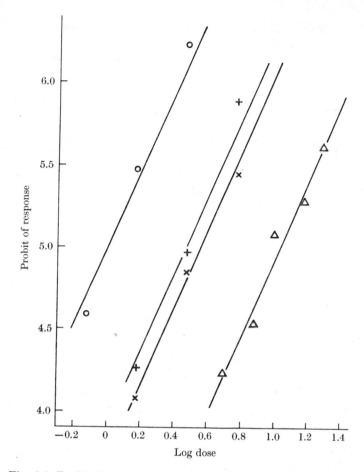

Fig. 6.1. Probit diagram, with provisional regression lines, for effects of four analgesics on mice (Ex. 13).

×: morphine   ○: phenadoxone
+: amidone   △: pethidine

sponding values of S_{xx}, S_{xy}, S_{yy} are then summed over the four drugs; for example

$$\Sigma S_{xx} = 10.567\,46 + 7.395\,74 + 8.029\,60 + 8.149\,75$$

$$= 34.142\,55.$$

Table 6.2   Second stage of calculations for
relative potencies of analgesics

	$\dfrac{1}{Snw}$	$\bar{x}$	$\bar{y}$
Morphine	0.005 117 71	0.5152	4.8744
Amidone	0.006 447 45	0.5212	5.1372
Phenadoxone	0.007 153 08	0.1392	5.3166
Pethidine	0.004 803 07	1.0233	4.9605

	$Snwx^2$	$Snwxy$	$Snwy^2$	Subtract S_{xy}^2/S_{xx}
Morphine	62.424 36	514.3790	4696.202	
	51.856 90	490.6644	4642.614	
	10.567 46	23.7146	53.588 −	53.218 = 0.370
Amidone	49.528 44	435.3704	4148.044	
	42.132 70	415.2801	4093.199	
	7.395 74	20.0903	54.845 −	54.575 = 0.270
Phenadoxone	10.739 52	125.4223	4011.634	
	2.709 92	103.4814	3951.549	
	8.029 60	21.9409	60.085 −	59.954 = 0.131
Pethidine	226.146 36	1075.7563	5168.934	
	217.996 61	1056.7896	5123.035	
	8.149 75	18.9667	45.899 −	44.141 = 1.758
Total	34.142 55	84.7125	214.417 −	210.184 = 4.233

Hence an improved estimate of the regression coefficient, after
one iteration, is obtained by equation (6.5) as

$$b = 84.7125/34.142 55$$
$$= 2.4811. \tag{6.10}$$

This regression coefficient is used in equation (4.18) to give
improved estimates of the lines:

$$\left.\begin{aligned}
Y_M &= 4.8744 + 2.4811(x - 0.5152) \\
&= 3.596 + 2.4811x, \\
Y_A &= 3.844 + 2.4811x, \\
Y_{Ph} &= 4.971 + 2.4811x, \\
Y_P &= 2.422 + 2.4811x.
\end{aligned}\right\} \tag{6.11}$$

Equations (6.11) have been used to calculate a new set of values of Y, in a final column of Table 6.1, with which a second iteration could be begun. Agreement is good, and further calculation is scarcely necessary. If final values of Y do not differ from those at the start of an iteration by more than 0.1, and differences are not predominantly in one direction, further iteration is unlikely to make changes to estimates that are of practical importance, although many more cycles may be needed to secure close numerical approach to true maximum likelihood estimates.

From equations (6.11), estimates of relative potency are readily formed. They are valid only if the data do not controvert the hypothesis that the regression coefficient is the same for all preparations. Both this and heterogeneity can be examined by χ^2 tests. For the present experiment, inspection of the data or of Fig. 6.1 almost suffices to convince that parallelism and homogeneity are satisfactory. A full test procedure requires that lines be fitted independently to the four preparations without the constraint of parallelism, as a basis for calculating a total heterogeneity χ^2; the difference between this χ^2 and one calculated from the 'total' line of Table 6.2, being determined by the extent to which lines with different slopes fit the data better, is a χ^2 that tests parallelism. In practice, the labour of calculating separate lines is usually omitted and the corresponding calculations for each
$$S_{yy} - S_{xy}^2/S_{xx}$$
made on the appropriate lines of Table 6.2. When, as in this example, truly independent lines are very nearly parallel, the effect of this change is negligible. When deviations from parallelism are more marked, the two forms of test will usually agree in giving a large χ^2. Only in doubtful cases is there really need for the full calculations. Table 6.3 shows this analysis of χ^2. A column of mean squares has been included, although unnecessary here. Had the heterogeneity χ^2 been significant, the second mean square would be used as a heterogeneity factor if this were thought appropriate. The test of parallelism would then have been a variance ratio test of the two mean squares (Fisher and Yates, 1964, Table V) instead of χ^2. In the present instance, reference to Table VII shows that Table 6.3 implies no worries.

Now
$$g = (1.960/2.4811)^2 \div 34.14255$$
$$= 0.01828.$$

Table 6.3   Analysis of χ^2 for Example 12 (first cycle)

	d.f.	Sum of squares	Mean square
Parallelism of regressions	3	1.704	0.568
Heterogeneity	6	2.529	0.422
Total	9	4.233	

Values in Table 6.2 can be used in conjunction with g and equations (6.1), (6.4) to give M and its fiducial limits for each preparation. Thus for amidone is found

$$M = 0.0999$$

with 95 % limits at 0.1889, 0.0149. Results for the three potencies are summarized in Table 6.5.

Table 6.4   Analysis of χ^2 for Example 12
(computer calculations)

	d.f.	Sum of squares	Mean square
Parallelism of regressions	3	1.528	0.509
Heterogeneity	6	2.335	0.389
Total	9	3.863	

The data have also been analysed by the computer program. Table 6.4 may be compared with Table 6.3. The final line is the total χ^2 resulting from maximizing the likelihood with respect to the separate α-parameters for the preparations and a common β, so as to estimate five parameters simultaneously. The degrees of freedom are always the total number of dose groups minus the number of parameters estimated. The middle line comes similarly from fitting a separate α and β for each preparation and adding the four χ^2 values. Numerically, Table 6.4 may look rather different from Table 6.3, but the practical implications are the same. The potency estimates from the computer analysis, calculated exactly as before but now effectively having achieved maximum likelihood limits after many iterations, are also put in Table 6.5. Agreement between potency estimates after one cycle of iteration and in the limit is good for each material, in support of the opinion that a single iterative cycle will often suffice for practical purposes.

Table 6.5   Estimates of relative potency for Example 12

	First Cycle		Computer	
	Estimate	95 % limits	Estimate	95 % limits
Amidone	1.259	1.545, 1.035	1.248	1.535, 1.024
Phenadoxone	3.583	4.459, 2.923	3.554	4.431, 2.894
Pethidine	0.336	0.404, 0.280	0.333	0.401, 0.277

6.3   Combination and comparison of relative potencies

Sometimes the relative potency of two preparations has been estimated in two or more independent experiments, and the question of combining the information arises. If the probit regression coefficient, β, is the same in all experiments, a new analysis might obtain a composite estimate of this parameter; greater precision in b brings a smaller g, and some of the theoretical difficulties of combining information are lessened.

The simplest procedure is that appropriate when variances of M, equation (6.3), can safely be used. The weighted mean of several estimates can then be formed, using for weights the reciprocals of the variances of M (Cochran, 1938), these variances having incorporated any necessary heterogeneity factors (§ 4.6). A weighted sum of squares of the deviations of the separate M from their mean should be tested as a χ^2 (Example 14), as evidence of whether the M values do agree satisfactorily (Miller, Bliss and Braun, 1939). If this test discloses no disagreement, the weighted mean may be assigned a variance equal to the reciprocal of the sum of the weights.

Ex. 14. *Combination of relative potencies.* Cohen and Leppink (1956), in an experiment to be described more fully later (Example 21), obtained four independent estimates of the potency of a pertussis vaccine (here identified solely as 'no. 4') relative to a standard NIH reference vaccine. Batches of 15–20 mice were 'protected' at three different doses of each vaccine, and the death rates after intracerebral injection of *Hemophilus pertussis* were recorded.

Doses of the standard vaccine were measured in units of 10^{-4} ml, and those of vaccine no. 4 in millions of organisms used in preparing the vaccine. Hence a measure of relative potency will be in units of ml of standard per 10^{10} organisms. Values of M, from new computations for the whole experiment, are -1.733, -1.488, -1.831, -1.875, with asymptotic variances from equation (6.3) respectively 0.048 15, 0.024 66, 0.031 42, 0.076 41. Using reciprocals of the variances as weights, a mean value of M is obtained:

$$\bar{M} = -\frac{20.8 \times 1.733 + \ldots + 13.1 \times 1.875}{20.8 + 40.5 + 31.8 + 13.1} \qquad (6.12)$$

$$= -179.0987/106.2$$

$$= -1.6864.$$

Before using $\bar{M}$, the heterogeneity of the separate estimates should be tested. The appropriate χ^2 is

$$\chi^2_{[3]} = 20.8 \times (1.733)^2 + \ldots + 13.1 \times (1.875)^2$$

$$- (179.0987)^2/106.2 \qquad (6.13)$$

$$= 2.77.$$

In (6.13), 179.0987 and 106.2 are the numerator and denominator already calculated for $\bar{M}$; the general formula should be obvious. Evidently (Table VII) χ^2 shows no evidence of discrepancies in the four values of M.

Hence the variance of $\bar{M}$ is 1/106.2, or 0.009 416, and the standard error 0.0970 (see also Example 18 and Table 6.8).

When two preparations have been tested in different experiments, estimation of M as the difference in the estimated log ED 50's may be misleading (even if values of b were similar). The level of susceptibility of the subjects may have differed in the two experiments, or experimental conditions may have altered the absolute potency of a preparation, though relative potencies within an experiment may be unaffected. If each preparation can be compared with a third which has been tested in both experiments, relative potency may still be estimated independently of susceptibility changes. Estimates M_{12} and M_{13} are made 'within experiments'; the required M is then estimated as

$$M_{23} = M_{12} - M_{13}, \qquad (6.14)$$

and

$$R_{23} = R_{12}/R_{13}. \qquad (6.15)$$

The variance of M_{12} is the sum of the variances of M_{13} and M_{23}.

The method should be used only when conditions in the two experiments are sufficiently similar for the estimate to be relevant. Cochran (1938) suggested this procedure for testing the significance of a difference in potency of poisons used in separate experiments. The significance should be assessed from the normal distribution, unless the variances have been adjusted by a heterogeneity factor.

Ex. 15. *Comparisons of relative potencies.* Tattersfield and Martin (1938, Table I), using rotenone and a resin derived from a Sumatra-type derris root in spraying trials against *Aphis rumicis*, found

$$M_{rs} = -0.800 \pm 0.025.$$

In a second experiment with the same technique, using rotenone and a *Derris elliptica* resin, they found

$$M_{rd} = -0.552 \pm 0.031.$$

Now $(0.025)^2 + (0.031)^2 = 0.001\,586 = (0.040)^2;$

hence the relative dose metameter of the two resins is estimated as

$$M_{ds} = -0.248 \pm 0.040,$$

and the Sumatra-type resin is estimated to be 0.56 times as toxic as the *Derris elliptica*. Approximate fiducial limits can be calculated from the standard error.

Ex. 16. *Relative toxicities of seven derris roots.* Earlier experiments by Tattersfield and Martin (1935) provide a more complex example. Seven samples of ether extracts of derris root were tested for their toxicity to *Aphis rumicis*, a different pair of roots being compared on each of six occasions. The relation between proportion of insects responding (i.e. badly affected, moribund or dead) and logarithm of dose (i.e. concentration of ether extract in mg/l) has been recomputed for this Example.† Values of m and M were (suffices identifying the seven roots):

$$\text{Occasion 1} \quad m_2 = 1.560, \quad m_5 = 1.636, \quad M_{25} = -0.076,$$
$$2 \quad m_5 = 1.716, \quad m_6 = 1.473, \quad M_{56} = 0.243,$$
$$3 \quad m_5 = 1.621, \quad m_7 = 1.386, \quad M_{57} = 0.235,$$
$$4 \quad m_4 = 1.650, \quad m_5 = 1.691, \quad M_{45} = -0.041,$$
$$5 \quad m_1 = 1.494, \quad m_3 = 1.488, \quad M_{13} = 0.006,$$
$$6 \quad m_1 = 1.431, \quad m_7 = 1.417, \quad M_{17} = 0.014.$$

The relative potency of any two roots can be estimated from comparisons within experiments. For example, the relative potency of roots 1 and 4 may be estimated by

$$M_{14} = M_{17} - M_{57} - M_{45} \qquad (6.16)$$
$$= -0.180.$$

Because the number of experiments is one less than the number of roots, this estimation is unique. The variance of M_{14} would be assessed by adding variances computed for M_{17}, M_{57}, M_{45} on occasions 6, 3, 4, with a heterogeneity factor, if thought appropriate. The analyses showed indications of heterogeneous deviations from the regression lines and no systematic pattern for these; by totalling over the six experiments,

$$\chi^2 = 109.913 \quad \text{with} \quad 63 \, \text{d.f.}$$

After so many years, to discuss reasons for the large χ^2 is useless: it may have been a consequence of imperfections in the technique of early experiments.

A convenient way of summarizing the ED 50's is to take a mean value for Root no. 5, which was tested four times, and to place the others at the estimated distances from this; this mean might be obtained by appropriate weighting, as in Example 14, but the simple mean of the four values, 1.666, is almost equally good. The log ED 50's may then be written:

Root no. 1: 1.445   Root no. 5: 1.666

2: 1.590   6: 1.423

3: 1.439   7: 1.431

4: 1.625

† The method of Chapter 7 was used to take account of a natural response rate.

Had more trials of pairs of the seven roots been made, some of the relative potencies could have been estimated by more than one chain of pairs. In order to obtain the best estimates, the method of least squares would then have been needed (see Example 17), though data from carefully controlled trials might be sufficiently consistent for satisfactory estimation by simple averaging.

Ex. 17. *Combination of relative potencies by the method of least squares.* Experiments by Martin (1940) on four different derris roots as poisons for *A. rumicis* provide a more complex example of the estimation of relative potencies. Of these roots, W.211, W.212, W.213, and W.214, two sets of three were tested on different occasions, and on a third occasion W.211 and W.213 were compared with rotenone. The estimated $\log ED50$'s are shown in Table 6.6, together with the weights (reciprocals of the variances). For each occasion, three values of m were obtained from a common regresssion coefficient as in Example 13, except that a modified analysis (§ 7.3) was used in order to take account of mortality in control batches of insects. All values of g were satisfactorily small (0.04 or less for 0.95 limits). The estimated $\log ED50$'s for any one occasion are not independent, since they are based on the same regression coefficient. Hence the variance of M for two poisons on one occasion is not simply the sum of two variances of the form of equation (3.12), but is an expression like equation (6.3). This complication has been ignored here, because the second term of equation (3.12) was always small relative to the first, as may be expected when experiments can be so planned that each $\bar{y}$ is near to 5.

Table 6.6   Estimated log ED 50's and weights for rotenone and four derris roots

Date	Rotenone	W.211	W.212	W.213	W.214
15.vi.38	—	2.433 (1710)	2.231 (3350)	—	2.019 (4500)
27.vi.38	—	2.429 (3190)	2.137 (2140)	2.268 (2770)	—
18.vii.39	1.031 (1390)	2.238 (1110)	—	2.226 (1170)	—

In order to estimate relative potencies under comparable conditions, account must be taken of differences in insect susceptibility on the three occasions. If relative potencies remain the same on all occasions, each log ED 50 must be expressible as the sum of three components, the general mean, a preparation (or column) constant, and an occasion (or row) constant. The constants can be suitably defined and then estimated by the method of least squares. For example, m_r, m_1, m_2, m_3, m_4 might denote average log ED 50's for rotenone and the four roots; d_1, d_2 can be constants for the first two dates and $-(d_1 + d_2)$ that for the third date. (This makes the average deviation for the three dates zero, whereas to introduce d_3 for the third occasion would create a redundancy and prevent unique determination of a solution). Thus for W.211 on the second occasion, the expected log ED 50 is

$$m_1 + d_2$$

and for rotenone on the third occasion the expected log ED 50 is

$$m_r - d_1 - d_2.$$

The 7 constants are estimated by minimizing the sum of the 'weighted' squares of differences between these expectations and the numerical values in Table 6.6: each square is multiplied by the weight in Table 6.6 before summation. Many standard schemes of calculation are available for this minimization, and computer techniques are particularly convenient. The estimates are obtainable from a set of 7 equations, each constructed by equating the observed and expected totals for one of the columns or one of the first two rows of Table 6.6. For example, the column for W.213 gives the equation

$$2770(m_3 + d_2) + 1170(m_3 - d_1 - d_2) = 2770 \times 2.268 + 1170 \times 2.226$$

$$= 8886.780. \qquad (6.17)$$

The solutions in respect of the m are summarized in Table 6.7; values for the occasion constants are $d_1 = 0.0770$, $d_2 = 0.0164$, indicating rather low susceptibility on the first occasion and rather high on the third. If each of the 7 estimates is multiplied by the right-hand side of the corresponding equation (e.g. $8886.780m_3$) and the total of these products subtracted from the weighted sum of squares of the nine log ED 50's in Table 6.6, the result is approximately a χ^2 for testing whether the hypothesis represented by the additive constants adequately describes Table 6.6. However, numerical accuracy to several more decimal places is essential in order to eliminate errors of rounding. Here

$$\chi^2_{[2]} = 12.22, \qquad (6.18)$$

the degrees of freedom being the difference between the number of entries in Table 6.6 and the number of constants estimated $(9 - 7)$. The statistical significance of χ^2 indicates some heterogeneity, although the relative potencies in Table 6.7 are a good enough representation of Table 6.6 for many practical purposes.

Table 6.7   Comparative potencies of rotenone
and four derris roots

Material	Analysis of Table 6.6			Full 9-parameter analysis		
	log ED50	ED50 (mg/l)	R	ED50 (mg/l)	R	Fiducial limits
Rotenone	1.1243	13.3	1.0000	13.1	1.0000	—
W.211	2.3815	240.7	0.0553	240.0	0.0546	0.0654, 0.0455
W.212	2.1410	138.4	0.0962	137.0	0.0957	0.1170, 0.0779
W.213	2.2717	187.0	0.0712	186.1	0.0705	0.0842, 0.0588
W.214	1.9420	87.5	0.1522	87.0	0.1508	0.1880, 0.1201

The most useful way of solving the set of equations of which (6.17) is the fourth is to obtain the inverse matrix of coefficients. The equations are written in order, corresponding with their formation from m_r, m_1, m_2, m_3, m_4, d_1, d_2 on the left. The totals on the right-hand sides are replaced by 1, 0, 0, 0, 0, 0, 0 and the new equations solved; this operation is repeated with 0, 1, 0, 0, 0, 0, 0 on the right, and so seven times in all until equations with 0, 0, 0, 0, 0, 0, 1 on the right have been solved. The sets of solutions are written in seven parallel columns, which will be found to form a 7×7 array symmetrical about the diagonal from northwest to south-east: this is the inverse matrix. The totals on the right of the original equations (8886.780 and the like) are then multiplied in order by the entries in the first row of the matrix and the products summed: the result is m_r. Repetition with the second row in place of the first gives m_1, and so on.

One trouble is that the elements of the inverse matrix must be evaluated to many decimal places, because otherwise multiplication by large numbers may cause rounding errors to be serious. However the whole operation is now a standard computer process. The merit is that these same elements give the variances and covariances of the estimates of the parameters. Heterogeneity apart, one might expect the variances of the estimated $\log \text{ED}\,50$'s to be approximately the reciprocals of the sums of the weights in the columns of Table 6.6 (1390, 6010, 5490, 3940, 4500 respectively). In fact, the lack of symmetry in Table 6.6 makes the variances appreciably higher. The first five rows and columns of the inverse matrix (omitting the elements relating to d_1, d_2) are, to an accuracy sufficient for illustration here,

$$V =$$

$$\begin{pmatrix} 0.000\,995 & -0.000\,051 & -0.000\,147 & 0.000\,015 & -0.000\,181 \\ -0.000\,051 & 0.000\,181 & 0.000\,022 & 0.000\,014 & 0.000\,010 \\ -0.000\,147 & 0.000\,022 & 0.000\,265 & -0.000\,023 & 0.000\,117 \\ 0.000\,015 & 0.000\,014 & -0.000\,023 & 0.000\,301 & -0.000\,076 \\ -0.000\,181 & 0.000\,010 & 0.000\,117 & -0.000\,076 & 0.000\,435 \end{pmatrix}.$$

$$(6.19)$$

At least three more places of decimals would be needed before the elements could safely be used as described in the previous paragraph. These variances could then be used in fiducial limit or similar calculations for $\text{ED}\,50$'s and relative potencies.

In the present example, the large χ^2 in equation (6.18) demands that account be taken of heterogeneity, whether by accepting estimates only as rough guides or by incorporating a heterogeneity factor into the analysis. Here, unfortunately, the degrees of freedom for χ^2 are too few for any satisfying judgement of what should be done.

An alternative is to make a new and deeper analysis of all the data. The regression coefficients for the three days were very similar, being between 5.38 and 5.72 with standard errors about 0.4. Consequently to fit probit regressions to the whole of the data simultaneously, with the constraint that relative potencies of the five preparations remain constant, seems sensible. This requires that the α-parameters in the regression equation shall have the same structure as was used above for the $\log \text{ED}\,50$'s.

If α_r, α_1, α_2, α_3, α_4 refer to the preparations, the equation for W.211 on the third occasion will be

$$Y = \alpha_1 - \delta_1 - \delta_2 + \beta x, \tag{6.20}$$

where δ_1, δ_2 relate to the first and second occasions. With the inclusion of a parameter for natural mortality (again see Chapter 7), this means a total of 9 parameters, and an estimation problem that a few years ago would have been truly intimidating. To-day, computer maximization of the likelihood over all the 53 dose groups simultaneously is practicable, each P being obtained from equation (3.1) with the appropriate version of Y.

This solution has been completed. For heterogeneity, it gave

$$\chi^2_{[44]} = 66.203 \tag{6.21}$$

with no indication of breakdown in linearity or parallelism. Therefore, the heterogeneity factor

$$h = 1.5046 \tag{6.22}$$

should be used. For estimates of ED 50 and the relative potency, the standard formulae based upon the estimates of the α-parameters are required, the major aim of this complicated analysis having been to eliminate the disturbances due to the δ_1, δ_2. An inverse matrix, analogous to (4.7), (5.1), and (6.19), but of size 9×9, gives variances and covariances except for the extra factor h. The general form of Fieller's theorem is used to calculate fiducial limits to such quantities as $(a_1 - a_r)/b$, and antilogarithms of these finally give the limits to the relative potency estimates. Results have been included in Table 6.7 for comparison with the previous analysis.

6.4 When approximations fail

The methods illustrated in §6.3 are satisfactory under the best conditions. If one or more of the assays whose results are to be combined show heterogeneity of deviations from regression lines, and if adjustment by heterogeneity factors is judged appropriate, the validity of the variances for forming weighted means is upset. The weight associated with a value of M might then be based on very few degrees of freedom, and results should be distrusted if this M appeared to have a relatively high weight. One possibility would be to omit ED 50 estimates that require hetero-geneity factors. Another would be to form a composite hetero-geneity factor for all the experiments, by adding the appropriate χ^2 values, so that all weights are similarly affected by the hetero-geneity. The χ^2 corresponding to equation (6.13) is now no longer valid, and must be replaced by a variance ratio test (Example 13). No general recommendations for this situation can safely be made.

Another trouble arises if values of b are markedly unequal on different occasions. Although this does not of itself disturb the discussion of the ED 50 for various materials, it does mean that conclusions might be substantially changed if the ED 90 or the ED 25 were used instead of the ED 50. In certain circumstances, such as the strict conditions of analytical bioassay (Finney, 1964), belief in the constancy of M for all experiments on the same pair of materials may be so strong as to dispose of worries arising from changes in β from day to day. Without this belief, interpre-tation of such calculations as were made in § 6.3 should be cautious in the face of any indications that β is not constant.

Yet a third type of difficulty enters if g is large in some ex-periments. This also destroys the legitimacy and trustworthiness of conclusions based on weighted means of M. Again no entirely satisfying general approach seems possible, unless constancy of β permits an analysis such as that at the end of § 6.3. A method that invokes a single composite value of g was suggested else-where (Finney, 1964, §§ 14.3, 21.2), and a slight modification is presented here.

Equation (6.1) may be written

$$M_{12} = (a_2 - a_1)/b, \qquad (6.23)$$

where for each material

$$a = \bar{y} - b\bar{x}. \qquad (6.24)$$

Now suppose that the numerator and denominator of (6.23) are available for each of several experiments, and with each ex-periment associate an arbitrary positive quantity ω, which may vary from experiment to experiment. Write

$$M^* = \Sigma\omega(a_2 - a_1)/\Sigma\omega b, \qquad (6.25)$$

where as usual Σ denotes summation over the experiments; whatever the set of values of ω, M^* can be regarded as estimating the relative dose metameter. The next step is to seek an optimal choice of ω.

The variances and covariance of the numerator and denominator of (6.23) are

$$
\left.
\begin{aligned}
v_{11} &= \frac{1}{{}_1Snw} + \frac{1}{{}_2Snw} + \frac{(\bar{x}_1 - \bar{x}_2)^2}{{}_1S_{xx} + {}_2S_{xx}}, \\
v_{22} &= 1/({}_1S_{xx} + {}_2S_{xx}), \\
v_{12} &= (\bar{x}_1 - \bar{x}_2)/({}_1S_{xx} + {}_2S_{xx}),
\end{aligned}
\right\}
\qquad (6.26)
$$

where the notation is that of equation (6.5). Hence the numerator and denominator of M^* will have as their variances and covariance

$$
\left.
\begin{aligned}
v_{11}^* &= \Sigma\omega^2 v_{11}, \\
v_{22}^* &= \Sigma\omega^2 v_{22}, \\
v_{12}^* &= \Sigma\omega^2 v_{12}.
\end{aligned}
\right\}
\qquad (6.27)
$$

The general form of Fieller's theorem, equation (4.35), can then be used to determine limits for M^*.

Very commonly a good choice for ω will be

$$
\omega = 1/v_{11}, \qquad (6.28)
$$

which will be near the optimal if all experiments were similar in doses tested, numbers of subjects per dose, and values of b. Although constancy of the true regression coefficient (β) or of σ is not essential to the validity of the estimate in (6.25), marked inequalities add to the difficulty of finding a satisfactory weighting system. An alternative worth considering is

$$
\omega = 1/(v_{11} - 2Mv_{12} + M^2 v_{22}), \qquad (6.29)
$$

but this is open to greater theoretical objection because of its explicit use of the separate values of M in the determination of ω. Example 18 makes the pattern of calculation clearer.

Ex. 18. *Improved combination of relative potencies and fiducial limits.* The four bioassays of a pertussis vaccine used in Example 14 can illustrate the method of the present section. Table 6.8 summarizes quantities abstracted from the computer output or easily formed by equations (6.26); values of g (at the usual 0.95 probability) have been included as indication that, in at least two of the 'blocks' yielding estimates for vaccine no. 4, g is not negligible. Equation (6.28) has been used for ω.

Table 6.8   Quantities required in weighted combination of
four assays of pertussis vaccine

| | Block of data | | | |
Quantity	2	6	7	12
M	-1.733	-1.488	-1.831	-1.875
$V(M)$	0.04815	0.02466	0.03142	0.07641
$1/V(M)$	20.8	40.5	31.8	13.1
$a_4 - a_S$	-2.5167	-2.7894	-2.9581	-1.8756
b	1.4526	1.8748	1.6155	1.0005
v_{11}	0.205335	0.180937	0.173670	0.129236
v_{12}	-0.078025	-0.067785	-0.060301	-0.040823
v_{22}	0.055507	0.048531	0.038522	0.028543
g	0.1011	0.0530	0.0567	0.1095
$\omega = 1/v_{11}$	4.87	5.53	5.76	7.74
ω^2	23.7169	30.5809	33.1776	59.9076

By summation over blocks:

$$\Sigma\omega b = 34.4910,$$

$$\Sigma\omega(a_4 - a_S) = -59.2375,$$

and therefore equation (6.25) gives

$$M^* = -1.7175.$$

From equations (6.26),

$$v_{11}^* = 23.90727,$$
$$v_{22}^* = 5.78857,$$
$$v_{12}^* = -8.36967.$$

For Fieller's theorem now,

$$g = \frac{t^2 v_{22}^*}{(\Sigma\omega b)^2} \tag{6.30}$$

$$= 0.01869.$$

Thus equation (4.35) gives

$$\text{Limits} = -1.7175 + \frac{0.01869}{0.98131}\left(-1.7175 + \frac{8.36967}{23.90727}\right)$$

$$\pm \frac{1.960}{34.4910 \times 0.98131}(12.2326 - 0.01869 \times 11.8056)^{\frac{1}{2}}$$

$$= -1.5428, \ -1.9442.$$

The potency, R, of vaccine no. 4 is therefore estimated as 0.0192, with fiducial limits at 0.0287, 0.0114; corresponding values from Example 14 are 0.0206 as the estimate, with limits (adding and subtracting 1.960 times the standard error of $\bar{M}$) at 0.0319, 0.0133. Evidently the difference is not great, but as usual proper allowance for g makes the fiducial range relatively a little wider. For practical purposes, Example 14 was adequate on the pertussis data, but had individual values of g been larger the inaccuracy in that method could have been serious. As so often, the only sure way of determining whether or not an approximate method suffices is also to complete more exact calculations, and then the approximation is pointless.

A very general experience with weighted means of any kind is that extreme numerical accuracy in weights is unimportant. For this reason, only 3 digits were retained in ω. If equation (6.29) were to replace (6.28), each ω would be approximately doubled without much change in their ratios, and so in this instance there is nothing to choose between the formulae.

6.5 Unequal tolerance variances

No law of nature requires that probit regression lines for different materials included in one experiment shall be parallel: if the tolerance distributions happen to have different variances, the lines cannot be parallel. In practice, a group of materials closely related chemically or in their mode of biological action commonly have very nearly equal β's. An extreme situation is that in which materials are identical in their biologically active principles but may differ in the concentration of these in the solutions or other preparations to be used; they must then have log tolerance distributions with identical variances. This is the basis of biological dilution assays (Finney, 1964).

When two probit regressions are not parallel, interpretation of any comparison between them is difficult. A relative dose metameter can be estimated at any stated level of response, but will not be constant at all levels. If regressions have been independently fitted, an asymptotic variance for such a value can be obtained as the sum of two expressions from equation (3.12). No result corresponding to Fieller's theorem is available. The logical difficulty is more serious than that of statistical technique.

One cannot speak of drug B being 7 times as potent as drug A if the ratio of equally effective doses varies with the percentage effect. What is needed before progress can be made is a clear definition of the character of any measure of comparison that is to be estimated. Cornfield (1964) and Finney (1965) have expressed contrasting points of view on this problem, neither of which is likely to be final.

Lack of parallelism may sometimes disappear if a more suitable dose metameter is chosen. This is unlikely unless some departure from linearity is also apparent on the logarithmic scale. Moreover, parallelism of linear regressions on any dose scale other than the logarithmic is no help to simple expression of a comparison between two preparations, because the constant difference in x at equal levels of response cannot be interpreted as a constant ratio of equally effective doses. Complications of parallelism may also arise when mixtures of biologically active substances are used, knowingly or unknowingly. This has led to further important developments of theory, especially with reference to insecticides (Chapter 11).

6.6   Illustration of computer output

In the development of the computer program described in Chapter 5, the printed output was planned with considerable care. For some of the problems of later chapters, recourse was had to other general-purpose programs, but facilities for these could easily be added to the special probit program if they were wanted often. Output from computer analysis of the bioassay of analgesics presented in Example 13 is reproduced below as an example of the kind of information that needs to be extracted for inspection by the user of the program. A different programmer might write a very different and possibly more efficient program, though it would have to contain essentially the same elements of input as were listed in § 5.3.

Analysis by computer is useless unless the output is easily identifiable and fully informative. Obviously in the kind of experiment under discussion the output must include estimates of the parameters and appropriate variances. Many additional items of output are useful either to aid identification or to contribute to interpretation. The choice of these depends upon the purpose of the experiments. Commercial requirements in the standardization of drugs or insecticides may suggest something quite different from scientific research. Two things must be remembered. First, the program should deal with all the supplementary calculations needed for the final summaries of results, and should not leave loose ends from which relative potencies, fiducial limits, and so on must be subsequently computed. Secondly, the temptation to include in the output every step of the analysis that is conceivably of interest must be resisted, lest the user be submerged beneath stacks of line-printer paper!

In the illustration that follows, the first part of the output is a piece of text as a heading. Here this has been written as a lengthy prose description, but it can be of any length. Next come the data, as further confirmation that the analysis relates to the right observations. The summary of the iterations can be as here merely a statement of the 'configuration' of the data (the words 'untreated' and 'maximal dose' refer to generaliza tions of the program required for the problems of Chapter 7), and a listing of the data to confirm that the right observations are being analysed. The details of a converging sequence of iterations for minimizing the negative log-likelihood need not be described here; for routine use of the program, the printing of these can be suppressed, but the information is valuable when poor data or a very complex problem make convergence slow and analysis difficult.

The summaries that follow begin with a classification of the problem and a reference number to the model used (because the program can be used for a wide range of mathematical formulations). The table of parameters and their standard errors is self-explanatory. In comparing with the calculations shown in Example 13, remember that the computer has operated in terms of normal equivalent deviates: agreement with equations (6.11) is good. The variance–covariance matrix is the generalization of equation (5.1) to five parameters; because this matrix is necessarily symmetric about its diagonal, only the lower left corner is shown and the upper right is used to record correlation coefficients between the error variations in the estimates of pairs of parameters (these last are not particularly interesting).

A further self-explanatory table is that showing expected frequencies and probabilities of response, leading to the χ^2 for the heterogeneity test. Next are tables of median effective doses and relative potencies, with fiducial limits computed for three different probabilities. Had they been wanted, these could have been followed by tables of each value of m and M with its asymptotic variance. Finally a short table of the log-likelihood function in the neighbourhood of the maximum is shown. For a problem with five parameters, far more complicated tabulation is possible, and can be valuable as reassurance that the maximum has been properly located, but in this instance it is not really needed.

Change of a single digit in the steering instructions causes all computations to be made in terms of the logit instead of the probit model. More important in practice is the possibility of dealing with situations in which the first attempt fails to give satisfactorily converging iterations. The program can be initiated several times, using different initial values, different step lengths, and different numbers of cycles of iteration, at relatively small cost; unless the data are very poor, eventually the estimation will be completed. With a little experience of handling problems, most sets of data will be analysed satisfactorily at the first attempt.

7

Adjustments for natural responsiveness

7.1 'Abbott's formula'

Numerical examples discussed in detail in earlier Chapters (e.g. Examples 6, 8, 13) have assumed responses of test subjects to be due solely to the applied stimuli. Often this is justifiable, and certainly these examples accord well with it. Some experiments, on the other hand, show clear evidence that responses can occur even at zero dose: either control batches of subjects have received zero dose or a sequence of low doses indicates a minimal response rate greater than zero. In an insecticidal test, some insects might die during the experimental period, by accident during handling or from natural causes, even though they had not been sprayed. In an experiment such as Grewal's (Example 13), some mice might show no evidence of discomfort from the applied shock and thus be classed as responders even without an analgesic (Grewal's data in fact did not suggest any such phenomenon under his conditions).

If a proportion C of subjects would respond without the applied stimulus, a dose sufficient to induce a response rate P in the remainder will have an expected rate of observed responses P^*, where
$$P^* = C + P(1 - C), \tag{7.1}$$
provided that the two sources of response operate independently. This may be rewritten
$$P = \frac{P^* - C}{1 - C}, \tag{7.2}$$

to express the response attributable to the dose as a function of the total observed response. Equation (7.2) is commonly known as *Abbott's formula*, though Tattersfield and Morris (1924) used it for the same purpose before Abbott (1925) published it. Indeed, equation (7.1) is simply the standard rule for the combination of independent probabilities. Healy (1952) has presented a table to expedite calculation of P from P^*.

Occasionally account must be taken of natural immunity as well as of natural responsiveness. A proportion D of subjects may be incapable of responding even at very high doses. Equation (7.1) becomes

$$P^* = C + P(1 - C - D),\qquad(7.3)$$

whence
$$P = \frac{P^* - C}{1 - C - D}.\qquad(7.4)$$

Van Soestbergen (1956) gave examples of the occurence and statistical analysis of this problem in estimating the *Toxoplasma* antibody content of sera from persons and animals infected with the parasite *Toxoplasma*.

Whether (7.1) or the more general (7.3) is appropriate, the probability that r subjects respond in a batch of n will still be determined by the binomial distribution. The difference is that P in equation (4.1) is replaced by P^*.

7.2   Approximate estimation of parameters

If C, D were known, equations (7.2), (7.4) could be used to calculate a p corresponding to each observed

$$p^* = r/n\qquad(7.5)$$

in an experiment. For long, this constituted the only recognition of the adjustment needed to statistical analysis. Evidently the effective number of subjects is reduced, the average number available for manifesting response to the dose being $n(1 - C - D)$. In discussing an ovicidal experiment, Bliss (1939a) stated: 'Both the number of eggs exposed and the percentage kill have been corrected for mortality in the untreated controls'. He reduced the number of eggs tested at each concentration by the percentage mortality in the controls; thus the weights attached to each probit were proportionately reduced. Estimates of potency were not altered, but the precision claimed was less than if no adjustment had been required.

Even when C, D are known exactly, this is not the only alteration required in the analysis. In equation (3.10), the product PQ arises as the variance of a binomial frequency distribution; the relevant distribution is now that defined by the total proportions

P^* and Q^*. The weighting coefficient in fact becomes (Finney, 1944a, 1949d):

$$w = \frac{Z^2}{\left(P + \dfrac{C}{1 - C - D}\right)\left(Q + \dfrac{D}{1 - C - D}\right)}. \qquad (7.6)$$

For the most important case, $D = 0$, this function,

$$w = \frac{Z^2}{Q\left(P + \dfrac{C}{1 - C}\right)}, \qquad (7.7)$$

has been tabulated (Table II).

When $C = 0$, (7.7) reduces to equation (3.10); for any other C, these coefficients must be multiplied by

$$P \Big/ \left(P + \frac{C}{1 - C}\right).$$

Even when C is no greater than 5 %, the reduction in w may be considerable, especially if the expected probit is small. Except for $C = 0$, w decreases more rapidly for small values of Y than for large. For practical purposes, interpolation in Table II is seldom needed. When $D \neq 0$, w must be calculated by equation (7.7), if necessary using Table I to obtain P for specified Y.

Usually C is not known exactly, but must itself be estimated from the data. Many experiments include a control batch of subjects at zero dose; if this batch is large relative to those used for other doses, C may be estimated satisfactorily from it alone. The observed rate among the controls, c, is itself subject to sampling variation, and, though it is an unbiased estimate of C, a better estimate may often be obtained by inclusion of additional information, especially that from any low non-zero doses. If the empirical response rate at a low dose is less than c, it suggests that c is an overestimate of the true natural rate, C. Similarly, a series of response rates at small doses that are nearly equal to one another but much in excess of c, indicates that c is an underestimate. A rough but improved estimate of C may be obtained from inspection of the data or from a freehand sketch of the sigmoid response curve. Similar remarks apply to D. An experiment having only a small control batch or lacking one entirely, cannot be used in this way unless low doses point the way very clearly, and the method of § 7.3 must be adopted.

When no control batches of subjects or other features of an experiment have suggested that $C > 0$ or $D > 0$, and experience of similar experiments does not seriously conflict with $C = D = 0$, the standard methods of analysis described in Chapters 3–5 may be adopted. Evidence for a small positive C (say less than 10 %), coming from c or from an estimate modified as proposed in the last paragraph, may be incorporated by using equation (7.2) to determine p for each observed p^*, and then applying the techniques of Chapters 4 and 5 with n as in the experiment and w modified to the expression in (7.7). The estimated C may exceed some of the p^* so that Abbott's formula yields negative p: the correct procedure is still to use working probits defined by equation (4.15), as only so can each observation exercise its right influence in the analysis. Working probits corresponding to negative values of p are not often required but may be calculated from Table III. No example need be given, as subsequent calculations are exactly as customary when $C = 0$. A similar procedure could be used with a small positive D.

7.3   Estimation by maximum likelihood

If C is evidently high, or if the control batch is too small to give a trustworthy estimate of C, recourse is had to generalizing the maximum likelihood process. The same applies to D. The theory for estimating C, D simultaneously is accessible (Finney, 1949d, 1964); that for C alone (Finney, 1949b) is an instructive example of development from § 4.2. If equation (7.1) obtains, P^* and Q^* must replace P and Q in equations (4.1), (4.2). Equations (4.12) will be replaced by:

$$\left. \begin{aligned} \frac{\partial P^*}{\partial \alpha} &= \frac{\partial P^*}{\partial P} \frac{\partial P}{\partial \alpha} = (1-C)\,Z, \\[2mm] \frac{\partial P^*}{\partial \beta} &= (1-C)\,Zx, \\[2mm] \frac{\partial P^*}{\partial C} &= Q. \end{aligned} \right\} \qquad (7.8)$$

Moreover, with p^* defined in equation (7.5) as the empirical response rate, and

$$p = \frac{p^* - C}{1 - C}, \qquad (7.9)$$

it follows that $p^* - P^* = (p - P)(1 - C).$ (7.10)

If S relates to summation over non-zero doses, and c is defined by

$$c = r_c/n_c,$$ (7.11)

the new version of equation (4.2) produces

$$\frac{\partial L}{\partial \alpha} = S \frac{nZ^2(1-C)}{Q\{C + P(1-C)\}} \left(\frac{p - P}{Z} \right) = 0,$$

$$\frac{\partial L}{\partial \beta} = S \frac{nZ^2 x(1-C)}{Q\{C + P(1-C)\}} \left(\frac{p - P}{Z} \right) = 0,$$ (7.12)

$$\frac{\partial L}{\partial C} = \frac{n_c(c - C)}{C(1-C)} + S \frac{n(p - P)}{C + P(1-C)} = 0.$$

As in § 4.2, a second differentiation can be followed by writing $p = P$. With the aid of the auxiliary variate

$$x' = Q/Z,$$ (7.13)

the coefficients can be written:

$$-\frac{\partial^2 L}{\partial \alpha^2} = Snw,$$

$$-\frac{\partial^2 L}{\partial \alpha \partial \beta} = Snwx,$$

$$-\frac{\partial^2 L}{\partial \beta^2} = Snwx^2,$$ (7.14)

$$-\frac{\partial^2 L}{\partial \alpha \partial C} = \frac{1}{1 - C} Snwx',$$

$$-\frac{\partial^2 L}{\partial \beta \partial C} = \frac{1}{1 - C} Snwxx',$$

$$-\frac{\partial^2 L}{\partial C^2} = \frac{n_c}{C(1-C)} + \frac{1}{(1-C)^2} Snwx'^2,$$

where w is as in equation (7.7). If the working probit, equation (4.15), is introduced, a system of iteration can be constructed that is essentially the formation of a weighted linear regression of y on x, x'. Equations (4.17), (4.18) are generalized to

$$bS_{xx} + \frac{\delta C}{1-C} S_{xx'} = S_{xy}, \tag{7.15}$$

$$bS_{xx'} + \frac{\delta C}{1-C} \left[\frac{n_c(1-C)}{C} + S_{x'x'} \right] = \frac{n_c(c-C)}{C} + S_{x'y}, \tag{7.16}$$

$$a = \bar{y} - b\bar{x} - \frac{\delta C}{1-C} \bar{x}'. \tag{7.17}$$

Equations (7.15)–(7.17) need explanation. A first approximation to C (§7.2 may help) gives p from equation (7.9). A provisional probit line, weights, and working probits are obtained, in the usual way except for use of the appropriate column of Table II; for each Y, Q/Z is also read from Table II. Weighted sums of squares and products are formed, such as

$$S_{xx'} = Snw(x-\bar{x})(x'-\bar{x}')$$

$$= Snwxx' - (Snwx)(Snwx')/Snw. \tag{7.18}$$

From n_c and c (if $n_c \neq 0$) are calculated

$$\frac{n_c(1-C)}{C} \tag{7.19}$$

and $\qquad \dfrac{n_c(c-C)}{C}, \tag{7.20}$

the contributions from information provided by the control subjects. Solution of equations (7.15)–(7.17) then gives the new regression equation $\qquad Y = a + bx,$

and δC is added to the previous estimate of C.

If the revised estimates differ appreciably from those on which the calculations were based, a new cycle must be computed. Iteration must continue until differences between successive cycles become unimportant. At that stage, δC will be practically zero, so that equation (7.15) will give b as the total regression coefficient of y on x, and equation (7.17) will give the constant term in the regression equation as $(\bar{y} - b\bar{x})$. Herein lies justification for the approximate analysis described in §7.2: if a value of C

close to the maximum likelihood estimate can be chosen by inspection of the data, the method of Chapter 4 with modified weighting coefficients may estimate the other parameters.

When the maximum likelihood estimates have been closely approached,

$$\chi^2 = S_{yy} - bS_{xy} - \frac{\delta C}{1-C} S_{x'y} \qquad (7.21)$$

tests heterogeneity; in assigning degrees of freedom to χ^2, a group of subjects at zero dose counts as one dose level, and one degree of freedom must be subtracted for the additional parameter, C. This χ^2 is subject to the usual limitations, in that it is strictly valid asymptotically and at the limit of iteration (at which point, the last term is zero).

Equations (7.15), (7.16) should be solved by first obtaining the inverse matrix of the coefficients, from which both the estimates of the parameters and their variances may be formed. Example 19 illustrates an application to the comparison of two sets of records with probit regression equations constrained to be parallel.

Ex. 19. *Maximum likelihood estimation with a natural response rate.* Martin (1940, Table 5) reported tests of the toxicity of two derris roots, W.213 and W.214, to the grain beetle *Oryzaephilus surinamensis*. Of 129 control insects, sprayed with the alcohol-sulphonated lorol medium but no derris, 21 were affected; hence

$$c = 0.163. \qquad (7.22)$$

The first column of Table 7.1 contains the log concentration (mg dry root per litre), and this is followed by n, r, p^*. All mortality percentages substantially exceeded 16.3 %, so that most of the information on C comes from c.

A provisional value of C was taken as 17.0 %. The response rates due to the poisons alone are then estimated: at the lowest concentration of root W.213, equation (7.2) gives

$$p = (0.460 - 0.170)/(1 - 0.170) = 0.349.$$

The empirical probits of p plotted against x allow two parallel regression lines to be placed by eye so as to give expected probits, Y. The weighting coefficients, from Table II in the column for 17 %, are multiplied by n to give the column nw in Table 4.1: for $Y = 7.6$, the weighting coefficient is 0.032 98, and multiplica-

Table 7.1   Toxicity of derris roots W.213 and W.214 to *Oryzæphilus surinamensis*

x	n	r	p^*	p ($C = 17.0$)	Empirical probit	Y	nw	$x' = Q/Z$	y	nwx	nwx'	nwy	Y (eqn. (7.26))
					Root W.213								
2.17	142	142	100.0	100.0	∞	7.6	4.7	0.343	7.94	10.199	1.6121	37.318	7.64
2.00	127	126	99.2	99.0	7.33	7.2	9.7	0.392	7.31	19.400	3.8024	70.907	7.16
1.68	128	115	89.8	87.7	6.16	6.3	35.0	0.565	6.15	58.800	19.7750	215.250	6.26
1.08	126	58	46.0	34.9	4.61	4.6	47.5	1.780	4.61	51.300	84.5500	218.975	4.59
							96.9			139.699	109.7395	542.450	
					Root W.214								
1.79	125	125	100.0	100.0	∞	7.2	9.5	0.392	7.59	17.005	3.7240	72.105	7.25
1.66	117	115	98.3	98.0	7.05	6.9	14.9	0.438	7.03	24.734	6.5262	104.747	6.89
1.49	127	114	89.8	87.7	6.16	6.4	31.4	0.539	6.12	46.786	16.9246	192.168	6.41
1.17	51	40	78.4	74.0	5.64	5.5	22.9	0.876	5.64	26.793	20.0604	129.156	5.52
0.57	132	37	28.0	13.3	3.89	3.8	17.6	4.557	3.89	10.032	80.2032	68.464	3.84
							96.3			125.350	127.4384	566.640	
					Controls								
—	129	21	16.3	—	—	—	—	—	—	—	—	—	

W.213: $1/_1 Snw = 0.01031992$ $\bar{x}_1 = 1.4417$ $\bar{x}'_1 = 1.1325$ $\bar{y}_1 = 5.5980$

W.214: $1/_2 Snw = 0.01038422$ $\bar{x}_2 = 1.3017$ $\bar{x}'_2 = 1.3233$ $\bar{y}_2 = 5.8841$

	$Snwx^2$	$Snwxx'$	$Snwx'^2$	$Snwxy$	$Snwx'y$	$Snwy^2$
W.213	215.1198	135.6391	163.7154	820.907	551.987	3147.90
	201.4016	158.2095	124.2803	782.040	614.326	3036.66
	13.7182	−22.5704	39.4351	38.867	−62.339	111.24
W.214	178.2746	111.9036	396.4995	779.415	602.854	3454.48
	163.1633	165.8817	168.6453	737.573	749.862	3334.17
	15.1113	−53.9781	227.8542	41.842	−147.008	120.31
Controls			$n_c(1-C)/C = 629.3235$		$n_c(c-C)/C = -5.312$	
Total	28.8295	−76.5485	397.1128	80.709	−214.659	231.55

tion by 142 gives 4.7. The auxiliary variate, x', is read from Table II for each Y, and the working probits, y, are obtained as usual.

Product columns, nwx, nwx', nwy, are next formed, the totals of which lead to the weighted means $\bar{x}$, $\bar{x}'$, $\bar{y}$, for each root. Sums of squares and products of deviations are calculated, and the contributions from the two roots added, with the further addition of (7.19):

$$129 \times 0.83/0.17 = 629.8235$$

to $S_{x'x'}$, and (7.20):

$$129 \times (0.163 - 0.17)/0.17 = -5.312$$

to $S_{x'y}$. Thus the combined evidence of the two roots gives equations (7.15), (7.16) in the form

$$\left.\begin{array}{r} 28.8295b - 76.5485\left(\dfrac{\delta C}{1-C}\right) = 80.709, \\[2ex] -76.5485b + 897.1128\left(\dfrac{\delta C}{1-C}\right) = -214.659. \end{array}\right\} \quad (7.23)$$

The first row of V, the inverse matrix of coefficients, is found by replacing the right-hand sides of equations (7.23) by 1, 0 and solving; similarly, the second row is found by solving with 0, 1 on the right. The result is

$$V = \begin{pmatrix} v_{bb} & v_{bc} \\ v_{bc} & v_{cc} \end{pmatrix} = \begin{pmatrix} 0.044\,847\,5 & 0.003\,826\,7 \\ 0.003\,826\,7 & 0.001\,441\,2 \end{pmatrix}. \quad (7.24)$$

Equations (7.23) are solved by adding the products of their right-hand sides with each row of V in turn:

$$b = 80.709 \times 0.044\,847\,5 - 214.659 \times 0.003\,826\,7$$

$$= 2.7982,$$

$$\frac{\delta C}{1-C} = 80.709 \times 0.003\,826\,7 - 214.659 \times 0.001\,441\,2$$

$$= -0.000\,52.$$

By substitution of the provisional value of C,

$$\delta C = -0.000\,52 \times 0.83$$

$$= -0.000\,43,$$

whence the revised estimate is

$$C = 0.17 - 0.0004$$

$$= 0.1696. \quad (7.25)$$

The mean values in Table 7.1 are then used in equation (7.17):

$$a_1 = 5.5980 - 2.7982 \times 1.4417 + 0.0005 \times 1.1325 = 1.564,$$

$$a_2 = 5.8841 - 2.7982 \times 1.3017 + 0.0005 \times 1.3233 = 2.242.$$

Hence the revised regression equations are

$$\left. \begin{array}{l} Y_1 = 1.564 + 2.798x, \\ Y_2 = 2.242 + 2.798x. \end{array} \right\} \tag{7.26}$$

The provisional and revised estimates of C, equations (7.22) and (7.25), are almost identical. Expected probits calculated from equations (7.26), in the last column of Table 7.1, agree closely with those used in the first cycle: no second cycle of calculations is needed. By equation (7.21),

$$\chi^2_{[6]} = 5.60; \tag{7.27}$$

ten groups of subjects were tested, and four parameters have been estimated, leaving 6 degrees of freedom for χ^2. Had χ^2 been large, the possibility that one or two classes with very small expectations were making unreasonably large contributions would have needed consideration. The expected probits from equations (7.26) would be converted into values of P, these used in equation (7.1) with $C = 16.96\%$ to give P^*, and expected frequencies calculated in each group (cf. Example 8).

In this example, the smallness of χ^2 suffices to show that there can be no serious deviation from parallelism. Formal test requires a new set of computations to fit separate regression coefficients to the two materials. Equations (7.23) must be reconstructed so as to keep distinct the information on β for each root while still pooling information on C. The new equations can be formed from totals in the lower part of Table 7.1 in an obvious manner:

$$\left. \begin{array}{rcl} 13.7182b_1 \quad -22.5704 \left(\dfrac{\delta C}{1-C} \right) &=& 38.867, \\[2mm] 15.1113b_2 \ -53.9781 \left(\dfrac{\delta C}{1-C} \right) &=& 41.842, \\[2mm] -22.5704b_1 - 53.9781b_2 + 897.1128 \left(\dfrac{\delta C}{1-C} \right) &=& -214.659. \end{array} \right\}$$

$$\tag{7.28}$$

The solutions (evaluation of an inverse matrix is unnecessary) are

$$b_1 = 2.8302,$$

$$b_2 = 2.7622,$$

$$\frac{\delta C}{1-C} = -0.00187.$$

By equation (7.21), but including terms for b_1 and b_2,

$$\chi^2_{[5]} = 5.57,$$

1 degree of freedom less than in equation (7.27) because an extra parameter has been estimated. The analysis of χ^2 in Table 7.2 shows that both the component for deviations from parallelism and that for residual heterogeneity are small enough to be attributed to random variation in the data.

Table 7.2   Analysis of χ^2 for derris–*Oryzaephilus surinamensis* experiment

Source of variation	d.f.	Sum of squares	Mean square
Parallelism of regressions	1	0.03	0.03
Residual heterogeneity	5	5.57	1.11
Total	6	5.60	

The matrix V gives the variances and covariances of b and $\delta C/(1-C)$. From the first diagonal element

$$V(b) = 0.04485,$$

whence $b = 2.798 \pm 0.212;$

from the second diagonal element

$$V(C) = 0.0014412 \times (0.83)^2,$$

whence $C = 16.96\% \pm 3.15\%.$

The sizes of these standard errors, relative to the changes from the provisional C and expected probits to the values after the first cycle, again show a second cycle to be unnecessary. The remaining element of V, the covariance, is required in any consideration of the precision of estimates of ED 50 or relative potency.

For 95 % fiducial limits,

$$g = t^2 V(b)/b^2$$

$$= 0.0220.$$

Variance formulae could therefore be used, but calculation according to Fieller's theorem will be illustrated here. The log ED 50 for either root is estimated from the appropriate member of equations (7.26), and has the algebraic form

$$m = \bar{x} + \left(5 - \bar{y} + \frac{\delta C}{1 - C} \bar{x}'\right) \Big/ b; \qquad (7.29)$$

this relates to 50 % mortality caused by the poison, and does not include natural mortality. Hence the estimated relative dose metameter, a modification of equation (6.1), is

$$M - m_1 \quad m_2 = \mathcal{X}_1 - x_2 - A/b, \qquad (7.30)$$

where
$$A = \bar{y}_1 - y_2 - \frac{\delta C}{1 - C} (\bar{x}_1' - \bar{x}_2') \qquad (7.31)$$

$$= -0.2862,$$

and therefore $M = 0.1400 + 0.2862/2.7982$

$$= 0.2423.$$

Now
$$V(A) = \frac{1}{{}_1 Snw} + \frac{1}{{}_2 Snw} + (\bar{x}_1' - \bar{x}_2')^2 v_{cc} \qquad (7.32)$$

$$= 0.010320 + 0.010384 + 0.000052$$

$$= 0.020756.$$

Similarly,
$$Cov(A, b) = -(\bar{x}_1' - \bar{x}_2') v_{bc} \qquad (7.33)$$

$$= 0.000730.$$

Note that components arising from estimation of C enter equations (7.32), (7.33) even though in the limit δC is zero. These equations are of standard statistical form for variances and covariances of linear functions of $\bar{y}_1, \bar{y}_2, b, \delta C/(1 - C)$. They can now be used in association with $V(b)$, to give fiducial limits to the ratio A/b. By equation (4.35), the limits are

$$-0.1023 + \frac{0.0220}{0.9780}\left(-0.1023 - \frac{0.000730}{0.044848}\right)$$

$$\pm \frac{1.960}{2.7982 \times 0.9780}\sqrt{[0.020756 + 2 \times 0.1023 \times 0.000730}$$

$$+ (0.1023)^2 \times 0.044848$$

$$- 0.0220 \times 0.020744]$$

$$= -0.1023 - 0.0027 \pm 0.7162\sqrt{(0.02092)}$$

$$= -0.0014, \quad -0.2086.$$

Insertion of these limits for A/b in equation (7.30) then gives the numerical values 0.3486, 0.1414. From the antilogarithms of M and its limits, root W.214 is estimated to have a potency 1.75 times that of W.213, with 95 % fiducial limits at 2.23, 1.38.

The asymptotic variance formula could have been safely used here, g being small. This would have been equivalent to writing $g = 0$ in the above calculations. Had formulae relating to the ED 50 for root W.213 been required, they could have been obtained by noting that equation (7.30) becomes the same as equation (7.29) after insertion of $\bar{x}_2 = 0$, $\bar{x}_2' = 0$, $\bar{y}_2 = 5$. These substitutions, and omission of the term in $_2Snw$ in equation (7.32), lead to the limits.

7.4   Analysis by generalized computer program

The computer program described in Chapter 5 is easily generalized to the problem of the present chapter. Indeed, the essential procedure remains that for maximizing L in equation (5.4), except that P^*, Q^* replace P, Q. Here P is as defined by equation (5.5) and P^* may have either of the generalized definitions in equations (7.1), (7.3).

The maximization routine described in § 5.4 applies without change, except that the matrix in equation (5.1) becomes of size 3×3 or 4×4 so as to accommodate the parameters C, D. Both the simplex and the quadratic phase of maximization are used as before, and estimates of the parameters α, β, C, D, are obtained after appropriate iteration. By comparison with § 7.3, extension of the computer analysis to the more complicated problem is easy. The special features of the auxiliary variate and the anomalous estimation of $\delta C/(1-C)$ disappear. The method leads directly to the estimates of the parameters, and almost as a by-product turns out the equivalent of equation (5.1), the elements of the matrix being the full set of variances and covariances for the parameters. The writing of the program involves no new problems; its execution will take a little longer, since more parameters must be estimated and more different values need to be tried.

With little extra trouble, of course, the program can complete the remaining tasks of the statistical analysis. In particular, expected frequencies are readily calculated and χ^2 found. Similarly, the ED 50's, relative potencies if two or more preparations have been tested, and variances or fiducial limits, are obtainable. No details need be presented: entirely standard programming can closely follow calculations such as those in the latter part of Example 19.

Ex. 20. *Computer analysis for Example 19.* The further generalization of a computer program so as to deal with two or more sets of observations tied to common values of b, C comes readily. Equation (5.4) still applies, with the understanding that the summations extend over all the data with P changing its α-parameter from one material or preparation to the next.

The general program has been applied to the data of Example 19. Naturally the numerical values obtained are close to those stated earlier, but they correspond more nearly to the true maximum likelihood. Agreement between observations and the estimated parameters is confirmed by

$$\chi^2_{[6]} = 5.990,$$

instead of the value in equation (7.27). A simple device enables the $\chi^2_{[5]}$ appropriate to separate regression equations to be obtained by essentially the same program. The variate x is replaced by two variates x_1, x_2. For root W.213, x_1 is log dose and x_2 always zero; for root W.214, x_2 is log dose and x_1 always zero. Multiple regression of probits on x_1, x_2 (in the manner of § 8.6, but with inclusion of C) then gives b_1, b_2 as the two regression coefficients and the required residual

$$\chi^2_{[5]} = 5.912.$$

The new version of Table 7.2 is easily formed, and of course the interpretation is the same.

The computer analysis leads to

$$C = 16.95\% \pm 3.17\%$$

and
$$b = 2.793 \pm 0.209,$$

very close to the values in Example 19. The regression lines are

$$Y_1 = 1.576 + 2.793x,$$

$$Y_2 = 2.243 + 2.793x.$$

The potency of W.214 relative to W.213 is now estimated as 1.734, with 95 % limits at 2.209, 1.375.

8

Experimental design

8.1 Planning for precision

If the prime object of an experiment is to estimate the dose corresponding to a particular percentage response, the constituent tests should be planned to estimate that dose as precisely as possible. The experimenter would achieve greater precision by increasing his total number of subjects, but often this course is not open to him. His problem is how to allocate a limited number of subjects to different doses in order to utilize them most effectively. No complete solution can be given, but some general principles can be established.

The commonest situation is that in which the ED 50 is to be estimated, but similar recommendations are applicable to neighbouring points such as the ED 45 or the ED 60. When nothing is known about the location of the ED 50 beyond a reasonable confidence that it lies between two widely separated limits, little can be done except to divide the N subjects equally between a large number of doses extending over a range rather wider than these limits. If N is large, a pilot investigation with only a few of the subjects (say $N/5$ or $N/10$) might be used to give some idea of the ED 50, the remaining subjects being assigned to an experiment planned as described below. At the other extreme, an experimenter who can afford only ten animals and wishes to estimate the mean toxic dose of a new poison or the mean curative dose of a new therapeutic agent is asking for the impossible: discussion of whether he ought to use two groups of five animals or to give every animal a different dose is often vigorous but always unprofitable.

The precision with which the value of x corresponding to a $P\%$ response is estimated depends very much upon the position of the doses tested relative to the ED 50. Consequently, any information on the ED 50 can be used to advantage in the selection of a set of doses for the experiment; this knowledge will be inexact, for if it were good no experiment would be needed, but the better it is the better will be the results of a carefully planned experiment. The interval between doses is also very important. If doses are close together, S_{xx} for any material will be small and b therefore of low precision; on the other hand, if doses are wide apart some of the weighting coefficients, w, must be small so that again estimation is unsatisfactory. A compromise must be made between seeking large values of w so that Snw is large and seeking intermediate values that will contribute well to S_{xx}. As will be seen, not only does the optimal compromise depend on the parameters themselves but it also needs to take account of the number of subjects.

Suppose that N subjects are to be used in estimating the ED 50 for a particular material. A convenient measure of the closeness of the estimate to the truth is I, the quarter-square of the fiducial interval for m, obtained from (4.37) as

$$I = \frac{t^2}{b^2(1-g)^2}\left[\frac{1-g}{Snw} + \frac{(m-\bar{x})^2}{S_{xx}}\right];\qquad (8.1)$$

here the n must satisfy the constraint

$$Sn = N. \qquad (8.2)$$

Suppose also that attention is restricted to experiments in which N is equally divided between k different doses, so that

$$n = N/k \qquad (8.3)$$

for each x tested.

The behaviour of I can be more usefully examined in relation to values of y, since these can be approximately controlled by an experimenter who chooses doses that he guesses will give specified responses. Substitutions from equation (4.26) lead to

$$Nb^2I = \frac{kt^2}{(1-g)^2}\left[\frac{1-g}{Sw} + \frac{\bar{y}^2}{Sw(Y-\bar{y})^2}\right] \qquad (8.4)$$

and

$$Ng = kt^2/Sw(Y-\bar{y})^2, \qquad (8.5)$$

where S denotes summation over the k doses and Y is the expected response at a dose; in the numerator of the second term of (8.4), $\bar{y}^2$ must be replaced by $(\bar{y}-5)^2$ when probits rather than N.E.D.'s are in use. The relative values of I for experiments that aim at different sets of values of Y can now be examined. If attention is restricted to equal spacing of doses, these should be symmetrically placed on either side of the $\log \text{ED}\,50$, so that $\bar{y}$ (or $\bar{y}-5$) is zero and (8.4) becomes

$$Nb^2I = \frac{kt^2}{(1-g)\,Sw}. \qquad (8.6)$$

A sufficiently good understanding can be gained by discussion of this idealized situation in which the choice of doses is assumed so successful that the second term in (8.4) can be neglected.

Table 8.1 presents results from these calculations. For each of $N = 18,\ 90,\ 450$ (chosen for divisibility by 2 and by 3), Nb^2I has been evaluated at probabilities 0.95 and 0.99 with $k = 2,\ 3$. Hypothetical experiments have been considered, in which the aim is to obtain responses of P, $(1-P)$ for $k = 2$ or P, 0.5, $(1-P)$ for $k = 3$. The numerical values shown for Nb^2I assume the aim to be so closely achieved that equation (8.6) computed for the intended responses approximates adequately to equation (8.4) computed for the observed responses. Thus the table will be a somewhat optimistic guide to the planning of an experiment.

Of course, responses more extreme than 0.1, 0.9 are scarcely meaningful when N is as small as 18, but entries have been included in order to emphasize the trend. One obvious feature of Table 8.1 is the decrease in Nb^2I as N is increased. Evidently the fiducial range narrows more rapidly than in proportion to $1/\sqrt{N}$, in consequence of the lessening importance of g for large N. A more interesting point is the minimum of each column, indicating approximately the responses at which the experimenter should aim in choosing his doses. In each triad of columns, the minimum moves down the column as N increases. This also is the effect of g: in equation (8.6), Sw will be maximized by having all responses close to 0.5, but as this limit is approached g becomes disastrously large. A broad conclusion is that for small N an experimental design that aims at responses of 0.1, 0.9 or 0.1, 0.5, 0.9 will be about optimal; for moderate N, these should be changed to 0.2, 0.8 or 0.2, 0.5, 0.8, and for large N they should be about 0.3, 0.7, or 0.3, 0.5, 0.7. An experiment with small N is necessarily almost useless for assessment of 99 % limits. The Table shows that a misjudgement causing the actual responses to be a little closer to 0.5 than was intended may be catastrophic, whereas responses a little wider apart than intended will usually have less serious consequences.

Table 8.1   Values of Nb^2I for estimation of ED50 with various true response rates

Subjects equally divided between doses with % responses shown below	95 % limits			99 % limits		
	$N = 18$	$N = 90$	$N = 450$	$N = 18$	$N = 90$	$N = 450$
3, 97	38.9	26.1	24.5	121.	48.0	42.9
4, 96	31.1	21.4	20.2	90.9	39.2	35.2
5, 95	26.5	18.5	17.4	75.7	33.7	30.4
10, 90	18.1	12.2	11.4	*56.3*	22.3	19.9
15, 85	*16.9*	9.93	9.18	80.0	18.5	16.1
20, 80	20.3	8.94	8.04	—	*17.2*	14.1
25, 75	55.8	*8.64*	7.39	—	17.6	13.1
30, 70	—	9.14	7.05	—	21.6	*12.7*
35, 65	—	12.2	*7.04*	—	62.7	13.2
40, 60	—	—	7.86	—	—	16.9
3, 50, 97	28.0	13.6	12.4	1211.	26.0	21.7
4, 50, 96	24.5	12.6	11.5	292.	23.9	20.2
5, 50, 95	22.5	11.9	10.9	*210.*	22.5	19.1
10, 50, 90	*20.3*	9.85	8.93	904.	18.8	15.7
15, 50, 85	25.7	9.00	7.96	—	*17.6*	14.0
20, 50, 80	91.0	*8.74*	7.40	—	18.1	13.2
25, 50, 75	—	9.11	7.10	—	21.2	*12.8*
30, 50, 70	—	10.8	*7.01*	—	37.1	12.9
35, 50, 65	—	22.1	7.30	—	—	14.4
40, 50, 60	—	—	9.04	—	—	23.8

Italics indicate tabulated value nearest to minimum; a dash indicates $g \geqslant 1.0$.

Table 8.1 also illustrates the gravely misleading conclusions that may be reached if $V(m)$, as defined by equation (3.12), is used in calculation of limits. Such a practice corresponds to the limiting values that Table 8.1 would show if N became very large. For example, at 95 % with $k = 2$, the values would decline steadily from 53.5 to 7.85 in the experiment aiming at responses of 0.2, 0.8, and to 6.07 at the foot of the column.

The table can easily be extended to any k. One major reason for using $k > 2$ is in order that deviations from the model can manifest themselves as non-linearity of the probit regression. A second reason is as a safeguard against a poor choice of doses. To some extent, Table 8.1 illustrates this by the less extreme changes in the columns for $k = 3$ as compared with those for $k = 2$. A small increase in Nb^2I near the optimal response rates is offset by a much greater decrease away from these rates, so that successful choice of doses has a somewhat greater latitude. This trend would be more marked for $k = 4$, but of course for large k the increase in Nb^2I near the optimal will dominate the situation.

8.2 Precision of assays

Many experiments are undertaken for estimating the relative potency of two materials. In planning one, attention should be given to the aim of obtaining an estimate as close as possible to ρ. Of course this cannot be guaranteed, but action can be directed at minimizing $V(M)$ or, better, at minimizing the interval between the fiducial limits of M.

Miller, Bliss and Braun (1939) discussed methods for increasing the precision of estimates of M. As for the estimation of a particular percentage point, consideration must usually be given to the most economical utilization of a limited number of subjects. If some information on the relative potency already exists, doses of the test material and the standard may be chosen so as to bear this ratio to one another. The final term in (6.3) or (6.4) will then be small, and alternative choices of dose can be compared almost exactly as in § 8.1. Again the need is to compromise between small values of w at extreme responses and small values of S_{xx} if all responses are near to 0.5. A modification of Table 8.1, based upon (6.4), has been presented elsewhere (Finney, 1964); Healy (1950) considered the same problem slightly differently. The doses in any experiment should as far as possible take account of all information and reasonable guesses available in advance. If practically nothing is known about the parameters, at least 4 or 5 doses of each material, sufficiently widely spaced for bracketting of ED 50's to be almost certain, must be used. If moderately good information on the ED 50's already exists, perhaps 3 doses will suffice and these should not go far outside the range corresponding to responses of 0.15, 0.85. The parameter β, the reciprocal of the standard deviation of the log tolerance distribution, may have intrinsic interest for an investigator; for its estimation, a somewhat wider range of doses is desirable. The maximum contribution to S_{yy} comes from subjects at doses that maximize $Z^2 Y^2 / PQ$, and a little numerical investigation shows this to happen at response rates about 0.06, 0.94. An experiment that concentrates on responses of 0.15, 0.85 will reduce S_{yy} about 25 % below the maximum.

Other things being equal, the fiducial range of M will be made narrower by any procedure that increases the slope of the regression lines. The parameter β, however, is not under the control of the experimenter in the same way as are the dose levels, but is an inherent property of the materials, test subjects, and experimental conditions. Although the precision of an assay of relative potency may be increased by a change in the test subject or in the experimental conditions, there may then be some doubt of whether the same quantity is being estimated. Bliss and Cattell (1943) refer to drug assays with animals in which estimates of relative potency were very different from the values found for man, in some cases even the order of potency being reversed. The relative potencies of insecticides may be altered by a change in spray medium or in method of application, and may be entirely different for different species of insect.

One might hope that precision could be increased by comparing responses to the substance under test with those found for the standard in all previous tests, rather than with the small amount of data for the standard obtained in the course of the current assay. The danger is that, though conditions can usually be so controlled that *relative* potency remains reasonably constant in tests made over a period of time, *absolute* potency often varies very considerably from day to day. Thus the data examined in Example 17 indicate that the poisons on the third occasion averaged 50 % higher potency than on the first. Bliss and Packard (1941) reported that exposure of eggs of *Drosophila melanogaster* to röntgen rays gave the same probit regression relation between kill and intensity of irradiation on a number of occasions over a period of seven years. Whitlock and Bliss (1943) described a series of antihelminthic tests in which the position of the regression line remained the same, within the limits of experimental error. Nevertheless, this state of affairs appears to be exceptional, and in any new field of study the investigator should be prepared for much greater instability in the mean log tolerance, μ, than in the slope, β, until experience shows otherwise.

8.3 The design of experiments

When several materials are to be compared, to test all on one day or all on one homogeneous stock of subjects may be impossible. Precautions must then be taken to avoid bias in conclusions from inequalities in the susceptibility of subjects on different days or in different stocks. A similar problem may arise when tests have to be made in several laboratories, or by several workers in the same laboratory: even under standardized methods of testing, variations may occur in the effectiveness of the same poison used by different workers. Examples 16 and 17 illustrate methods of combining results from groups of tests where each group contains a different selection of poisons from all those under investigation and the general level of potency or effectiveness may vary from one group to another. These, however, are not ideal examples of how to deal with the general problem: such lack of symmetry complicates the analysis and can lead to final estimates of log ED 50's that differ widely in precision.

In an investigation of materials which are too many for all to be tested on a single occasion, or which for some other reason have to be divided into groups for testing, an element of balance in the arrangement improves the quality of unbiased and reliable comparisons between every pair of materials. The doses and numbers of subjects should be chosen, in the light of existing knowledge, with the aim that all estimates of log ED 50 shall have about the same precision. Careful planning before an investigation is begun is always to be preferred to the haphazard accumulation of results: statistical analysis cannot extract satisfactory answers to the questions propounded if the experiments themselves were ill designed.

Moore and Bliss (1942) described an experiment in which seven different organic compounds were tested for toxicity to *Aphis rumicis*. Sets of three compounds, to be tested under comparable conditions on each of seven days, were so chosen that each was used on three days and occurred once and once only on the same day as each of the others. Moore and Bliss discussed the analysis of variance of the 21 values of m, ignoring the differential weighting of the observations in the usual probit analysis. They obtained unbiased comparisons between the poisons, though still more precise estimates might possibly be made from the data. The experiment showed a heavy mortality among insects sprayed with the spreader only, but the general methods in Chapter 7 have not been applied.

This experiment is said to have a *balanced incomplete block* design. If the seven insecticides are denoted by A, B, ..., G, the seven blocks or sets of three may be written:

Day:	1	2	3	4	5	6	7
	A	A	A	B	B	C	C
	B	D	F	D	E	D	E
	C	E	G	F	G	G	F

No general presentation of the extensive subject of experimental design can be attempted here. Books such as Cochran and Cox (1957), Cox (1958), and Finney (1960) can be understood by a beginner with some knowledge of analysis of variance; Fisher and Yates (1964) include tables of many standard designs and condensed rules for statistical analysis. The important step in the statistical analysis is to use properly the 'within day' comparisons, so avoiding contamination of estimates of relative potency with day-to-day changes in susceptibility of subjects. For example, insecticides D, F occur together on day 4; however, days 2 and 6 permit D to be compared with A, C, E, G by within-day comparisons and days 3, 7 permit F to be compared with these four. Statistical analysis combines this information into an optimal estimate of the relative dose metameter for D, F, and simultaneously deals with all other pairs.

The merit of this design comes solely from the expectation that 'blocks' (or days) will differ. Experience of a field of research will help the investigator to recognize circumstances in which control of variation by some form of balancing over blocks will help him. The blocks may be groupings according to days of experiment, sources of subjects, alternative pieces of apparatus, different laboratories, or any other characteristics of the experiment. Many illustrations of good experimental design in biological assay have been published; only a few of these have related to quantal responses.

Ex. 21. *Comparison of nine vaccines with a standard.* An experiment reported by Cohen and Leppink (1956) provides a useful contrast to the unsymmetric pattern of Example 17. The authors wished to compare 9 different pertussis vaccines with the standard 'NIH reference vaccine no. 5'. The experiment was arranged in 12 blocks, presumably corresponding to different days of beginning. A block included tests of the standard and 3 of the 9 new vaccines, each at three levels of dose in ratios 25:5:1. The blocks were so constructed that every pair from the 9 vaccines occurred in exactly one block. Thus the 9 vaccines were placed in a balanced incomplete block arrangement (constructed with the aid of what is known as a *balanced lattice*), but each block was augmented by the standard vaccine, S (Table 8.2).

Table 8.2   Sets of four vaccines in the twelve blocks of Cohen and Leppink's experiment

Block	T:	S, 1, 2, 3	Block	VII:	S, 2, 4, 9
,,	II:	S, 4, 5, 6	,,	VIII:	S, 3, 5, 7
,,	III:	S, 7, 8, 9	,,	IX:	S, 1, 6, 8
,,	IV:	S, 3, 6, 9	,,	X:	S, 2, 6, 7
,,	V:	S, 2, 5, 8	,,	XI:	S, 1, 5, 9
,,	VI:	S, 1, 4, 7	,,	XII:	S, 3, 4, 8

Cohen and Leppink have described fully the experimental procedure. Here all that need be said is that for each block about 240 mice were divided into groups of about 20 for vaccination with the twelve test doses. After 10–14 days, all mice were injected intracerebrally with a standard strain of *Hemophilus pertussis* and observed for 14 days longer. Some mice from the groups of 20 were lost from the experiment, possibly by death before challenge with the bacteria. Statistical analysis was based on 660 mice for S and 210–230 mice for each of the other vaccines. Probit methods were applied to the proportion of mice surviving among those that were injected, but numbers and proportions are not shown here. The first step, a little different from Cohen and Leppink, was to use the standard computer program (cf. § 6.6) for each block in turn, so estimating four values of m and their asymptotic variances from parallel regressions. Of course the program could give three relative potencies for each block, but the method used is a simpler way of taking account of the design. No attempt was made to force the regression lines of different blocks to be parallel: b ranged from 1.875 in block 6 to 1.001 in block 12.

Table 8.3 shows the 48 values of the log ED 50. The asymptotic variances of m ranged from 0.0419 to 0.0112; m was sufficiently close to $\bar{x}$ for the final term in equation (3.12) to be relatively small every time, so that for any relative potency $V(M)$ was not very different from the sum of the two values of $V(m)$. For the next step, differences of variance were ignored and the standard type of analysis for planned experiments was applied.

Table 8.3   Estimates of log ED50 for experiment of Table 8.2 (units: 10^{-4} ml for S, 10^6 organisms for other vaccines)

Block	S	1	2	3	4	5	6	7	8	9
I	1.988	3.171	2.868	2.990	—	—	—	—	—	—
II	2.143	—	—	—	3.875	3.837	3.677	—	—	—
III	1.793	—	—	—	—	—	—	2.579	2.971	2.758
IV	2.163	—	—	3.101	—	—	3.131	—	—	2.983
V	1.775	—	2.767	—	—	3.413	—	—	3.459	—
VI	2.032	3.017	—	—	3.519	—	—	3.203	—	—
VII	1.656	—	3.141	—	3.488	—	—	—	—	2.726
VIII	1.620	—	—	2.821	—	3.549	—	2.404	—	—
IX	1.938	2.861	—	—	—	—	2.983	—	3.095	—
X	2.240	—	3.152	—	—	—	3.358	2.976	—	—
XI	1.946	3.046	—	—	—	3.877	—	—	—	3.008
XII	1.270	—	—	2.736	3.144	—	—	—	2.842	—

The details of the analysis were slight modifications of those published in many standard texts for balanced incomplete blocks. As in Example 17, each m was equated to the sum of a parameter corresponding to the particular preparation and a parameter for the block plus an error component. The parameters were then estimated by minimizing the sum of squares of deviations of the 48 values from the corresponding totals of preparation and block parameters. Calculations of a pattern familiar to all users of incomplete block designs lead to the analysis of variance in Table 8.4.

Table 8.4   Analysis of variance for Table 8.3

| Adjustment for mean | | 380.1939 | |
Source of variation	d.f.	Sum of squares	Mean square
Blocks	11	2.5068	
Vaccines, eliminating blocks	9	16.1229	
Block and vaccine parameters	20	18.6297	
Error	27	0.6610	0.02448
Total	47	19.2907	

The analysis of variance shows clearly that differences between treatments are large relative to the error mean square. The error mean square itself may be compared with the average of the 48 values of $V(m)$, 0.022618; the close agreement suggests that practically all variability is attributable to the binomial distributions within the blocks. Table 8.5 contains the estimates of the preparation parameters, that is to say revised means for the log ED 50's for the ten preparations. The algebra of the least squares analysis leads to formulae for the variances of differences between the parameters; these are 8/13 of the error mean square for differences between two of the nine vaccines, 5/13 of this for differences of any one of the nine from S. Thus the potency of vaccine 1 relative to S is finally estimated by

$$M = 1.8803 - 2.9608$$
$$= -1.0805.$$

Addition and subtraction of t times the standard error (0.0970 with 27 degrees of freedom), followed by the taking of antilogarithms, gives a relative potency 0.0831 with 95 % limits at 0.1314 and 0.0525, in terms of the units stated earlier. This step can be repeated for each vaccine. The limits are substantially narrower than those appropriate to the data on any one block, and in general a little better than would be obtained by combining the four values of M from the blocks in which a specified vaccine was tested.

The above is only an outline of the analysis of this complicated experiment. The reader may wonder why one of the methods of

Table 8.5   Final estimates of log ED 50 for Table 8.3

S:	1.8803	5:	3.6261
1:	2.9608	6:	3.1335
2:	2.9382	7:	2.7930
3:	3.0357	8:	3.2874
4:	3.4962	9:	2.8604

Example 17 was not followed, either minimizing a weighted sum of squares for the log ED 50's or undertaking a single large probit analysis for simultaneous estimation of all parameters (22 including a common β). These are certainly more general approaches, but they are also computationally much heavier, and Example 21 has been chosen to illustrate the standard unweighted procedure for a symmetric design. Had all $V(m)$ been equal, the first method of Example 17 would have given exactly the same results: the range of less than four-fold variation here is unlikely to make much difference in so symmetrical an experiment. The single comprehensive analysis is preferable in theory. The regression coefficients are not particularly well determined for the 12 separate blocks, and a few values of g are quite large. If the hypothesis of a common β were found to be tenable, the much greater precision of its estimation would be a real gain, and some theoretical objections to assumptions inherent in the above analysis would be removed. Simultaneous estimation of so many parameters is not to be lightly undertaken, for even on a modern computer this requires a substantial amount of time. The computations have been completed for this assay of pertussis vaccine, and gave essentially the same results as did the analysis presented above.

8.4   Factorial design

Recognition of the factorial principle is essential for full exploration of the causes underlying even the simplest biological phenomena. Fisher (1966) succinctly stated the case for factorial design:

In expositions of the scientific use of experimentation it is frequent to find an excessive stress laid on the importance of varying the essential conditions *only one at a time*. . . In the state of knowledge or ignorance in which genuine research, intended to advance knowledge, has to be carried on, this simple formula is not very helpful. We are usually ignorant which, out of innumerable possible factors, may prove ultimately to be the most important, though we may have strong presuppositions that some few of them are particularly worthy of study. We have usually no knowledge that any one factor will exert its effects independently of all others that can be varied, or that its effects are particularly simply related to variation in these other factors. On the contrary, when factors are chosen for investigation, it is not because we anticipate that the laws of nature can be expressed with any particular simplicity in terms of these variables, but because they are variables which can be controlled or measured with comparative ease. If the investigator, in these circumstances, confines his attention to any single factor we may infer either that he is the unfortunate victim of a doctrinaire theory as to how experimentation should proceed, or that the time, material or equipment at his disposal are too limited to allow him to give attention to more than one narrow aspect of his problem.

Factorial design was first widely employed for agricultural field trials, but its value in the logical structure and interpretation of experiments is as great in the laboratory as in the field. Problems of quantal responses have often been investigated by varying the level of a single factor in the conditions defining the stimulus, others being held as nearly constant as was practicable. Factorial design entails: first, a selection of the more important factors relating to the stimulus or the subject; secondly, the adoption for the experiment of a convenient number of states or levels of each factor selected; thirdly, the making of tests on batches of subjects under the conditions defined by various combinations of levels of these factors, non-experimental factors being held as nearly constant as possible. In this way the virtues of carefully standardized conditions are combined with the obtaining of information on the effects of variations in these conditions. The measurements of all factors constituting the stimulus may be referred to collectively as the *dose*.

In considering adoption of a factorial set of treatments, the different needs of an assay and an investigation into the laws determining the reaction of the subject to the stimulus should be borne in mind. The purpose of an assay is to assess the value of an arbitrary unit of the stimulus under test in units of a standard stimulus; provided that the test stimulus can be fully measured in these units, there will generally be no advantage in using more than one factor for the assay. If a vaccine is to be assayed relative to an accepted standard vaccine, the result ought to be unaffected (except for sampling variation) by age, sex, or diet of the test animals, and to include different levels of these or other factors would have little merit. Indeed, any indication that the relative potency depended on other conditions (such as sex, diet, or type of cage) would contradict the central assumption of the assay, and would suggest that the vaccine could not be measured uniquely in units of the standard. For an assay, the choice of conditions in respect of these other factors should depend solely upon practical convenience and experience of which gives the more precise results, unless indeed challenging or confirming the central assumption is itself an objective.

When the relation between the reaction of the subject and measures of the stimulus is the object of study, a factorial experiment has many advantages. Before the action of an insecticide can be fully understood, the direct effects of various factors defining the dose and its method of application, and the interactions between these, must be investigated in detail. By comparison with agricultural experiments, laboratory tests of insecticides take only a short time, and, at least in the preliminary stages of research, a series of experiments on single factors may give better returns than one experiment including many factors. Information gained from these simple trials, however, aids plans for the more extensive factorial experiments essential to the elucidation of relations between the factors.

The discussion in § 8.5 again relates to insecticidal studies, but the principles readily apply to other fields. Laboratory research on insecticidal potencies has been primarily directed at the effect of concentration of the toxic substance on the mortality of the insects. To a lesser extent, duration of exposure to the poison has been examined, though few published experiments have included both factors. Other factors, such as temperature or quantity of poison used, have received even less attention.

8.5   Qualitative factors

Among the complex of factors measuring dose will usually be one, such as concentration, that takes different values on a continuous scale. Methods of statistical analysis applicable for two or more quantitative factors are discussed in § 8.6 and subsequently. There may also be factors of a purely qualitative nature, such as differences in the medium in which the poison is applied (e.g. oil or water; Martin, 1943), or differences in the method of application of an insecticide (e.g. spray or film; Tattersfield and Potter, 1943). Other factors, though capable of quantitative measurement, may simply be recorded descriptively, as in comparisons between 'warm' and 'cold' conditions during spraying, between 'old' and 'new' stocks of insecticide, or between different strains of insect.

The problem of §§ 6.1, 6.2 is that of one quantitative and one qualitative factor. The method may be extended to experiments with one quantitative and several qualitative factors in all their combinations. Provided that the data can be adequately described by parallel regression lines, all effects of the qualitative factors can be evaluated in terms of comparisons between ED 50's. The essential features of the analysis are then small modifications of standard procedures for factorial experiments (Cochran and Cox, 1957). The unequal precisions associated with the different regression lines cause some complication, but this is negligible if similar numbers of subjects have been used for all tests. If the lines are not parallel, interpretation is more difficult, but nothing need be added to what has been said in § 6.5.

Ex. 22. *A factorial experiment on the insecticidal potency of pyrethrins.* Potter and Gillham (1946; Finney, 1946) tested the toxicity of pyrethrins to adult *Tribolium castaneum* Hbst., under conditions described fully in their paper. All that need be noted here is that the insects were stored in either a hot (H) or a cool (C) room before spraying, and in either a hot or a cool room after spraying. Tests were made for each of the four possible combinations of storage treatments, and also, for each combination, with a spray to which 1% terpineol (T) has been added. The eight series of tests form a 2^3 factorial system, three factors (storage of insects before spraying, storage after spraying, terpineol) each of which has two alternative states (hot or cool for the first two, absence or presence for the third). The series are identified by code symbols: $C.H.T.$ refers to the tests in which terpineol was added to the spray for insects stored cool before, hot after spraying. In each series, about seventy insects were tested at each of three doses, 0.0195%, 0.0130%, and 0.0065% total pyrethrins. For statistical analysis, doses were taken in units of 0.001%, in order to avoid negative values of x, a transformation of scale that makes no difference to measures of relative potency.

Eight parallel regression lines were fitted exactly as described in § 6.2, except that allowance was made for natural mortality. Of 517 insects receiving a spray containing no pyrethrins, 7 died: with this large number tested, and evidently a very small true natural mortality, 7/517 has been taken as the true value of C and the approximation of § 7.2 used. The full calculations as in § 7.3 have been put through the computer, but the differences were negligible as the estimate obtained for C was 1.39%. Table 8.6, constructed as was Table 6.3, shows no serious departure from parallelism but significant heterogeneity. Careful inspection of the lines, and of differences between observed and expected frequencies, discloses no systematic pattern of non-linearity, and for illustrative purposes here a heterogeneity factor is taken as the composite mean square from Table 8.6:

$$32.95/15 = 2.197.$$

Table 8.6   Heterogeneity and parallelism of regressions
for factorial experiment

Nature of variation	d.f.	Sum of squares	Mean square
Parallelism	7	17.28	2.47
Residual heterogeneity	8	15.67	1.96
Total	15	32.95	

The regression coefficient is
$$b = 4.0285 \pm 0.3784,$$
whence for a probability of 0.95
$$g = 0.033\,89.$$

Table 8.7 contains the values of a for the 8 lines, in an order to be explained. Chief interest attaches to comparisons between median lethal doses, rather than to the separate values, in assessment of the relative potencies of different conditions. In particular, the *main effect* of each factor must be studied: Is the average potency of all series stored hot before spraying different from that of all series stored cool before spraying? Similar questions can be put for the other two factors. Secondly, the *interactions* between these factors, the extent to which the effect of one is affected by the state of another, must be studied. A convenient method of examining main effects and interactions, customary in other uses of factorial design, is fully described in the usual texts. The main computations are shown in Table 8.7. The values of a have been arranged in order, starting with 'hot before, hot after, no terpineol' and changing each factor in turn according to the system shown in the first column. The first four entries of column (1) are the sums of the four successive pairs of values of a, and the last four entries are the differences of these pairs; the first number must always be subtracted from the second and signs maintained correctly. Thus

$$1.1038 + 0.7083 = 1.8121,$$

$$1.1038 - 0.7083 = 0.3955.$$

Column (2) is derived from column (1), and column (3) from column (2) by repetition of this process.

Table 8.7   Calculations for main effects and interactions

Series	a	(1)	(2)	(3)
H H	0.7083	1.8121	6.4173	12.2665
C H	1.1038	4.6052	5.8492	2.2517
H C	2.2207	0.3223	0.5593	7.9977
C C	2.3845	5.5269	1.6924	−0.2827
H H T	−0.2747	0.3955	2.7931	−0.5681
C H T	0.5970	0.1638	5.2046	1.1331
H C T	2.3531	0.8717	−0.2317	2.4115
C C T	3.1738	0.8207	−0.0510	0.1807

The first entry in column (3), the total of the eight values of a, is of no particular interest. The second, which can be seen to be the difference between totals of the four 'cool before spraying' and the four 'hot before spraying' values of a, is used in comparing the potency for these two conditions. In fact, for the average potency of cool relative to hot,

$$M = T_a/4b, \qquad (8.7)$$

where T_a is the second entry in Table 8.7. The third entry, used similarly in equation (8.7), gives the main effect of cool versus hot after spraying, and the fifth gives the main effect of the addition of terpineol. The remaining entries give interactions. The fourth, for example, contains the difference between 'cool versus hot before spraying, cool after' and 'cool versus hot before spraying, hot after', that is to say between

$$(2.3845 + 3.1738) - (2.2207 + 2.3531) = 0.9845$$

and $(1.1038 + 0.5970) - (0.7083 - 0.2747) = 1.2672.$

Equation (8.7) then gives M for this interaction of conditions before and after spraying. The sixth and seventh entries give interactions of the two storage conditions separately with terpineol, and the last entry gives the interaction of all three factors. Note that the meaning of each entry in column (3) corresponds with the number of changes between the factor code for the series and the first factor code, HH. For each calculation,

$$T_a = T_y - bT_x, \qquad (8.8)$$

where T_x, T_y are combinations of $\bar{x}$, $\bar{y}$ formed in the same way as T_a. Hence

$$M = \frac{T_y}{4b} - \frac{T_x}{4}. \qquad (8.9)$$

Now $$V(T_y) = \Sigma \frac{1}{Snw}, \qquad (8.10)$$

multiplied by a heterogeneity factor if necessary (as in this example), irrespective of which line of Table 8.7 is under discussion, and T_y is uncorrelated with b because each $\bar{y}$ is uncorrelated with b. Therefore, fiducial limits to M can be found by applying Fieller's theorem to the ratio of T_y and $4b$, afterwards subtracting $T_x/4$ from each limit. The asymptotic variance

$$V(M) = \frac{2.197}{16b^2} \left\{ \Sigma \frac{1}{Snw} + \frac{(4M + T_x)^2}{\Sigma S_{xx}} \right\} \qquad (8.11)$$

(2.197 is the heterogeneity factor) has analogy with equation (6.3); the factors 4 and 16 enter because M is based on average differences between two sets of four conditions. Table 8.8 summarizes results from equations (8.7), (8.11), the small g ensuring that the standard errors give a fair representation.

Table 8.8　Summary of effects, expressed as relative potencies

Effect	M	R	(R_L, R_U)
1. Cool v. hot before spray	0.140 ± 0.036	1.38	(1.16, 1.67)
2. Cool v. hot after spray	0.496 ± 0.045	3.13	(2.56, 4.02)
3. Terpineol	-0.035 ± 0.035	0.92	(0.77, 1.10)
4. Interaction of 1, 2	-0.018 ± 0.035	0.96	(0.80, 1.14)
5. Interaction of 1, 3	0.070 ± 0.036	1.17	(0.99, 1.41)
6. Interaction of 2, 3	0.150 ± 0.036	1.41	(1.19, 1.70)
7. Interaction of 1, 2, 3	0.011 ± 0.035	1.03	(0.86, 1.23)

Table 8.8 shows cool storage to have increased the potency of pyrethrins substantially and significantly, especially after spraying. On average, terpineol reduced the potency slightly, but not significantly. Some positive interaction of the terpineol effect with storage is evident, the after-spraying interaction being significant: although terpineol had no average effect on potency, it augmented the benefits of cool storage. Addition of lines 2 and 6 in Table 8.8 gives M for cool versus hot storage after spraying *in the presence* of terpineol, subtraction of line 2 from line 6 gives the value *in the absence* of terpineol. Just as M in line 2 of Table 8.8 is the average of these two, so M in line 6 is half their difference. The two quantities could be obtained by direct calculation from the first or second four entries in the first column of Table 8.7, using essentially the same methods but now for a 2^2 factorial system.

An easily understood summary for factorial experiments is shown in Table 8.9. The relative potency for each effect is stated as an average for all series, and also separately averaged for the four series at each level of the other factors in turn. Calculation of the second pair of entries in the last column has been described, and others follow the same pattern.

Table 8.9   Relative potencies of pyrethrins
under various conditions

Relative potency for	Mean	Before spraying		After spraying		Terpineol	
		Hot	Cool	Hot	Cool	Absent	Present
Before spraying, cool instead of hot	1.38	—	—	1.44	1.32	1.17	1.62
After spraying, cool instead of hot	3.13	3.27	3.01	—	—	2.22	4.43
Terpineol added	0.92	0.79	1.08	0.65	1.30	—	—

8.6   Probit planes

If two quantitative factors, such as the time of exposure to an insecticide and its concentration, are to be studied simultaneously, batches of subjects must be tested at various combinations of values of the two factors. The test conditions may be chosen as various concentrations for each of several exposure times, or as various exposure times for each of several concentrations. Usually a better plan is to take all combinations of a set of concentrations and a set of exposure times (say 4 concentrations and 5 times used in their 20 combinations); the more extreme combinations – low concentrations for short times, or high concentrations for long times – may be omitted if they seem unlikely to give useful results.

The probit of the response may be linearly related to the logarithm of the measure of each factor when the other is fixed. A plane may then represent the joint effect of time and concentration: the equation

$$Y = \alpha + \beta_1 x_1 + \beta_2 x_2 \qquad (8.12)$$

gives the probit in terms of log concentration (x_1) and log time (x_2). Bliss (1940) expressed this by saying that the logarithms of the exposure time and concentration that produce a specified response are linearly related. Maximum likelihood estimation of the parameters (Finney, 1943a) is a natural generalization of the one-factor analysis. The computations, using multiple regression analysis for estimation of the coefficients β_1 and β_2, are easily systematized and may be further extended to three or more factors with no difficulty other than increase in labour.

For generalization of the classical probit regression technique, a provisional plane must first be obtained, with the aid of empirical probits read from the standard table. If the experiment tested several concentrations at each of a series of times, but not necessarily the same set of concentrations for every time, lines would be drawn (by eye) relating the empirical probit to the log concentration for each time, with the restriction that the lines are parallel and at distances apart proportional to the differences in the log times. If the experiment tested a series of concentrations and a series of times in all, or nearly all, their combinations, a method adapted from Richards (1941) may be used. The probits are plotted against $(x_1 + x_2)$, the sum of log concentration and log time; points with constant x_2 should then follow one set of parallel straight lines with slope β_1, and points with constant x_1 a second set of parallel lines with slope β_2. The intervals between pairs of lines of either set will be proportional to the differences between the corresponding values of x_2 or x_1. Thus a plane representation of the three dimensional figure relating probits to x_1 and x_2 is obtained, and provisional probits can be read from two intersecting sets of parallel lines, drawn by eye, in the diagram. Care at this stage may make a satisfactory approximation to the maximum likelihood solution obtainable in one or two cycles of computation.

The working probit, y, and its weight, nw, may be derived from the provisional probit and the percentage kill exactly as when only one factor is involved. Using the technique of weighted multiple regression to derive a, b_1, and b_2 as estimates of the parameters, equation (8.12) is estimated just as was the one-factor probit equation in § 4.4. If this equation differs substantially from the provisional one, it may be used to determine a new series of provisional probits, with which the cycle of computations is repeated. When there is an appreciable natural mortality, the methods of Chapter 7 may be used; the full analysis (§ 7.3) may easily be extended to allow for the additional independent variate.

The statistical technique was first elaborated for the analysis of data obtained by Tattersfield and Potter (1943). Example 23 is a revised version of the computations appropriate to the

6

results of one of their experiments; as previously published (Finney, 1942b, 1943a), weighting coefficients were not modified to allow for the control mortality.

Ex. 23. *The effect of concentration and deposit on the toxicity of a pyrethrum preparation to* Tribolium castaneum. Tattersfield and Potter (1943) described experiments on the toxicity of a pyrethrum extract in a heavy oil to the beetle *T. castaneum*; they used combinations of several concentrations of the pyrethrum extraçt and several weights of spray deposit on the glass disk (covered with fabric) on which the insects were placed. In the first experiment, four concentrations and three deposits were tested; each combination of concentration and deposit was used, on separate batches of insects, both as a direct spray and as a film on which the insects were afterwards placed. Batches of about ten insects were used for each spraying, and all treatments had three-fold replication.

Other insects were exposed to different deposits of the base oil alone, applied as a spray or as a film. The mortalities gave no indication of difference from that of unsprayed controls and all control batches have been added: of 311 beetles, 12 were 'badly affected, moribund, or dead', giving a control rate of 3.86 %. The data (*loc. cit.* Table 2) are presented in Table 8.10.

Table 8.10   Responses of *Tribolium castaneum* when exposed to a pyrethrum spray or film (responses/insects exposed)

Pyrethrin I Concentration (mg/ml)	Deposit (mg/cm²)					
	Direct spray			Film		
	0.29	0.57	1.08	0.29	0.57	1.08
0.5	1/27	4/29	6/30	3/29	4/27	8/28
1.0	15/29	19/29	15/24	10/30	14/28	17/28
2.0	27/30	26/27	31/31	24/29	27/28	26/28
4.0	28/28	30/30	19/19	29/29	29/29	17/17

An extension of the ideas of Chapter 6 suggests the fitting of two parallel probit planes. Table 8.11 displays the lengthy, but straightforward, computations for one cycle, in which C has been assumed known to be 4 %. In order to avoid negative numbers, concentrations and deposits have been multiplied by 10, so that x_1 and x_2 are each 1.00 in excess of the true log concentrations and log deposits. In Table 8.11, these x_1, x_2 are followed by n, the number of insects under any one treatment, r, the total killed, and p^*, the proportionate mortality. Because the control mortality is estimated with considerable precision from 311 insects, and at 4 % is not in conflict with the remainder of the data, the approximate method of § 7.2 has been used. The column of adjusted mortalities, p, has therefore been obtained with $C = 0.04$ in equation (7.2), and the empirical probits of p have been entered from Table 3.2.

In Fig. 8.1, empirical probits are plotted against $(x_1 + x_2)$. For both the spray and the film technique, two intersecting sets of parallel lines were drawn, representing the regressions of probits on x_1 for fixed x_2 and on x_2 for fixed x_1. The lines were drawn by eye (remembering zero and 100 % responses at certain doses) in such a way as to intersect vertically above or below the plotted points. Each intersection gives the provisional probit, Y, for a combination of x_1 and x_2, from which weights and working probits were calculated in the usual way; the weighting coefficients, w, were taken from the 4 % column in Table II.

For either technique of application of insecticide, the weighted regression equation of y on x_1 and x_2 is an approximation to the maximum likelihood estimate of the probit plane. In order to obtain two parallel planes, the regression coefficients must be calculated from the totals of corresponding sums of squares and sums of products of deviations for the technique separately (cf. § 6.1). To anyone familiar with multiple regression, the next stage of calculation is standard. Products nwx_1, nwx_2, and nwy are formed independently for spray and film; for nwx_2, subtotals of nw for each deposit can be multiplied by x_2, but for nwx_1 and nwy individual products must be entered. Column totals Snw, $Snwx_1$, $Snwx_2$, and $Snwy$ are entered for each technique. Summation of products such as nwx_1 by x_1 or nwx_1 by x_2 then leads to sums of squares and products of deviations; for example

$$S_{x_1 x_2} = Snwx_1 x_2 - (Snwx_1)(Snwx_2)/Snw. \qquad (8.13)$$

Thus all the quantities at the foot of Table 8.11 have been calculated, the values for the two techniques being added to give the last line.

Table 8.11 Computations for first cycle of probit plane estimation for toxicity of pyrethrum to *Tribolium castaneum*

x_1	x_2	n	r	$p*$	$p(C = 4)$	Empirical probit	Y	nw	y	nwx_1	nwx_2	nwy
						Exposure to direct spray						
0.70	0.47	27	1	3.7	0	$-\infty$	3.4	3.6	2.91	2.520		10.476
1.00	0.47	29	15	51.7	50	5.00	4.8	16.6	5.00	16.600		83.000
1.30	0.47	30	27	90.0	90	6.28	6.2	10.6	6.28	13.780		66.568
1.60	0.47	28	28	100.0	100	∞	7.6	1.1	7.94	1.760		8.734
								31.9		34.660	14.993	168.778
0.70	0.75	29	4	13.8	10	3.72	3.8	7.9	3.72	5.530		29.388
1.00	0.75	29	19	65.5	64	5.36	5.2	17.0	5.36	17.000		91.120
1.30	0.75	27	26	96.3	96	6.75	6.6	6.1	6.73	7.930		41.053
1.60	0.75	30	30	100.0	100	∞	8.0	0.4	8.30	0.640		3.320
								31.4		31.100	23.550	164.881
0.70	1.04	30	6	20.0	17	4.05	4.2	12.6	4.06	8.820		51.156
1.00	1.04	24	15	62.5	61	5.28	5.6	12.7	5.25	12.700		66.675
1.30	1.04	31	31	100.0	100	∞	7.0	3.9	7.42	5.070		28.938
1.60	1.04	19	19	100.0	100	∞	8.4	0.1	8.67	0.160		0.867
								29.3		26.750	30.472	147.636
								92.6		92.510	69.015	481.295
						Exposure to film						
0.70	0.47	29	3	10.3	7	3.52	3.3	3.1	3.57	2.170		11.067
1.00	0.47	30	10	33.3	31	4.50	4.7	16.7	4.51	16.700		75.317
1.30	0.47	29	24	82.8	82	5.92	6.1	11.2	5.90	14.560		66.080
1.60	0.47	29	29	100.0	100	∞	7.5	1.4	7.85	2.240		10.990
								32.4		35.670	15.228	163.454

0.70	0.75	27	4	14.8	11	3.77	3.6	5.4	3.80	3.780	20.520
1.00	0.75	28	14	50.0	48	4.95	5.0	16.5	4.95	16.500	81.675
1.30	0.75	28	27	96.4	96	6.75	6.4	8.1	6.67	10.530	54.027
1.60	0.75	29	29	100.0	100	∞	7.8	0.7	8.12	1.120	5.684
							30.7			31.930	23.025 161.906
0.70	1.04	28	8	28.6	26	4.36	4.0	9.7	4.42	6.790	42.874
1.00	1.04	28	17	60.7	59	5.23	5.4	13.8	5.22	15.800	82.476
1.30	1.04	28	26	92.9	93	6.48	6.8	4.8	6.37	6.240	30.576
1.60	1.04	17	17	100.0	100	∞	8.2	0.1	8.49	0.160	0.849
							30.4			28.990	31.616 156.775
							93.5			96.590	69.869 482.135

Spray $\bar{x}_1 = 0.9990$ $\bar{x}_2 = 0.7453$ $\bar{y} = 5.1976$

Film $\bar{x}_1 = 1.0330$ $\bar{x}_2 = 0.7473$ $\bar{y} = 5.1565$

	$Snwx_1^2$	$Snwx_1x_2$	$S\text{-}wx_2^2$	$Snwx_1y$	$Snwx_2y$	$Snwy^2$
Spray	97.01900	67.43520	56.40009	502.7093	356.5278	2614.423
	92.42009	68.94792	51.43704	480.8272	358.7103	2501.565
	4.59891	-1.51272	4.96305	21.8821	-2.1825	$112.858 - 109.75 = 3.11$
Film	104.27900	70.86200	57.30655	515.5154	361.2989	2566.119
	99.78212	72.17804	52.21045	498.0687	360.2812	2486.141
	4.49688	-1.31604	5.09610	17.4467	1.0177	$79.978 - 75.65 = 4.33$
Total	9.09579	-2.82876	10.05915	39.3288	-1.1648	$192.836 - 183.40 = 9.44$

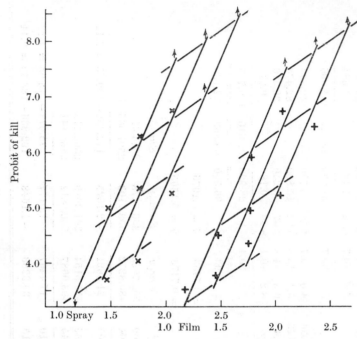

Fig. 8.1. Diagrammatic representation of probit planes for toxicity of a pyrethrum spray to *T. castaneum* at various concentrations and deposits (Ex. 23).

× : spray   + : film

Continuous lines: fixed deposit, change in concentration.
Broken lines: fixed concentration, change in deposit.

The regression coefficients for the parallel planes are the solutions of

$$9.096b_1 - 2.829b_2 = 39.329, \\ -2.829b_1 + 10.059b_2 = -1.165. \Big\} \qquad (8.14)$$

These should be solved by the inverse matrix method, as in Example 19, because the elements of the inverse are needed in variance calculation. Taking first 1, 0 and then 0, 1 on the right-hand side, this matrix was obtained as

$$V = \begin{pmatrix} 0.120\,48 & 0.033\,88 \\ 0.033\,88 & 0.108\,94 \end{pmatrix}, \qquad (8.15)$$

whence

$$b_1 = 39.329 \times 0.120\,48 - 1.165 \times 0.033\,88$$
$$= 4.6989,$$
$$b_2 = 1.2056.$$

(8.16)

The fitting of the regression planes accounts for a portion

$$b_1 S_{x_1 y} + b_2 S_{x_2 y} = 183.40 \qquad (8.17)$$

of the sum of squares of deviations of y, $S_{yy} = 192.84$. Since S_{yy} is based on two sets of 11 degrees of freedom, and the fitted parameters remove 2, the residual

$$\chi^2_{[20]} = 9.44 \qquad (8.18)$$

tests goodness of fit. Agreement is good, with no fear of non-parallelism.

Had the question of whether the planes for the two techniques are parallel needed examination, the procedure would have been as usual: fit planes to the two parts of the data independently and compare the reductions in S_{yy} when 4 parameters are estimated instead of 2. In fact, this gives (approximately) $\chi^2_{[2]} = 2.0$ for differences between β_1, β_2 for the two planes. Larger values of χ^2 would have demanded examination of contributions from extreme doses, as in Example 8.

The equation to either probit plane is of the form

$$Y = \bar{y} + b_1(x_1 - \bar{x}_1) + b_2(x_2 - \bar{x}_2). \qquad (8.19)$$

Substituting the appropriate values of $\bar{y}$, $\bar{x}_1$, and $\bar{x}_2$, the estimated equations are

$$Y_S = -0.395 + 4.699x_1 + 1.206x_2,$$
$$Y_F = -0.598 + 4.699x_1 + 1.206x_2.$$

(8.20)

(Replacement of x_1, x_2 by $(x_1 + 1)$, $(x_2 + 1)$ respectively would make the equations relate to logarithms of the actual concentrations and deposits listed in Table 8.10). Expected probits calculated from the equations agree so well with the provisional values (Table 8.11) that further iteration is unnecessary, but they could be used to initiate a new cycle of calculation. The diagonal elements of V estimate the variances of b_1, b_2 and the remaining element is their covariance.

The above calculations are as in previous editions of this book. To-day, the alternative of analysis by computer is available. The general program needs only the extension of defining Y by equation (8.12) instead of equation (3.6); the likelihood function remains unchanged, but maximization must include the additional parameter. Opportunity was taken also to estimate C exactly as in § 7.3, of course including the evidence of the 12/311 response rate observed for the controls. This led to

$$C = 3.96\% \pm 1.09\%, \tag{8.21}$$

almost exactly as taken for Table 8.11. The residual deviations from the fitted plane had

$$\chi^2_{[20]} = 9.60, \tag{8.22}$$

in close agreement with equation (8.18).

The regression equations were found to be

$$\left.\begin{aligned} Y_S &= -0.4145 + 4.7040x_1 + 1.2230x_2, \\ Y_F &= -0.6183 + 4.7040x_1 + 1.2230x_2, \end{aligned}\right\} \tag{8.23}$$

and for the regression coefficients

$$V = \begin{pmatrix} 0.129\,203 & 0.034\,455 \\ 0.034\,455 & 0.109\,041 \end{pmatrix}, \tag{8.24}$$

showing further good agreement with (8.20) and (8.15) respectively.

One might guess that response would depend primarily on total amount of pyrethrin (the product of concentration and deposit) rather than on the components separately. This would imply that b_1 and b_2 differ only by an amount consistent with sampling variation, so that equal proportionate increases in concentration or deposit would have equal effects on the kill. Such equality is clearly contradicted by the results, for the variance of the difference between b_1 and b_2 is

$$V(b_1 - b_2) = 0.1292 - 2 \times 0.0345 + 0.1090$$

$$= 0.1692, \tag{8.25}$$

and therefore $b_1 - b_2 = 3.481 \pm 0.411.$ (8.26)

In fact a doubling of concentration was as effective as an increase of $(b_1 \log 2)/b_2$ in log deposit, or a multiplication of the deposit by 14.4; the precision of this ratio is low, and 95 % fiducial limits determined by formula (4.35) from the variances and covariances in equation (8.24) are 5.8 to 238. Absorptive properties of the substratum undoubtedly influence the relative importance of concentration and deposit: in further experiments, Tattersfield and Potter again found concentration to have the greater effect, but, when a hardened filter paper replaced the fabric covering the disk, deposit became the more important factor.

In multifactorial experiments there is no unique median lethal dose; for example, with the direct spray any pair of values of x_1 and x_2 satisfying

$$4.704x_1 + 1.223x_2 = 5.415 \qquad (8.27)$$

is estimated to give a 50 % kill. Similarly the inequality of b_1 and b_2 implies a difference in relative potencies in respect of concentration and of deposit. The difference between equally effective log doses may be taken as any pair of M_1, M_2 satisfying

$$4.704M_1 + 1.223M_2 = 0.2038. \qquad (8.28)$$

If equal concentrations were used, log deposits 0.167 less for the spray than for the film would be expected to show equal kills, and if equal deposits were used, equal kills would be expected when the log concentration was 0.043 less for the spray than for the film.

8.7 Ostwald's equation

If equation (8.12) be considered as a relation between values of x_1 and x_2 that produce a stated response, it may conveniently be expressed

$$z_1^{\beta_1} z_2^{\beta_2} = \text{constant}, \qquad (8.29)$$

z_1 and z_2 being the absolute (not logarithmic) measures of dose, or

$$z_1^{\beta_1/\beta_2} z_2 = \text{constant}.$$

This equation is essentially that used by Busvine (1938) and others:

$$z^n t = k, \qquad (8.30)$$

where z is the concentration, t the time of exposure, n and k constants, as an empirical law relating the concentration and time required for a specified toxic effect. Bliss (1940) pointed out that this was a particular case of the equation

$$(z - z_0)^n t = k, \qquad (8.31)$$

used by Ostwald and Dernoschek (1910) to represent the relation between adsorption and toxic effect (z_0 being a threshold concentration below which no effect occurs).

Bliss discussed experiments in which the time taken to reach 100 % kill at selected concentrations was measured. Results by such a procedure are unreliable, since the time is determined by the most extreme member of each batch. A more satisfactory method is to expose batches of subjects to concentrations and for times chosen by the experimenter in a two-factor design. The probit of the response rate can then be related to the two dose factors by the regression equation

$$Y = \alpha + \beta_1 \log (z - z_0) + \beta_2 \log t. \qquad (8.32)$$

To fit this equation by generalization of the classical probit calculations would be tedious, but substitution of equation (8.32) for equation (8.12) in a standard computer program is simple and z_0 can then be estimated as a fourth parameter.

Fortunately, such allowance for a threshold appears seldom to be needed. If the threshold concentration commonly differed from zero to an important extent in concentration–time tests, this would also affect tests carried out for a fixed time. Hence adjustment would be needed in one-factor experiments such as have been considered in earlier chapters: x in equation (3.6) would have to be taken as $\log (z - z_0)$ instead of $\log z$, with z_0 estimated from the data. The fact that in so many experiments equation (3.6) is adequate seems evidence against non-zero threshold concentrations. Possibly some data that show a curved relation between probits and log concentrations might be better fitted by

$$Y = \alpha + \beta \log (z - z_0). \qquad (8.33)$$

8.8   Extensions to several factors and interactions

The method of § 8.6 is easily extended to experiments involving three or more dose factors. For example, if various combinations of concentration, deposit, and time of exposure to an insecticide were tested, equation (8.12) could be generalized to

$$Y = \alpha + \beta_1 x_1 + \beta_2 x_2 + \beta_3 x_3. \qquad (8.34)$$

Standard probit regression techniques or computer programs can be adapted in an obvious manner to equation (8.34) or extensions to even more factors. The price to be paid is that, as the number of parameters increases, the amount of computation increases at least as rapidly as the square of this number. Moreover, forming a satisfactory provisional equation with which to begin iterations is more difficult when several factors are included, and therefore the number of iterative cycles required may be greater than that for unifactorial experiments.

Equation (8.12) implies that the effects of the two factors on the response probit are independent and additive. If a concentration–time experiment were conducted at several fixed concentrations with various times of exposure, the regression of probit on log time might be linear for each concentration yet the lines for the different concentrations need not be parallel. Such a situation might indicate heterogeneity of the material, changes in experimental conditions between the tests of different concentrations, or a true dependence of slope of the lines upon concentration. Bliss (1940) suggested that, as a first approximation, the standard deviation of the log-time tolerances (the reciprocal of the slope) might be expressed as a linear function of the log concentration, which leads to a combined regression equation of Y on x_1 (log concentration) and x_2 (log time):

$$Y = \frac{\alpha + \beta x_1 + x_2}{\alpha' + \beta' x_1}. \tag{8.35}$$

Equation (8.35) is linear in x_2, so that for any fixed concentration the relation between Y and x_2 is a straight line, but it is not linear in x_1. It therefore differs fundamentally from the usual findings in single-factor experiments that Y is linearly related to the log concentration. The equation

$$Y = \alpha + \beta_1 x_1 + \beta_2 x_2 + \beta_{12} x_1 x_2 \tag{8.36}$$

seems preferable to (8.35), being linear in each of x_1, x_2 when the other is held constant. The slope of the regression of probit on either x_1 or x_2 increases (decreases if β_{12} is negative) as the other increases. The coefficient β_{12} measures the *interaction* between the two factors, or the extent to which the increase in Y for unit increase in x_1 or x_2 exceeds that predicted by the purely additive equation (8.12).

Estimation of parameters for equation (8.36) presents no new difficulties. The product $x_1 x_2$ is used exactly as though it were the measure of a third factor, so that the computations are of the same form as for equation (8.34); the mean value of $x_1 x_2$ must be defined by

$$x_1 x_2 = Snwx_1 x_2 / Snw,$$

which is not the same as $\bar{x}_1 . \bar{x}_2$.

The idea of equation (8.36) can be extended to more factors. For example, with three factors, one obtains the equation

$$Y = \alpha + \beta_1 x_1 + \beta_2 x_2 + \beta_3 x_3 + \beta_{12} x_1 x_2 + \beta_{13} x_1 x_3$$
$$+ \beta_{23} x_2 x_3 + \beta_{123} x_1 x_2 x_3, \quad (8.37)$$

in which the regression coefficient β_{12} measures the interaction between the first two factors, and β_{123} measures the three-factor interaction. To fit such an equation to experimental results, the four products $x_1 x_2$, $x_1 x_3$, $x_2 x_3$ and $x_1 x_2 x_3$ are treated as though they measured separate factors, and a seven-variate regression equation is calculated. Tests of heterogeneity and of parallelism are easily made, and standard errors of parameters are derived from the inverse matrix used in solving the equations for the estimates. The notion of relative potency, so useful in tests of one factor, has no simple multifactorial analogue, although one can still seek to represent different series of results by parallel equations; 'parallel' here means that equations such as (8.34) or (8.37) adequately describe the data, with the restriction that corresponding regression coefficients, β, are the same for every series and thus that the equations for the several series differ only in α.

Ex. 24. *Distribution of ovulation times.* One important use of probit methods is in the study of frequency distributions of phenomena that can be determined only by destruction of subjects (e.g. Leslie, Perry and Watson, 1945). Parsons, Hunter and Rayner (1967) wished to study whether the presence of a ram continuously during the oestrus of ewes affects the time of ovulation. By careful observation, the beginning and end of oestrus could be recorded for each ewe; the time of ovulation, however, could not be observed, and information was obtainable only by examination of reproductive tracts after slaughter.

The experiment consisted in establishing two sets of about 100 Merino ewes, and recording duration of oestrus (x_2) for each. Those of one set were continuously with their rams after mating, those of the other were removed from the rams except for short periods. In each set, 100 ewes were slaughtered at various times (x_1) after oestrus ended, and each was classified as ovulated or not ovulated after dissection of the reproductive tract. The probit regression equation for proportion ovulated on x_1 was then used to investigate characteristics of the frequency distribution of ovulation time (i.e. time from start of oestrus to ovulation).

The authors analysed in terms of x_1, x_2 as defined, and not their logarithms; they did not state a reason, and whether this fitted the data better is unknown. If ovulation time were independent of length of oestrus, a regression of probit on x_1 should suffice to describe the data. In fact, for the ewes running continuously with the ram, inclusion of x_2 in the regression significantly improved the fit; for the other set of ewes, a similar effect appeared although the regression coefficient was not significantly greater than zero.

Some suggestion appeared that the variance of ovulation times might vary with length of oestrus. Equation (8.36) was therefore tried. The estimate b_{12} was negative for the ewes with the rams, positive for the others, although neither differed significantly from zero: an interesting idea on a possible effect of the ram is not clearly demonstrated by these data but may repay further research.

8.9 Variances and fiducial limits

The methods of §§ 8.6, 8.8 will always give a variance–covariance matrix for the several regression coefficients, in the usual way as the inverse matrix of sums of squares and products of deviations of the x-variate. For example, associated with equation (8.34) would be

$$V = \begin{pmatrix} v_{11} & v_{12} & v_{13} \\ v_{12} & v_{22} & v_{23} \\ v_{13} & v_{23} & v_{33} \end{pmatrix}. \tag{8.38}$$

From this and Snw can be formed the variances and covariances of other functions required in connexion with a problem.

The estimate of equation (8.34) could be written

$$Y = \bar{y} + b_1(x_1 - \bar{x}_1) + b_2(x_2 - \bar{x}_2) + b_3(x_3 - \bar{x}_3). \tag{8.39}$$

One might wish to discuss the value of x_2 estimated to give 50 %

response for specified values, X_1 and X_3, of x_1 and x_3. The estimate is evidently

$$m = \bar{x}_2 + \{5 - \bar{y} - b_1(X_1 - \bar{x}_1) - b_3(X_3 - \bar{x}_3)\}/b_2. \quad (8.40)$$

Fiducial limits are found by applying Fieller's theorem to the ratio that estimates $m - \bar{x}_2$. The variances of numerator and denominator, formed in the usual way as variances of linear functions, are

$$\frac{1}{Snw} + v_{11}(X_1 - \bar{x}_1)^2 + 2v_{13}(X_1 - \bar{x}_1)(X_3 - \bar{x}_3) + v_{33}(X_3 - \bar{x}_3)^2$$

$$(8.41)$$

and $\qquad\qquad\qquad v_{22}; \qquad\qquad\qquad (8.42)$

the covariance is

$$- v_{12}(X_1 - \bar{x}_1) - v_{23}(X_3 - \bar{x}_3). \quad (8.43)$$

By writing, approximately, $g = 0$, a formula equivalent to use of an asymptotic variance is obtained.

Essentially the same procedures would be followed with equation (8.36).

9

Extensions of the standard model

9.1 Variation between batches

Analyses in earlier chapters have tacitly assumed that all subjects at a dose were in one batch tested at one time. In many experiments, several distinct batches of subjects are tested at a dose. For example, an apparatus for testing insecticidal sprays might take a maximum of twenty insects; if the precision of larger numbers is wanted, the apparatus will be used two or more times at each dose. The variability between mortalities in batches given the same dose then provides a measure of heterogeneity of behaviour of the batches, including both biological differences between batches and variation from batch to batch in experimental technique. A comparison of this variability with the residual variation *between doses* (after removal of the probit regression component) can test whether that residual is explicable as due solely to the natural batch variation, and thus of whether the regression line is an adequate description of the relation between dose and response.

The usual procedure is to add the values of n (number of subjects) and of r (number responding), and to analyse as if these totals referred to a single batch. Very often this is satisfactory. In general, maximum likelihood estimates are the same whether all subjects for one dose are regarded as a single batch or the identity of each batch is retained throughout calculation. Provided that the same provisional line is used, the estimates are the same at the end of each cycle of computations. This remains true for batches of unequal size if the variance of the response at every dose is inversely proportional to n. Even should heterogeneity introduce a component of batch variance independent of n, if the batches are of nearly the same size, the less onerous computations needed when batches are combined often make that the preferred method. The number of batches need not be the same at every dose.

When the residual variation between doses shows a significant

χ^2 (§ 4.6), this may indicate either a real departure from linearity or a non-independence of the responses of individuals of the same batch. If the latter explanation is adopted, all variances have to be multiplied by a heterogeneity factor, which often is estimated with low precision because the degrees of freedom available are very few; fiducial limits will vary irregularly in repeated determinations, and will tend to be widely spaced on account of the increased value of t for few degrees of freedom. Using the data for each batch separately not only permits a test of linearity but, supposing that there is heterogeneity between batches though no significant departure from linearity, also gives a heterogeneity factor based on more degrees of freedom.

Ex. 25. *The toxicity of ammonia to* Tribolium confusum. Strand (1930) tested ammonia as a fumigant for *T. confusum*, using two batches of insects at each of eight concentrations. The tests may have constituted two distinct experiments, one batch in each, but for present purposes the pairs of batches will be assumed to be replicates of the type just discussed. Table 9.1 contains the results.

The computations are not difficult. The easiest way to organize them is first to combine the batches (by totalling corresponding n and r) and to fit a regression equation to these, iterating as often as necessary. The final cycle of iteration is then repeated with the same initial values of Y but retaining the separate identities of the batches. In this way, two values of $Snw(y - Y)^2$ are obtained, with different degrees of freedom; subtraction of one from the other gives a sum of squares 'between batches within doses', with degrees of freedom equal to the total number of batches minus the number of doses.

From the output of the standard computer program applied to Strand's data, the residual that would ordinarily be tested as χ^2 is

$$18.384 \quad \text{with} \quad 6 \, \text{d.f.}$$

when the batches are combined, and

$$30.794 \quad \text{with} \quad 14 \, \text{d.f.}$$

when the batches are separate. Table 9.2 summarizes the situation.

Table 9.1   Results of exposure of *Tribolium confusum*
to ammonia (Strand, 1930, Table I)

Log concentration (mg/l)	No. of subjects (n)	No. responding (r)	% response p
0.72	29	2	7
	29	1	3
0.80	30	7	23
	31	12	39
0.87	31	12	39
	32	4	12
0.93	28	19	68
	31	18	58
0.98	26	24	92
	31	25	81
1.02	27	27	100
	28	27	96
1.07	26	26	100
	31	29	94
1.10	30	30	100
	31	30	97

Table 9.2   Heterogeneity and batch variation for Table 9.1

	d.f.	Sum of squares	Mean square
Residual between doses	6	18.384	3.064
Between batches, within doses	8	12.410	1.551
Total residual	14	30.794	

Although the variation between batches, tested as a χ^2 with
8 degrees of freedom, is not statistically significant, neither are
the mean squares between and within doses significantly different. Inspection of a probit diagram (not reproduced here)
discloses no systematic deviation from linearity. Both mean
squares are appreciably greater than unity, and the most reasonable conclusion seems to be that heterogeneity between batches

has increased both. (After 40 years, one cannot hope to investigate all possibilities: one might perhaps wonder whether by accident records of the second batches for the second and third doses have been interchanged!) Further analyses might therefore use a heterogeneity factor from the last line of Table 9.2:

$$30.794/14 = 2.200 \quad \text{with} \quad 14 \,\text{d.f.}$$

The estimates of the parameters, the same whether or not batches are kept separate, lead to the equation

$$Y = -3.988 + 10.149x.$$

9.2   Randomization

At this point, a brief digression on the order in which component parts of an experiment should be performed seems appropriate. Strictly speaking, different batches of subjects should be chosen and then the numbers of batches required for each dose should be taken entirely at random from these. Moreover, the various doses should be tested in random order, or in a randomization restricted in accordance with a previously planned experimental design. In practice, order of testing is often held to be unimportant in a well-controlled laboratory experiment provided that the whole is completed within a reasonably short time (one day, perhaps). The theoretical requirements of randomization are therefore frequently sacrificed to the practical convenience of testing all batches at one dose consecutively, and completing all tests on one material before those on a second are begun.

The perils of any departure from the strict canons of randomness established in many fields of applied statistics must not be underestimated. Many possibilities of bias exist, in respect of the parameters themselves and of assessments of their precision. In the circumstances of a particular experiment, an investigator may be able to dispose of some of these as inherently unlikely, but rarely can he eliminate all. The insidious danger of these biases is that they are seldom detectable or even suspected from the internal evidence of an experiment, at least on statistical grounds. The investigator who takes such a risk is in effect backing his own judgement that biases will be negligible: statistical techniques cannot distinguish consistent and nearly constant biases, however large, from the phenomena that are under estimation.

For example, if subjects are not grouped into batches at random, the action of forming batches may be producing unwanted differences, such as a preponderance of one sex, of the less agile, or of the more readily visible subjects (Murray, 1937; Bliss, 1939b). Any consequent difference in susceptibility of insects tested at different times in the day might be interpreted as a difference in potency of two poisons (McLeod, 1944). If doses of a single material are tested in ascending or in descending order, any steady trend in susceptibility of subjects or any increase in the dose actually received (through incomplete cleaning of the apparatus after the preceding dose) might manifest itself by an increase or decrease in the estimated slope of the regression line or even by a departure from linearity; when doses are given in random order, such a trend would properly increase the heterogeneity between batches but would not bias estimates of potency.

Procedures for random allocation of subjects to batches or for random ordering of a sequence of tests are fully described in many standard texts. Usually a table of random numbers is a convenient aid; Fisher and Yates (1964) give such a table and explain how to work with it easily.

9.3   Records of individuals

In some experiments, doses may vary from subject to subject and may be individually recorded. One possibility is that dose can be roughly controlled in advance but measured accurately only after administration. An early example was Campbell's poison-sandwich technique for testing stomach insecticides (Campbell and Filmer, 1929; Campbell, 1930), which involved feeding separate amounts of poison to each insect and measuring actual dose after ingestion. The results consist of a series of doses with, for each, a record of whether or not a single insect was killed. Such techniques tend to be more troublesome than those commonly employed for testing batches of insects at selected doses, and the series may be comparatively short. Tests on fifty insects in all cannot be expected to yield the precision obtainable when batches of fifty are tested at each of several doses. Nevertheless, the parameters can still be estimated by the probit method. Little modification of the instructions in preceding chapters is required, but the analysis presents some new features; Bliss (1938) gave a similar but fuller discussion.

The data cannot be plotted directly as probits, because each dose has a zero or 100 % response. The provisional regression line for initiating computations must therefore be obtained by some grouping of records. Results for several consecutive doses might be combined to give response rates based on ten or more individuals, the probits of which can then be plotted against log dose. In some experiments the doses themselves may indicate a convenient grouping, the experimenter having deliberately aimed at certain values. In others, doses may be spread fairly uniformly over the whole range so that grouping has to be entirely arbitrary; non-independent overlapping groups of about ten might be used in order to obtain more points, taking the first to the tenth doses as one group, the sixth to the fifteenth as a second, the eleventh to the twentieth as a third, and so on. A diagram showing the empirical probit of each group plotted against the mean log dose allows a provisional line to be drawn in the usual manner; the grouped data tend to underestimate the slope of the line, which should therefore be drawn so as apparently to err slightly on the side of steepness.

Improvement of the line by iteration may proceed exactly as in § 4.4, each dose contributing a line to a table like Table 4.1, in which every working probit is a maximum or a minimum. Rather many cycles of iteration may be needed. Early cycles, however, might be computed from small independent (non-overlapping) groups of four or five consecutive doses, despite the underestimation of slope. One or two final cycles could then be based on individual doses.

Unfortunately, trouble with the routine χ^2 test for homogeneity (§ 4.6) arises most acutely from individual response records. A single instance of response at a low dose or non-response at a high may inflate the value of χ^2 to an undue extent: an observation of 1 in a class with expectation 0.1 is not unduly remarkable, yet it will contribute 8.1 to χ^2. Such a χ^2 has a sampling distribution very different from that represented by Table VII; the expectation remains equal to the degrees of freedom, but the dispersion is greater, increased probabilities of very high values of χ^2 being balanced by a greater probability of abnormally low values. If the computed χ^2 is small, no further step is needed.

Certainly this is no evidence against homogeneity, unless it is so small as to suggest that a neglected feature of the experiment has caused unusual regularity or even that the data have been faked! If χ^2 is large, an extreme adaptation of the method of Example 8 might be tried, records being grouped sufficiently to give reasonably large expectations of responses and non-responses in every group. When the total number of subjects in the experiment is less than fifty, only a few groups can be formed and the test of homogeneity will not be very sensitive. No satisfactory alternative test is available. Indeed, even if the tolerances of these few subjects could be measured individually, a significance test on the departure from normality of the distribution of their logarithms would not be very sensitive; when the only results obtainable are quantal responses, a sensitive test is even less to be expected.

Bliss (1938) gave a good example of the computations, with data on the toxicity of sodium fluoride to grasshoppers. However, as must often happen when so few subjects are involved, the estimate of the regression coefficient was imprecise ($y = 0.44$), and the full formula for fiducial limits should have been used. He suggested adjustment of the χ^2 test, not by altering the calculation of the statistic but by reducing the numbers of degrees of freedom allotted to it, using an empirical rule for the expectations in terminal dose groups. This method not only lacks theoretical basis but seems liable to have an effect opposite to that intended: it will attribute significance to high values of χ^2 even more readily than will the unadjusted test.

A common practice in tests of this nature is to express each dose as amount per unit weight of the test subject, in order to allow for differing sizes and consequent probable differences in resistance. Thus doses of sodium fluoride in Bliss's example were recorded as mg/g of body weight. As an approximate adjustment this is not unreasonable, but a better analysis would be to obtain a probit plane for dependence on logarithm of the actual dose and logarithm of body weight. The assumption that resistance to the poison is directly proportional to body weight is thereby avoided, and the influence of body weight is estimated from the data. If desired, the effect of body weight may subsequently be averaged out (§ 9.5).

Ex. 26. *Dose factors in clinical radiotherapy.* Cohen (1951) reported the results of irradiation of about 100 cases of epidermoid cancer. For each case, the dose was expressed by two variates, here called z and t. These are respectively the measured radiation (in 10^3 röntgens) standardized by a 'reciprocal biological efficiency' factor to an equivalent in terms of γ-radiation, and the time (in days).

Theory suggested that an *equivalent clinical dose* can be measured by z/t^ν, the logarithm of which will show a normal tolerance distribution in respect of the classification of the tumour as cured or recurred in the following three years. The value of ν, the *recovery exponent*, was believed to be about 0.25. If x_1, x_2 are written for $\log z$, $\log t$ respectively, the probability of recurrence should conform to the probit plane

$$Y = \alpha + \beta_1 x_1 + \beta_2 x_2, \tag{9.1}$$

where

$$\nu = -\beta_2/\beta_1. \tag{9.2}$$

More will be said about the computations for this type of analysis in Example 27. For Cohen's 98 cases, the standard computer program led to

$$b_1 = 12.809,$$

$$b_2 = -3.808,$$

and to

$$\chi^2_{[95]} = 223.93.$$

At first sight, this enormous value of χ^2 casts serious doubts on the model. Inspection of a table of expectations shows that one subject showed a recurrence of tumour despite values of z, t that estimate the probability of recurrence as 0.0059, and that this subject therefore contributed over 169 to χ^2. Two tempting explanations are that the record is erroneous or that the particular patient was from a population differing from the other 97 in some important respect. A third possibility is that the unknown sampling distribution of the quantity labelled as χ^2 is responsible: perhaps a deviation of the most extreme member of a set of 98 to the extent represented by this large contribution to χ^2 is not as extraordinary as it first appears. Certainly the fact that the most extreme of 98 observations behaves in a way that for a randomly chosen observation has a probability 0.0059 is not a matter for great concern. Further inquiry about this patient might throw light on what happened, but for present purposes the question cannot easily be pursued.

Evidently the recovery exponent is estimated by

$$3.808/12.809 = 0.297.$$

The variances and covariance of b_1, b_2 can be used to compute fiducial limits, by equation (4.35). If no account is taken of the heterogeneity factor, 2.357, the 95 % fiducial limits ν are 0.383 and 0.240, so that clearly $\nu = 0.25$ is not seriously in conflict with the evidence. Omission of the questionable case reduces χ^2 (now with 94 degrees of freedom) to 80.16, confirming that all indications of heterogeneity arise from the one case. The regression coefficients are substantially changed; the recovery exponent is estimated as 0.284, much the same as before but now with fiducial limits 0.346 and 0.240.

Ex. 27. *Deep inspiration and vaso-constriction of a finger.* Under carefully controlled conditions, a deep breath may produce a transient reflex vaso-constriction in the skin of the fingers. Gilliatt (1947) reported tests of whether or not this phenomenon occurred in three subjects, in a number of tests for each of which the volume of air inspired and the average rate of inspiration were measured. Assessment of the degree of vaso-constriction was not practicable, and the record of response was therefore quantal.

Table 9.3 summarizes the thirty-nine tests; the first nine, the next eight, and the remaining twenty-two correspond with the three subjects. Had the tests been more numerous, differences between subjects would have needed investigation, but these few data conform well to a hypothesis that subjects do not differ in sensitivity. The tests are represented as points in Fig. 9.1.

Finney (1947b) described estimation of a probit plane for these data. The points in Fig. 9.1 suggest that the response will seldom occur if either the volume (V) or the rate (R) of inspiration is very small, and that each member of a family of curves

$$VR = \text{constant} \tag{9.3}$$

might approximately represent a fixed probability of response.

Table 9.3   Records of vaso-constriction after deep inspiration
(Gilliatt, 1947)

Volume of air (l)	Rate of inspira- tion (1/s)	Response	Volume of air (l)	Rate of inspira- tion (1/s)	Response
3.7	0.825	+	1.8	1.8	+
3.5	1.09	+	0.4	2.0	−
1.25	2.5	+	0.95	1.36	−
0.75	1.5	+	1.35	1.35	−
0.8	3.2	+	1.5	1.36	−
0.7	3.5	+	1.6	1.78	+
0.6	0.75	−	0.6	1.5	−
1.1	1.7	−	1.8	1.5	+
0.9	0.75	−	0.95	1.9	−
			1.9	0.95	+
0.9	0.45	−	1.6	0.4	−
0.8	0.57	−	2.7	0.75	+
0.55	2.75	−	2.35	0.3	−
0.6	3.0	−	1.1	1.83	−
1.4	2.33	+	1.1	2.2	+
0.75	3.75	+	1.2	2.0	+
2.3	1.64	+	0.8	3.33	+
3.2	1.6	+	0.95	1.9	−
			0.75	1.9	−
0.85	1.415	+	1.3	1.625	+
1.7	1.06	−			

Such an equation would follow from a probit regression of the
form
$$Y = \alpha + \beta(x_1 + x_2),\qquad(9.4)$$

where x_1, x_2 are the logarithms of V, R respectively. A more
general equation is
$$Y = \alpha + \beta_1 x_1 + \beta_2 x_2,\qquad(9.5)$$

leading to $V^{\beta_1}R^{\beta_2} = \text{constant},\qquad(9.6)$

as equations for fixed probability of response. Of course the
probit regressions could be replaced by regressions corresponding
to some non-normal underlying tolerance distribution.

In the original calculations, inspection of the probits of the
proportions responding in ranges of $(x_1 + x_2)$, say -0.3 to 0.1,

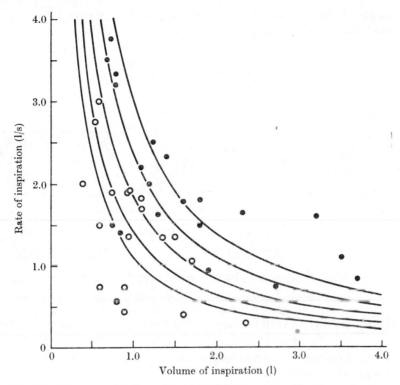

Fig. 9.1. Vaso-constriction of finger for different combinations of volume and rate of inspiration (Ex. 27).

○: no vaso-constriction
●: vaso-constriction

Contours of response surface are drawn for probabilities 0.1, 0.25, 0.5, 0.75, 0.9.

− 0.1 to 0.3, 0.1 to 0.5, and 0.3 to 0.7, suggested an approximation to equation (9.4)

$$Y = 4.2 + 3.0(x_1 + x_2). \qquad (9.7)$$

This was used to give working probits and initiate iteration for equation (9.5). Recalculation by the computer program has now led to

$$Y = 3.4674 + 6.6292x_1 + 5.8842x_2. \qquad (9.8)$$

Agreement with the data is excellent:

$$\chi^2_{[36]} = 30.72, \qquad (9.9)$$

which gives no sign of heterogeneity. The two regression co-efficients have a variance–covariance matrix

$$V = \begin{pmatrix} 4.58896 & 3.35692 \\ 3.35692 & 4.33482 \end{pmatrix}. \tag{9.10}$$

Evidence for any difference between β_1 and β_2 is slight: from (9.10),

$$V(b_1 - b_2) = v_{11} - 2v_{12} + v_{22}$$

$$= 2.2099, \tag{9.11}$$

and therefore $b_1 - b_2 = 0.745 \pm 1.486.$

If iteration is again begun from (9.7) but directed at a common regression coefficient and equation (9.4), the result is,

$$Y = 3.3839 + 6.3038(x_1 + x_2), \tag{9.12}$$

and $$\chi^2_{[37]} = 30.30. \tag{9.13}$$

Comparison of (9.9) and (9.13) illustrates the fact that maximizing likelihood is not the same as minimizing χ^2: a reduction in χ^2 when the number of parameters is *reduced* is unusual but not impossible. Certainly equation (9.12) is not to be judged any worse a fit than equation (9.8). The only good reason for basing the next calculations on equation (9.8) is to illustrate the greater complications.

Table 9.4   Comparison of observed and expected
frequencies of vaso-constriction

Estimated probability of response	Frequencies				
	Expected		Total	Observed	
	−	+		−	+
< 0.1	0.15	8.85	9	1	8
0.1–0.6	4.40	6.60	11	2	9
0.6–0.9	8.30	2.70	11	9	2
> 0.9	7.66	0.34	8	8	0
Total	20.51	18.49	39	20	19

Table 9.4 examines the expectations further, in a way that is scarcely needed here. The 39 tests have been grouped according

to which of four ranges of probability of response is given by equation (9.8). Had the agreement not been obviously so good, the quantities

$$(\text{observed} - \text{expected})^2/\text{expected}$$

would have been summed and tested as an approximate χ^2 with 1 degree of freedom (4 groups less 3 parameters estimated).

In Fig. 9.1, five contours have been drawn to indicate the behaviour of the probability of response. These were obtained by substituting probits of selected percentages for Y in equation (9.8) and then finding pairs of values of x_1, x_2 (and hence of V, R) that satisfy the relation. For example, substitution of 5.6745 for Y leads to the equation

$$V^{6.029}R^{5.884} = 161.1 \qquad (9.14)$$

as the estimated relation for a probability 0.75 that vaso constriction occurs.

Fiducial limits to any contours can be obtained. The variance of Y, the expected probit estimated for any combination of x_1, x_2, is

$$V(Y) = \frac{1}{Sw} + v_{11}(x_1 - \bar{x}_1)^2 + 2v_{12}(x_1 - \bar{x}_1)(x_2 - \bar{x}_2) + v_{22}(x_2 - \bar{x}_2)^2. \qquad (9.15)$$

By the argument that underlies equation (4.34), a quadratic equation corresponding to any desired limits can be formed. If Y_0 is the probit corresponding to a specified response rate (e.g. 5.6745 for 75%), and t is taken for the required probability, the fiducial region will consist of values of x_1, x_2 for which

$$(Y - Y_0)^2 \leqslant t^2 V(Y). \qquad (9.16)$$

Thus a pair of limit curves for any probability can be inserted by calculation from

$$(a + b_1 x_1 + b_2 x_2 - Y_0)^2 = t^2 V(Y); \qquad (9.17)$$

this equation can be regarded as a quadratic in x_1 for any chosen value of x_2, and the solutions give appropriate pairs (x_1, x_2) that can be transformed back to the V, R scale and plotted. Figs. 9.2, 9.3 give two illustrations of such limit curves.

These data can show how precision may be lost in respect of certain estimates if unnecessarily many parameters are included in a problem. Had equation (9.4) been judged adequate,

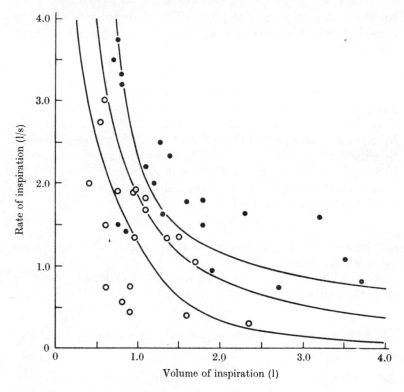

Fig. 9.2. Fiducial limit curves to 0.5 contour in Fig. 9.1, at probability 0.95.

○: no vaso-constriction

●: vaso-constriction

contours for the response rate and limit curves would be obtained in much the same way. Equation (9.12) replaces equation (9.8) as the estimate, and equation (9.15) is replaced by

$$V(Y) = \frac{1}{Sw} + \frac{(x - \bar{x})^2}{S_{xx}} \qquad (9.18)$$

where x is written for $(x_1 + x_2)$ as the variate in the probit regression calculations. In place of equation (9.17), the limit curves are now given by

$$(a + bx_1 + bx_2 - Y_0)^2 = t^2 V(Y). \qquad (9.19)$$

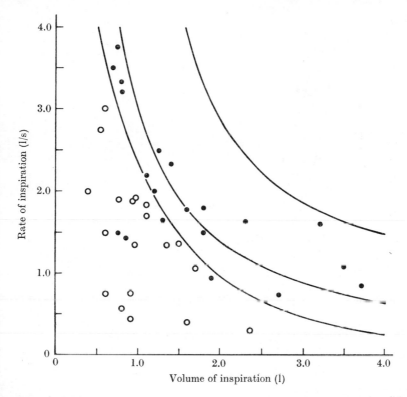

Fig. 9.3. Fiducial limit curves to 0.9 contour in Fig. 9.1, at probability 0.95.

○: no vaso-constriction
●: vaso-constriction

Fig. 9.4 shows how little the estimates are altered: over the interesting ranges of x_1, x_2, equations (9.8) and (9.12) are not very different. The limits, however, are changed. If one can safely assert that vaso-constriction occurs in accordance with equation (9.4), the contour of the response surface representing 50 % response is more precisely determined (the limits being narrower) than if the greater freedom of equation (9.5) is permitted, but changes for the limits of other contours are more complicated.

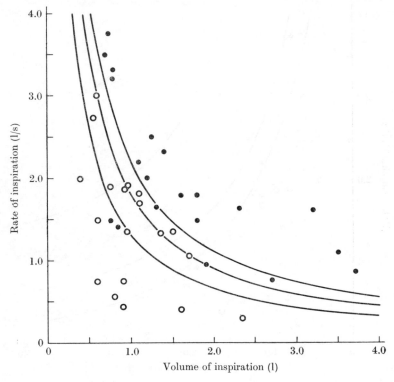

Fig. 9.4. Contour of response surface for probability 0.5, with 0.95 fiducial limits, with constraint that $\beta_1 = \beta_2$.

○: no vaso-constriction
●: vaso-constriction

9.4 Many-parameter problems

Consider again the analysis of Example 22. This legitimate statistical treatment of the data is not the best possible. The responses for the CCT series were all high, and consequently Snw is small, yet in the calculation of relative potencies it entered in exactly the same way as the CHT series for which Snw is much greater. Account was taken of this in forming $V(M)$ for the various potency estimates, but the unweighted mean of quantities whose variances are different will have a larger variance than a suitably weighted mean. The same criticism could be made of the summarizing of relative potencies in Example 21. On the other hand, the first analysis in Example 17 employed unequal weights in a general least squares estimation that was satisfactory because the separate estimates of log ED 50 were sufficiently precise. The second analysis in Example 17 provides the key to the more logically satisfying procedure that can nowadays more often be used because of increased computing power. This involves simultaneous estimation of all parameters, usually though not necessarily by maximum likelihood, despite the large number of parameters. With such a method, Example 22 requires simultaneous estimation of at least 9 parameters, and Example 21 at least 22 parameters: even to-day, such problems are large enough to make simpler approximations worth considering.

Difficulties arise in the most extreme form when an experiment provides only individual records, for example with litter-mate control in animal experiments. If the test subjects are, say, rats, experience of other quantitative properties suggests that members of the same litter are likely to vary less in their individual tolerances than do members of different litters. An experimenter testing six dose levels (perhaps three from each of two preparations whose relative potency was to be estimated) might hope to increase precision by assigning to each dose one animal from each of several litters of six. On the assumption that every intra-litter variance was the same, the method of § 9.3 might seem adaptable to the estimation of a series of probit lines (two for each litter if two preparations were involved) constrained to be parallel; the slope, the reciprocal of the standard deviation of log tolerances within litters, should be steeper than if litter-mate control were not used. An average value of M could then be formed, as in equations (8.7) and (8.9), from the common value of b and appropriate functions of the $\bar{x}$ and $\bar{y}$. The calculations would be more laborious than if the litter classifications were ignored and data for all subjects at one dose were pooled, but, for subjects showing real differences between litters, the precision of the relative potency would benefit. Moreover, the design of an experiment should always influence its analysis: to incorporate litter-mate restrictions into the design and to ignore them in the analysis would be improper and could be misleading.

Execution of this analysis will almost certainly encounter

trouble. Complications enter if the three rats of a litter on one preparation either all respond or all fail to respond (cf. § 5.9). Even more acute problems arise if all combinations of factors in the experiment are not fully replicated. Suppose that work with one rat takes 30 minutes, so that the experiment cannot be completed in one day. One possibility would be to complete two litters on each of six days, but the alternative of a Latin square design might be preferable. Table 9.5 illustrates an arrangement of twelve litters, each represented once on each day and with each dose represented twice on each day. For higher precision, the design might be extended by further sets of six litters with other Latin square arrangements. If the response were not death but some minor reaction that could be repeated indefinitely without permanent effect on the subject, each row might represent a single animal tested on six successive days.

Table 9.5   Latin square design for tests of two preparations, S and T, each at three doses

Litter	Days					
	1	2	3	4	5	6
I	T_2	S_3	T_1	T_3	S_2	S_1
II	T_1	S_2	S_1	T_2	T_3	S_3
III	S_2	T_3	T_2	S_1	S_3	T_1
IV	S_3	T_1	T_3	S_2	S_1	T_2
V	T_3	S_1	S_3	T_1	T_2	S_2
VI	S_1	T_2	S_2	S_3	T_1	T_3
VII	S_2	T_3	S_1	T_1	S_3	T_2
VIII	T_3	T_1	T_2	S_2	S_1	S_3
IX	S_3	S_1	S_2	T_2	T_1	T_3
X	T_2	S_3	T_1	S_1	T_3	S_2
XI	T_1	S_2	S_3	T_3	T_2	S_1
XII	S_1	T_2	T_3	S_3	S_2	T_1

The twelve-litter scheme will serve for illustration. Had a quantitative response been measured for each test, differences between litters and between days would be eliminated by a standard analysis of variance, and all comparisons between doses would be made with the same precision as if based entirely on differences within one litter and within one day. This ought still to be possible for quantal responses: the method is less obvious, but derives from the basic formulation of tolerance distributions. Each test gives information on the tolerance for a particular combination of litter and day, information inexact by comparison with a hypothetical direct measurement but still information of value. If direct measurements of log tolerance or some other quantitative character were available, the assumptions underlying the analysis of variance would be that the expected value for each combination of litter and day could be expressed as the sum of two parameters, one constant for the litter and the other constant for the day, and that the distribution of measurements about their expectations was normal with a variance constant for the whole experiment. Such assumptions are made regularly in the analysis of quantitative data, and an analysis for quantal data may equally well be based upon them. Unfortunately, the orthogonality of the Latin square disappears when unequally weighted probits replace quantitative measurements.

Somewhat similarly to Example 17, suppose the expected probit for a subject in litter i receiving a dose with logarithm x on day j to be

$$Y = \alpha_i + \delta_j + \beta x, \qquad (9.20)$$

where β corresponds to the constant variance of log tolerances and the first parameter is written α_{iS} or α_{iT} according to whether the standard or the test preparation is used. Thus the means of the log tolerance distributions, μ_{iS} and μ_{iT}, will be $-\alpha_{iS}/\beta$ and $-\alpha_{iT}/\beta$. Applied to Table 9.5, equation (9.20) implies a total of 30 parameters, of which one is redundant (two sets of twelve μ_i, but the six δ_j must be defined as deviations totalling to zero). One can scarcely hope to estimate so large a set of parameters satisfactorily from 72 individual quantal responses. The problem is simplified a little if the relative potency is constant for all litters:

$$\mu_{iT} - \mu_{iS} = \text{constant}, \qquad (9.21)$$

and therefore $\alpha_{iT} - \alpha_{iS} = \text{constant}$,

so that only 18 essential parameters remain.

A general maximum likelihood computer program can be employed, but experience of handling problems with so many parameters and rather few observations is small. Almost certainly,

7   FPA

iteration would progress slowly and might fail to converge, either totally or unless started from a sufficiently good first approximation. Alternatively, classical probit techniques can be adapted, analysis beginning with choice of a set of expected probits for the 72 combinations of litter and day. These can be read from provisional average regression lines, each value being modified by an addition representing the deviation of its litter from the general average and another representing the deviation of its day from the general average. Thus, in a litter which shows general evidence of low tolerances, all expected probits will be increased by the same amount. In the first cycle of iteration, these adjustments may be little better than guesses. Weights and working probits are found from the usual formulae, and weighted totals of y for each litter, each day, and each dose formed. By equation of these totals to similarly weighted totals of expected values, representing a true linear regression of probit on log dose modified by litter and day parameters, estimates of the parameters may be formed. This step is essentially the same as for the 'fitting of constants' in any non-orthogonal experimental design, except that the weights are no longer integral numbers of plots but values of nw. From the estimates of the parameters, a new set of expected probits is constructed, and iteration proceeds as in simpler problems.

A Latin square of k^2 cells is a fractional replicate of a k^3 factorial arrangement, since only $1/k$ of all combinations of three factors (in this instance litter, day, and dose) are used. Analysis is possible only if interactions between factors can be ignored. This condition, sometimes forgotten in other applications of Latin square designs, means here that tolerance differences between litters are assumed to remain the same from day to day. The fact that dose S_2 is tested only on 12 of the 72 possible combinations of litter and day then does not prevent comparison of the responses it produces with corresponding results from other doses. Full analysis is essential if risk of false inference from non-orthogonal comparisons is to be avoided. If all tests had been made simultaneously, so that the day classification did not occur, the same principles could be applied and would ensure that the final estimate of relative potency was the best weighted average from the several litters. The calculations would be simpler, but the method of Example 22, now applicable because of the complete replication of all combinations of litters and doses, might often be preferred, even at the price of loss of precision.

If the laborious probit transformation were replaced by the angle, equation (3.26), the work would be eased because the weighting coefficient is now independent of Y. Claringbold, Biggers and Emmens (1953) described and illustrated the simplifications that occur, though their examples are not of individual records. Strictly, iteration is still required though now with equal weighting coefficients, but even a single cycle should come close to the solution. Moreover, the constancy of w preserves orthogonality (if n is also constant), and in consequence the equations for estimating the parameters degenerate into evaluations of arithmetic means. One cannot seriously maintain that the angle transformation corresponds with a realistic tolerance distribution, but in many respects (such as estimating an ED 50) it is likely to be adequate. It will tend to be unsatisfactory if there are many zero or 100 % responses, and therefore is least suitable for individual records. Another useful possibility is to begin analysis with the angle transformation and then, when one or two cycles of iteration have given some stability, to convert expected angles to corresponding expected probits for continuation.

Claringbold (1955) has argued against 'the assumption...that a distribution of thresholds must be postulated before efficient analysis may be made', and has contrasted this with the use of a transformation simply as a different scale of measurement. His contrast is not easily understood; he refers specifically to Fisher (1954), whose paper outlines the general theory of transforms of binomial data but who emphasizes that the theory can be readily applied to whatever transformation is appropriate, not that choice should be dictated solely by convenience. That the tolerance distribution is rarely of known mathematical form is true, but it cannot be ignored: any transformation contains an assumption, explicit or implicit, about the form of that distribution, or the corresponding response regression, that is to be used as an approximation. None the less, Claringbold performed a valuable service in systematizing the angle-type calculations.

In a further paper (1956), Claringbold illustrated yet another analysis for individual responses, in a bio-assay experiment very like that of Table 9.5. He proposed the extreme simplification of analysing the untransformed responses, simply 0 and 1, as though equally weighted on a continuous scale. An exceedingly simple analysis of variance is obtained. Although indefensible theoretically, this may be adequate for experiments that are intrinsically so imprecise. Claringbold's argument is persuasive and his own experience encouraging, but further study is needed before the method is unreservedly recommended.

The general problem of analysing quantal response data from complex experiments may be regarded as solved, for the principles of analysis described here are adaptable to any design. Whether or not the gain in precision justifies the labour of computation must be decided by experience, as also must the extent to which approximations can replace the full scheme of analysis.

9.5   The mean response

Occasionally an estimate is required for the mean response when a population of subjects is associated at random with doses from a specified frequency distribution. Suppose that the true probit regression equation is

$$Y = \alpha + \beta x. \tag{9.22}$$

If subjects are selected at random and exposed to doses from a distribution for which ξ is the mean log dose, the mean response will not in general be obtained by substitution of ξ for x, but will depend upon the variance of the doses. In fact, if the doses are normally distributed with mean ξ and variance σ^2, the mean proportion responding can be proved to be that for which the probit is

$$Y^* = 5 + (\alpha - 5 + \beta\xi)/(1 + \beta^2\gamma^2)^{\frac{1}{2}}. \tag{9.23}$$

If ξ happens to be the ED50, the mean response will be 50 % ($Y^* = 5$) whatever the value of γ, but if ξ differs from the ED50 the mean response will be nearer to 50 % than is a probit $(\alpha + \beta\xi)$. For example, suppose that

$$Y = 3 + 2x, \tag{9.24}$$

and that $\xi = 2$. Then

$$Y^* = 5 + 2/(1 + 4\gamma^2)^{\frac{1}{2}}. \tag{9.25}$$

If γ, the standard deviation of log doses, is very small, the mean response is 97.7 % ($Y^* = 7$); if $\gamma = 0.5$, this mean is only 92.1 % ($Y^* = 6.41$), and as γ increases it approaches the limit of 50 %. In practice, of course, estimates of α, β must be used in equation (9.23): nothing is known of the effect of errors of estimation.

A slightly more complex problem is that of a response probit expressed as a regression on two dose factors, as in § 8.6:

$$Y = \alpha + \beta_1 x_1 + \beta_2 x_2. \tag{9.26}$$

If the doses to which a population is exposed are such that x_1 can be controlled but x_2 is normally distributed about ξ_2 with variance γ_2^2, the probit of the mean response is still linearly related to x_1 but the regression coefficient is reduced to

$$\beta_1^* = \beta_1/(1 + \beta_2^2 \gamma_2^2)^{\frac{1}{2}}. \tag{9.27}$$

The one-factor regression equation is then

$$Y^* = 5 + (\alpha - 5 + \beta_2 \xi_2)/(1 + \beta_2^2 \gamma_2^2)^{\frac{1}{2}} + \beta_1^* x_1. \tag{9.28}$$

Equation (9.28) is useful when x_2, though formally a dose factor, is in fact a measurable characteristic of the individuals tested. For example, the potency of a poison may depend not only on the concentration but also on the weight of the subject. By subdivision of the data into weight classes, or by using individual records (§ 9.3), the weight (or perhaps the logarithm of the weight) can be introduced as x_2 in the probit regression equation. In order to predict the effect of any chosen concentration on a random selection of subjects, the contribution of weight to the equation must be averaged by equation (9.28). For this purpose, ξ_2 and γ_2 can be estimated from the animals used for the toxicity test if these were randomly selected from the whole population; alternatively, the parameters of the weight distribution may be estimated from measurements of a subsidiary random sample on which no toxicity tests were made.

If equation (9.26) were known, equation (9.28) could be derived from it for any specified ξ_2, γ_2. A more realistic situation is that in which the probit regression of response on x_1 is being studied in the presence of uncontrolled variation in x_2, where x_2 itself is perhaps not of intrinsic interest. In the example mentioned above, subjects might be chosen at random for weight and used in a study of the dependence of response on concentration. The regression estimated is

$$Y^* = \alpha^* + \beta_1^* x_1, \qquad (9.29)$$

of the form of equation (9.28); note that β_1^* is necessarily less than β_1. Equations (9.27), (9.28) show that

$$\frac{5 - \alpha - \beta_2 \xi_2}{\beta_1} = \frac{5 - \alpha^*}{\beta_1^*}. \qquad (9.30)$$

Hence the ED 50 in respect of concentration for $x_2 = \xi_2$ is directly estimable from equation (9.29); the three parameters of equation (9.26), however, cannot be separately estimated even if ξ_2, γ_2 are exactly known.

9.6   The meaning of standard errors

Standard errors, fiducial limits, or other measures of precision, must not be uncritically accepted as though they indicated the variation that would be encountered in all repetitions of an experiment. Because they emphasize analytical technique rather than interpretation, text-books of statistics tend to leave the reader with the impression that an estimate of a parameter has only one standard error; any similar misunderstanding arising from §§ 3.4, 3.5 and later sections now needs correction. In reality, even though only one standard error may be assessable from a particular set of data, an estimate has different standard errors appropriate to the different kinds of comparison in which it may be used.

The temptation to regard a percentage effective dose estimated from one sample of subjects, whether animal or human, as a 'normal value' typical of the species, must not be bolstered with argument from a standard error based on the internal variation of the sample. If any impression has been given that, for example, the value obtained in Example 7 for the LD 50 of rotenone to *Macrosiphoniella sanborni*, is an estimate of a 'normal value' for this species, that impression must be corrected; it is necessarily only an indication of what would be found in repeated tests of the same stock of *M. sanborni* under the same conditions of testing. Moreover, the meaning of 'same' is open to discussion!

This limited interpretation must always be attached to statements about the ED 50 of a single material. Tests performed in a London laboratory may be used to estimate an ED 50 and its fiducial limits, and the results may be valid for future work in that laboratory; they should not be expected to apply to similar tests in New York, where the subjects may be obtained from a different source, may be genetically different, and will have different nutritional and environmental experience. Tolerances are no more to be supposed independent of these conditions than are weights or other measurements of size. Small differences in experimental technique between different laboratories may also affect results. A standard error calculated from internal evidence of one experiment relates only to repetition under the same conditions, using subjects and materials that can be regarded as random samples from the same populations as were those in the original experiment. The ED 50 may have one system of variances between determinations by one worker, another system of variances between workers using the same apparatus and stock of animals in one laboratory, and yet another between determinations in different laboratories. Comparison of estimates obtained from several experiments by one worker, or in one laboratory, will assess only the variations to be expected from day to day under those conditions and will still not be related to variations between laboratories or between different stocks of subjects. However extensive the experimentation on one population, any assertion that conclusions will apply to a different population must rest upon extra-statistical considerations.

At first sight, this argument might appear to deny the usefulness not merely of statistical analysis but of experimentation. The extent to which past experience can form a logical basis for future action is too great an issue for discussion here. Part of the difficulty seems to be removed by adoption of *comparative experiments*. Even though the absolute performance of a particular treatment varies erratically with changes in experimental conditions, the relative performance of two treatments may be much less disturbed by these changes (Anscombe, 1948). The relative toxicity of two chemically related insecticides, for example, at least for a particular species of insect, may be a characteristic of the materials, rather than of the circumstances of any one experiment. There is some evidence for this, but the truth – if truth it be – is empirical. Full study of the problem would involve lengthy experimentation into the effect of environment on the responses of insects to insecticides. In so far as these experiments demonstrated consistent differences, they could encourage inference relating to the future, while their inexplicable variations would warn the experimenter against any too confident application of absolute standards.

In one class of experiment to which the methods of this book are relevant, a theoretical reason exists for believing a numerical result to be applicable more widely than in the circumstances of the particular experiment that produced it. If two preparations contain the same active constituent, but this is diluted to different extents by constituents that do not affect the response, estimation of relative potency constitutes an *analytical dilution assay*. The subjects are used purely as an aid to the measurement of the relative amounts of the active constituent in the two preparations; though choice of subjects and experimental conditions will influence the precision, it should not limit the meaning of the estimate. The result is an estimate of a property which, in theory, is independent of the species of subject or other circumstances in its determination. Empirical verification of this statement is still desirable, but should be undertaken as a check on the validity of fundamental assumptions about the nature of the preparations rather than as a study of other sources of variation (Jerne and Wood, 1949; Finney, 1964).

9.7   Control charts

When a particular type of experiment is employed as a routine, quantitative features that might reasonably be expected to show some stability over time should be kept under continuous scrutiny. For example, a standard vaccine may be used regularly in the bioassay of new batches, and experimental conditions will naturally be held as constant as is practicable. Values of m and b for the standard might be abstracted from each assay and plotted against a horizontal time scale: standard errors or fiducial limits for each point can also be shown, so that the chart indicates whether (and to what extent) variation with time exceeds that internal to the experiments. Even χ^2 values for linearity and parallelism tests might be similarly scrutinized. Once such a 'control chart' has been established, any marked change in its form can draw attention to unsuspected influences on the quality of the experiments. Occasional unusually high or low points obviously suggest accidents or unstated changes of conditions. A steady trend, without necessarily achieving any extreme value, suggests general deterioration or improvement in the conduct of the assay. Another possibility is a pattern associated with time of day or day of week, suggestive of fatigue effects.

Bliss (1952) and Bliss and Pabst (1955) have described control chart techniques in connexion with biological assay. Cohen, van Ramshorst and Tasman (1959) illustrated this well from a series of 28 assays of a standard tetanus toxoid, in which the estimates varied erratically between 0.065 ml and 0.125 ml. They also presented an excellent control chart for 103 values of b in 39 consecutive assays involving one or more of four vaccines. Greenberg (1953) summarized results from many assays of diphtheria toxoid. Of 198 samples for which parallelism with a control toxoid was tested, 28 showed significant deviation from parallelism, 14 % instead of the 5 % to be expected if all assays were valid; some of these samples proved satisfactory on re-checking, so that possibly environmental conditions were at fault initially. Greenberg did not show his data in a form that permitted time trends to be examined.

Routine practitioners of biological assay or other users of quantal response techniques may have much to learn from the methods of quality control of statistical characteristics that are common in the manufacturing industries.

10

Various special techniques

10.1 Wadley's problem

In some quantal-response situations, the number of subjects exposed to a dose is unknown although the numbers at different doses are randomly selected from a definable distribution. For example, unit volumes of a bacterial suspension might be exposed to different doses of a bactericide, or to different temperatures, and the numbers surviving subsequently counted. If the original number of bacteria per unit volume follows a Poisson distribution with mean ν, and a treatment sufficient to kill a proportion P is applied, the number of survivors per unit volume will follow a Poisson distribution with mean $\nu(1-P)$. Analysis of counts of survivors, s, at various doses will involve estimation of ν as well as of the parameters of the tolerance distribution. A similar problem can arise in experiments on the control of immature stages of an insect pest of fruit. Samples of fruit may be treated at various doses, and counts made of the numbers of insects that survive and emerge. The number of larvae exposed to treatment in any sample, however, cannot be discovered (except perhaps by laborious counts on the dissected fruit). Direct information on the numbers exposed to treatment may be provided by the number of flies developing from a parallel sample of untreated fruit. If the distribution of initial numbers were Poisson, the problem would be formally the same as for the bacteria, except that the control sample estimates ν independently of any additional information accruing from comparisons between treated samples at different doses.

Wadley (1949) stated this problem and obtained the maximum likelihood solution. Finney (1949d) showed the close similarity to the technique of adjustment for natural mortality (Chapter 7). From a provisional estimate of ν, an empirical proportion killed at each dose,

$$p = 1 - \frac{s}{\nu}, \qquad (10.1)$$

may be assessed. On the assumption that the tolerance distribution is log normal, the usual probit calculations may be initiated. Working probits are still found as in § 4.3, but, because a Poisson distribution occurs instead of a binomial, the weighting coefficient becomes

$$w = Z^2/Q; \qquad (10.2)$$

this function is tabulated in Table V. Since ν is the same for all doses, it may be introduced into totals at the end instead of by calculation of νw for each dose. If the number of samples per dose varies, each w must be multiplied by the corresponding number. An auxiliary variate, that of equation (7.13) with sign reversed:

$$x' = -Q/Z, \qquad (10.3)$$

is required, and may be read from Table II. Revised estimates of the three parameters are then obtained from

$$bS_{xx} + \frac{\delta\nu}{\nu} S_{xx'} = S_{xy}, \qquad (10.4)$$

$$bS_{xx'} + \frac{\delta\nu}{\nu} S_{x'x'} = S_{x'y}, \qquad (10.5)$$

and
$$a = \bar{y} - b\bar{x} - \frac{\delta\nu}{\nu} \bar{x}', \qquad (10.6)$$

equations exactly analogous to (7.15), (7.16), (7.17), though the change in sign in the definition of x' should be noted. If a sample tested at zero dose shows s_0 survivors, so that s_0 is a direct estimate of ν, $S_{x'x'}$ must be increased by the provisional value of ν and $S_{x'y}$ by $(s_0 - \nu)$. The cycle is iterated as often as necessary to give a close approach to the maximum likelihood estimates. A test of heterogeneity may be based on

$$\chi^2 = S_{yy} - bS_{xy} - \frac{\delta\nu}{\nu} S_{x'y}. \qquad (10.7)$$

The variances and covariance of b, $\delta\nu/\nu$ are the elements of the inverse matrix of coefficients from equations (10.4), (10.5).

Other results, such as the ED 50 and its fiducial limits, are obtainable in a standard manner. As usual,

$$m = (5-a)/b \qquad (10.8)$$

estimates the log ED 50. The variance of b is already available in the inverse matrix. Also

$$V(5-a) = \frac{1}{Svw} + \bar{x}^2 V(b) + 2\bar{x}\bar{x}'Cov\left(b, \frac{\delta\nu}{\nu}\right) + \bar{x}'^2 V\left(\frac{\delta\nu}{\nu}\right) \quad (10.9)$$

and $\qquad Cov(5-a,b) = \bar{x}V(b) + \bar{x}'Cov\left(b, \frac{\delta\nu}{\nu}\right). \qquad (10.10)$

Fieller's theorem can be used to find fiducial limits.

To adapt the general computer method is very simple. The probability that s survivors are found at a dose for which P is the probability of response is

$$Pr(s) = e^{-\nu(1-P)}[\nu(1-P)]^s/s!, \qquad (10.11)$$

the standard Poisson formula. This replaces equation (2.7), and equation (5.4) must therefore be replaced by

$$L = -\nu S(Q) + \log \nu S(s) + S(s \log Q). \qquad (10.12)$$

A general program such as that discussed in Chapter 5 can readily permit insertion of the expression (10.12) when required, with maximization with respect to α, β, ν following automatically. Inclusion of a sample tested at zero dose brings no difficulty, but makes a contribution to L for which $Q = 1$.

Anscombe (1949) pointed out that, although possibly appropriate to a well-stirred fluid in which bacteria are suspended, a Poisson distribution is usually less satisfactory as a representation of an insect infestation in fruit. For the latter, a negative binomial distribution might be nearer to the truth; it also has the property that counts for treated samples at any one dose will follow the same type of distribution with modified parameters. Unfortunately, however, ν is now replaced by two parameters, both of which enter into the new weighting coefficient. Anscombe suggested a modification in experimental procedure that might help to overcome the difficulty. Once again, a general maximum likelihood or analogous estimation program would be used, but experience is needed of whether a reasonable amount of data can estimate the important features of the probability structure.

Ex. 28. *The exposure of bacteria to heat.* Wadley (1949) used two numerical examples to illustrate his method, unfortunately both

artificial though devised to simulate reality. Table 10.1 purports
to relate to suspensions of bacteria held at $107°C$ for times
ranging from 20 to 50 minutes. Wadley used minutes as the dose
metameter, but here x is taken as the logarithm of time.

The third column of the table contains values of p calculated
from a provisional estimate that $\nu = 300$. Empirical probits and
a provisional regression line are found in the usual manner; with
a value of ν as large as this, to take Y to 2 decimals seems desirable,
even at the cost of interpolation in the various standard tables.
For example, the value of x' for the second dose, corresponding
to $Y = 5.53$, was obtained by linear interpolation between
$Y = 5.5$ and $Y = 5.6$:

$$w = 0.7 \times 0.8764 + 0.3 \times 0.8230 = 0.8604,$$

and w and y were similarly calculated.

The calculations of νw, $\nu w x$, $\nu w x'$, $\nu w y$ present no special
features, and from them are formed S_{xx}, etc., with the usual
care for arithmetical checking. For these data, $s_0 = 280$, and
therefore $S_{x'x'}$, $S_{x'y}$ had to be increased by 300.0, -20.0 respectively. Hence equations (10.4), (10.5) become

$$3.0983b + 17.3295\frac{\delta\nu}{\nu} = 22.102,$$

$$17.3295b + 425.7585\frac{\delta\nu}{\nu} = 103.960.$$

The inverse matrix of the coefficients on the left is

$$V = \begin{pmatrix} 0.417\,895 & -0.017\,009\,5 \\ -0.017\,009\,5 & 0.003\,041\,08 \end{pmatrix};$$

multiplication of each row by the elements on the right of the
equations gives

$$b = 22.102 \times 0.417\,895 - 103.960 \times 0.017\,009\,5$$

$$= 7.4680,$$

$$\frac{\delta\nu}{\nu} = -0.059\,793.$$

Equation (10.6) can be expressed

$$a = (1483.332 - 7.4680 \times 387.830 - 0.059\,793 \times 240.900)/255.8$$

$$= -5.5801.$$

Table 10.1 Calculations for Wadley-type analysis on numbers of bacteria in samples from suspensions held at 107° C for various times

Log time (x)	No. of bacteria (s)	(ν = 300) p	Empirical probit	$-5.7+7.6x$ $= Y$	vw	x'	y	vwx	vwx'	vwy
1.301	233	22.3	4.24	4.19	31.4	-2.755	4.24	40.851	-86.507	133.136
1.477	85	71.7	5.57	5.53	120.8	-0.860	5.57	178.422	-103.888	672.856
1.602	29	90.3	6.30	6.48	76.9	-0.521	6.27	123.194	-40.065	482.163
1.699	3	99.0	7.33	7.21	26.7	-0.391	7.31	45.363	-10.440	195.177
Controls	280	—	—	—	—				—	—
					255.8			387.830	-240.900	1483.332

S_{xx}	$S_{xx'}$	$S_{x'x'}$	S_{xy}	$S_{x'y}$	S_{yy}
591.1050	-347.9099	352.6264	2271.049	-1272.970	8762.21
588.0067	-365.2394	226.8679	2248.947	-1396.930	8601.54
3.0983	17.3295	125.7585	22.102	123.960	160.67
		300.0000		-20.000	
		425.7585		103.960	

Also the revised estimate of ν is

$$300 + \delta\nu = 300 - 17.94$$
$$= 282.06.$$

Now a, b do not differ greatly from their provisional values (-5.7 and 7.6), and the revised ν is not very different from 300. A second cycle of calculation is certainly desirable, but even this first cycle gives a reasonable assessment of the parameters. From equation (10.7)

$$\chi^2_{[2]} = 160.67 - 7.4680 \times 22.102 + 0.059\,793 \times 103.960$$
$$= 1.83,$$

which indicates very satisfactory agreement with theory.

The elements of V give the variances and covariance of b, $\delta\nu/\nu$, and of course
$$V(\bar{y}) = 1/S_{xx}.$$

In particular, the standard error of b may be taken as the square root of $0.417\,895$; the standard error of the estimate of ν must be obtained by multiplying the square root of $0.003\,041\,08$ by 300, the provisional value of ν.

Details of further iterations will not be presented, because much greater accuracy is readily achieved by the computer program. If the likelihood function is expresssed directly in terms of ν, as in equation (10.12), the parameter ν will be estimated immediately without the intermediate stage of finding $\delta\nu/\nu$. Such analysis gave:

$$b = 7.5034 \pm 0.6673,$$
$$\nu = 282.02 \pm 15.93,$$

and $$\chi^2_{[2]} = 2.77.$$
The fitted equation is

$$Y = -5.633 + 7.503x;$$

the log ED 50 is estimated as 1.417, corresponding to estimation that 26.1 minutes exposure would destroy 50 % of the bacteria. Equations (10.9), (10.10) can be used to put limits on this estimate, by calculations of standard type.

Of course, if no untreated sample were available to give a value s_0, computation would proceed in the same manner but without the adjustments to $S_{x'x'}$ and $S_{x'y}$. For the classical probit problem, observations at zero dose often contribute relatively little information even when a natural response rate must be estimated (Chapter 7). Here, however, precision is likely to be increased greatly by observation of s_0, because this estimates a denominator for p and enters vitally into all the calculations. If the data of Table 10.1 are re-analysed without use of the s_0, precision is much reduced, the variances of b, ν being increased by factors of approximately 3 and 16.

10.2 Estimation of an extreme percentage point

The guidance given in § 8.1 is not very satisfactory if the dose corresponding to a very large or very small response rate is to be estimated. The user of an insecticide is likely to be more interested in the LD 99 than in the LD 50, and in a study of undesirable side effects of a drug the ED 5 may be important. The precision with which the ED 99 can be estimated using N subjects will be much less than the precision for the ED 50; if tests are made at doses near the ED 99, the weights will be low, yet if tests are restricted to the central region the second term in equation (3.12) will make a large contribution to the variance. The second procedure scarcely deserves serious consideration because it involves extrapolation beyond the range of the observations, an action to be avoided unless the regression relation used is known to be absolutely, or very nearly, correct in algebraic form. Though the normal distribution is usually adequate to explain the dose-response relation for a considerable range on either side of the 50 % response, it may not be good enough at extremes. If the true regression equation of probit on log dose were to have a slight curvature, extrapolatory estimation of the ED 99 might be grossly misleading.

Bartlett (1946) drew attention to the importance of estimating an extreme percentage point from observations made in its neighbourhood, and advocated an inverse sampling procedure. His method of obtaining maximum likelihood estimates, presented here in a slightly more general form, has analogies with § 10.1. It may be adopted for any investigation in which the subjects are tested one at a time and each can be classed as responding or not responding before the next is tested. The method is described for the estimation of a high percentage point, but will apply to a low percentage point if the rôles of 'response' and 'non-response' are interchanged.

Suppose that subjects are tested one by one at a dose x until s have failed to respond, at which stage a total of n has been tested. In this inverse procedure, n is a variable for fixed s, instead of the more usual direct sampling situation of s being a variable for fixed n. An experiment will involve tests of several doses, with record of n for each dose; s need not be the same for every dose. Bartlett suggested the inverse sampling method, using a series of doses towards the tail of the tolerance distribution, as a way of securing estimates of response rates on numbers of subjects that increase towards the extreme.

The probability that n subjects have been tested at a dose by the time that the quota of s non-responses is achieved can easily be proved to be

$$Pr(n|s) = \begin{cases} \binom{n-1}{s-1} P^{n-s} Q^s & \text{for} \quad n \geqslant s, \\ 0 & \text{for} \quad n < s. \end{cases} \quad (10.13)$$

With the understanding that s is fixed by the experimenter and n is now the variable observed, replacement of equation (2.7) by equation (10.13) leaves equation (5.4) essentially the same in form:

$$L = S(n-s)\log P + Ss \log Q. \quad (10.14)$$

Maximization of this by a general computer technique is indistinguishable from the standard probit method; as is well known, such modification in the stopping rule for sample size does not alter the behaviour of the likelihood function.

However, if an iterative routine is developed as in Chapter 4, the formulae look a little different because at each dose s is predetermined instead of n. An empirical response rate

$$p = 1 - q = 1 - \frac{s}{n} \quad (10.15)$$

leads to empirical probits, a provisional regression line, weighting coefficients

$$w = Z^2/PQ^2, \quad (10.16)$$

and working probits

$$y = Y - \frac{Q}{Z} + \frac{Q^2}{Z}\left(\frac{1}{q}\right). \quad (10.17)$$

The weighted regression of y on x, with sw (not nw) as the weight per observation, is calculated. Table VI shows the minimum working probit

$$y_0 = Y - \frac{Q}{Z}, \qquad (10.18)$$

the range, Q^2/Z, and w, from which the working probits and weights for an analysis may rapidly be constructed; the table has entries only for $Y = 5.0$ and upwards, as lower values are unlikely to be wanted. The final regression equation,

$$Y = a + bx, \qquad (10.19)$$

is used for estimation of the required percentage point, and fiducial limits are obtained with the aid of equation (4.35).

In a numerical study of precision analogous to that in § 8.1, Bartlett found instances of a smaller variance for inverse sampling. His investigation was scarcely fair to direct sampling, for which he considered only experiments with doses symmetrically distributed about the ED 50; an experimenter who wished to estimate the ED 90 would in any event avoid low doses and concentrate on the high ones. In fact, if a second experiment were performed by the ordinary direct method, with the same doses and with values of n chosen to be equal to those observed in the inverse experiment, the precision of an estimated percentage point would be about the same. This follows because sZ^2/PQ^2 and nZ^2/PQ are approximately equal, and will tend to equality in large samples.

The real argument for the inverse method is not one of precision but of relevance. If an estimated ED 90 is to be of practical use, it must have come from observations on doses that bracket it, and the weight of information from doses above the ED 90 should be comparable to that from lower doses: only so can the evils of extrapolation be avoided. Without foreknowledge of the dose-response relation, satisfactory allocation of N available subjects between a series of doses is impossible. If instead a series of doses almost certain to bracket the ED 90 is chosen for inverse tests, a reasonable balance of information from the different doses will be maintained; the ratio of the standard error of the rate of non-response, q, to q itself will be almost constant at high response rates (Finney, 1949a). The scheme of experimentation might be to make tests at each dose in turn until one subject has failed to respond, then with the remaining subjects to make further tests at each dose until $s = 2$, then to increase until $s = 3$, and so on until the N subjects have been used. If the last subject responds, and so does not complete a group of tests for one dose, the subjects for that dose can be regarded as relating to direct sampling with use of the ordinary formulae for w and y. Even this procedure is not ideal, for Z^2/PQ^2 increases steadily with P and a disproportionate amount of the total information may come from very high doses. The precision of the estimated ED 90 may be less than could be obtained in some other way; a more equal division of subjects between the same doses, for example, might give a much greater total weight to the observations, and despite an increase in $(m - \bar{x})$ equation (4.35) might give narrower limits. The requirement of relevance, however, must be regarded as more important than formal precision, and even the scheme last mentioned might involve undesirable extrapolation. In the recommended scheme, the regression line may not be the same as would be obtained from experimentation in the neighbourhood of the ED 50, but it will be an approximation determined from, and relevant to, observations around the percentage point to be studied. The ill effects of using an assumed tolerance distribution that may differ from the true will have been minimized, in a general rather than a rigorous mathematical sense, though deviations from strict normality are likely to be more important for estimation of an extreme percentage point than for the ED 50.

For many quantal response investigations, the inverse approach would be impracticable. Estimation of the LD 99 of an insecticide by spraying insects one at a time is unlikely to prove a popular technique! In other circumstances, for example in the application of destructive tests to certain industrial products, subjects are normally tested singly, and the inverse method could be adopted as easily as the direct.

10.3   Estimation by up-and-down techniques

Inverse sampling (§ 10.2) is useful mainly if the experimental technique involves testing subjects one at a time, in such a way that the result of any test is known before the next is begun. If also a change in the level of dose is so little trouble to the experimenter that he is prepared to test each subject at a dose different from that of its predecessor, other procedures for estimating an ED 50 deserve consideration.

In the *staircase method*, a series of equally spaced log doses is chosen, say $..., x_{-3}, x_{-2}, x_{-1}, x_0, x_1, x_2, ...$, where x_0 is believed to be about the log ED 50. The first subject is tested at x_0. Thereafter, the result of any test determines the dose for the next test: if the subject responds, the next subject is tested at a dose one step lower, if it fails to respond, the next is tested at a dose one step higher. Whether or not the first dose is successfully chosen, later doses tend to concentrate about the ED 50: the rule ensures that the further from the ED 50 any test dose happens to be the greater is the probability that the next dose is one step nearer.

This method was first suggested by Dixon and Mood (1948). Subsequent discussion has suffered from confusion between the design of the experiment and the possibility of a simple analysis. As with the methods of earlier chapters, the choice of doses is very important. If, through ignorance or carelessness, x_0 is chosen far from μ or the interval between successive log doses is small relative to σ, many subjects will be used before the neighbourhood of the ED 50 is approached; if the interval is large, successive doses may oscillate between extremes of response probabilities (say alternately $P = 0.99$, $P = 0.03$). However, this danger is less than for an experiment in which subjects are assigned at the start to a fixed set of doses, because now the early results enable an unwise start to be adjusted. Brownlee, Hodges and Rosenblatt (1953) reported much higher efficiency for estimation of the ED 50 than for the 'probit method'. This is scarcely surprising, for any procedure that enables some subjects to be tested at doses chosen on the evidence of preliminary trials with a few subjects is likely to aid precise estimation of the ED 50. Practical difficulties of experimentation, or the delays inherent in testing one subject at a time, may forbid the staircase procedure or modifications of it. If these objections do not apply, certainly the procedure (or even a simple sequence of a pilot followed by a definitive experiment) offers hopes of substantial gain in efficiency. The results obtained by Brownlee *et al.* must be primarily attributable to the improvement in design rather than to any inherent property of the staircase rule.

Dixon and Mood proposed a simple estimation of the ED 50, essentially as the mean of the values of x tested. This may seem surprising, but of course it arises from the concentration of doses produced by the sequential rule. More exactly, if the first subject is tested at the dose with logarithm x_0 and other log doses are denoted by x_i (i being positive, negative, or zero), the Dixon and Mood estimator is

$$ m = x_0 + d \left(\frac{S_r(i)}{n_r} - 0.5 \right) ; \qquad (10.20) $$

here d is the interval between successive log doses, S_r denotes summation over all subjects that respond, and n_r is the number of these. A similarity with the Spearman–Kärber estimator (§ 3.7) is evident. Dixon and Mood also obtained a simple approximate estimator for σ. These formulae can be improved slightly by a rule excluding the first few subjects, so as to avoid undue influence of a start far from the ED 50.

Dixon and Mood also described a somewhat complicated maximum likelihood procedure. In fact, a rule of experimentation that determines each dose from the previous experimental results gives exactly the same likelihood function as one in which the total numbers of subjects were assigned to doses in advance. The sequential rule does not affect the independence of successive responses. The standard probit technique can therefore be applied to the total numbers at the doses. The highest and lowest values of x used have the property that, because of the staircase rule, one must show only responses and the other must show only non-responses. This small-sample peculiarity becomes progressively less important as the total number of tests is increased, and the chance occurrence of long sequences of responses or of non-responses pushes the extreme values of x farther from the neighbourhood of the ED 50; moreover, maximum and minimum working probits make some allowance for these extreme responses. Maximum likelihood estimation has some theoretical flaws when applied to small samples (§ 5.9), but staircase experiments do not present any special obstacle to its applicability.

Dixon (1965) published a tabulation of maximum likelihood estimates for short experiments. If x_0, d are specified, his table can be used to read off the estimate appropriate to a sequence of responses (+) and non-responses (−) such as $+ + + - - + - +$.

Dixon also gave an approximate table for the errors of estimation. An experiment might be organized as several distinct staircase sequences, rather than one long sequence, in order to save time, and these tables would then be very useful; for a long sequence, either the original Dixon and Mood quick approximation or the ordinary probit technique is needed.

Ex. 29. *An illustration of staircase estimation.* Table 10.2 contains records generated artificially as an example of staircase methods. These 'data' were obtained by sampling from a log normal distribution of tolerances for which $\mu = 2.0$, $\sigma = 0.15$ and the dose interval is 0.2.

Equation (10.20) gives

$$m = 2.3 + 0.2 \left(\frac{-13 \times 1 - 7 \times 2 - 0 \times 3}{26} - 0.5 \right)$$

$$= 1.992.$$

Dixon and Mood's formulae, not quoted here, indicate an estimate of about 0.17 for σ and a standard error of 0.036 for m. The alternative of applying standard probit methods to the totals of responses and non-responses, the final columns of Table 10.2, presents no difficulty. It yields

$$m = 1.980,$$

and σ is estimated from $1/b$ as 0.164. The approximate standard error for m is 0.034, but g is large (0.25) and the 95 % fiducial limits are found to be 2.056, 1.900.

The two methods of estimation agree well, as does either with the known values for the parameters.

The effectiveness of the staircase method depends vitally upon the choice of x_0 and d. Dixon and Mood recommended that they should be as near to μ and σ as initial information permits; for good results, d should lie between $2\sigma/3$ and $3\sigma/2$. In Example 29, the estimate of ED 50 is certainly more precise than if the fifty subjects had been allocated in almost equal numbers to the four doses 1.7 ,1.9, 2.1, 2.3. Moreover, if nothing had been known originally, except perhaps that the ED 50 was believed to lie between 1.5 and 2.5, an even wider range of doses, and consequently still lower precision, might have resulted from an

Table 10.2 Example of the staircase method

(Artificial data, with $\mu = 2.0$, $\sigma = 0.15$)

i	x_i	Results of 50 tests	Totals +	Totals O
1	2.5	+	6	0
0	2.3	+ + +	13	5
−1	2.1	+ + O O O + + + O + O + + +	7	12
−2	1.9	+ + O O O O + + O + + O + + O O O + + +	0	7
−3	1.7	O O + + O O O O O + O O + O +		
−4	1.5	O O O		
Totals			26	24

+ = response;   O = non-response.

attempt to apply the principles suggested in § 8.1. In some situations, the staircase method is the ideal solution to the problem of the pilot investigation, for the first few tests play the part of the pilot in indicating roughly where the ED 50 lies, yet all observations can be combined in the final analysis. If the original uncertainty about the ED 50 is great, there is no reason why a start should not be made with a large dose interval which may be halved or quartered as soon as the ED 50 has been roughly located; probit methods apply without modification, and no doubt suitable approximations could be developed.

In a very detailed study of the staircase method and variations on it, Wetherill (1963) found the asymptotic efficiency to depend rather heavily on the choice of d, whereas that of the stochastic approximation method (see below) is less sensitive to the corresponding choice. Not surprisingly, the very features of the staircase that make it good for estimation of μ lead to the information on σ being poor, and a very different design would be wanted if σ were a primary object of study. Wetherill questioned the wisdom of a preponderance of interest in the ED 50, presumably having in mind that ED 99 or ED 10 could be more important. For a bioassay, conclusions should be independent of which response level is used, and high precision for the ED 50 is a dominant consideration. In other circumstances, a more extreme level may be wanted: Wetherill suggested some modified staircases that would estimate different levels, but the properties of these have not been extensively studied.

An alternative up-and-down technique has been based upon the stochastic approximation theory of Robbins and Monro (1951). Suppose that n_i subjects are tested at x_i and r_i respond, and that the log dose x_{i+1} is determined from x_i by the formula

$$x_{i+1} = x_i - d_i \left(\frac{r_i}{n_i} - P \right), \qquad (10.21)$$

where $0 < P < 1$ and d_i for $i = 1, 2, 3, \ldots$ is a predetermined sequence of constants. The theory shows that the d_i can be so defined as to ensure that the doses have a high probability of approaching the value for which P is the expected response rate. Considerable freedom of choice is permissible for the d_i, but

$$d_i = c/i, \quad \text{for constant } c, \qquad (10.22)$$

has some optimal properties and is easy to operate. Thus for the ED 50 a good rule is

$$x_{i+1} = x_i - c\left(\frac{r_i}{n_i} - 0.5\right)\Big/i. \qquad (10.23)$$

Note that n_i can vary from one dose to another, although in practice it is likely to be held constant; one can have $n_i = 1$, so that every r_i is 0 or 1.

The optimal value of c is the reciprocal of the ordinate (Z) at the ED 50: for a normal tolerance distribution, c should be chosen as close to $\sigma\sqrt{(2\pi)}$ (or about 2.5σ) as can be guessed. With this and a corresponding guess at the ED 50 used as x_1, an experiment of k consecutive doses might be conducted. Wetherill (1963) has shown that estimation of the ED 50 by x_{k+1} is closely related to maximum likelihood, although the two are not identical. As for the staircase method, maximum likelihood can properly be used, but the extreme simplicity of using x_{k+1} must always be a great attraction.

In a series of papers, Cochran and Davis (1963, 1964, 1965) examined the merits of variations in n_i (taken as constant, n, for all doses) and k. They were concerned with small total numbers, and looked at possible arrangements for $nk = 6$, 12, 24, and 48. For a particular nk, an arrangement with small k and large n has the appeal that it avoids delay in completing the experiment. If x_1 is close to μ, the effect of change in n for fixed nk is small; the mean square error is almost independent of $(x_1 - \mu)$ unless this deviation is appreciably greater than σ, but the value of $(x_1 - \mu)$ at which the mean square error begins to be seriously inflated by contributions from the bias of the estimate is reduced when n is chosen to be large. Evidently caution in the choice of n is important unless the investigator is confident that he can choose x_1 well. Wetherill concluded that knowledge of σ, on which choice of c is based, is not very important as the estimation process is fairly robust in respect of departure of c from the optimal. Cochran and Davis commented that the ranges studied by Wetherill were not very extreme, and that deviations of x_1 from μ or c from 2.5σ of a magnitude likely in biological applications could well enlarge the mean square error.

Advice presented by Cochran and Davis may be paraphrased as follows. If σ is known within a factor of 2:1, the geometric mean of its limits enables c to be between 0.7 and 1.4 times its optimal. If x_1 can be guessed within σ of the ED 50, the stochastic approximation process with $k = 6$ or more can be recommended. This will estimate the ED 50 with standard errors of about 0.38σ, 0.27σ, 0.20σ for $nk = 12$, 24, 48 respectively. If x_1 may deviate from the ED 50 by as much as 2σ, c should be chosen larger, almost equal to that value suggested by the upper limit of σ; the standard error of the estimated ED 50 will then be about 10 % greater. If σ is known only within a factor if 4:1, the process may be badly wrong unless x_1 can be chosen within 0.6σ of the true μ and this is seldom possible.

Cochran and Davis suggested variants on the stochastic approximation method. For example, the rule expressed by equation (10.22) can be modified for the first few subjects, d_i perhaps being held constant until x_i approaches μ, or the whole experiment can be conducted in two stages. In general, the protection afforded by such modifications is purchased by lesser precision when conditions are favourable.

One attraction of the stochastic approximation method is that equation (10.21) can be used with $P \neq 0.5$, so enabling other dose levels to be estimated. Wetherill reported unsatisfactory experience in attempts to estimate ED 75 and ED 95, primarily because if the sequence of doses overshoots the mark many subjects must be tested before steps downward compensate completely.

Like the staircase method, stochastic approximation does not give records that are very satisfactory for estimation of σ. If the design of an experiment is chosen to concentrate doses close to μ, it can scarcely be expected to provide good information on σ, and Wetherill's difficulties with the iterative maximum likelihood confirm this.

Up-and-down methods of estimating the ED 50, or other parameters of a tolerance distribution, have become an important topic of research in their own right. What has been said above is no more than an introduction. Further variations on the obvious staircase and stochastic approximation methods deserve study: they can scarcely lead to generally more precise estimates, but improvements in practical suitability may be achieved. In their publications, Cochran and Davis used the normal tolerance distributions, but Wetherill used the logistic equation (3.22).

This can scarcely have affected the qualitative relevance of Wetherill's findings. Tsutakawa (1967) and Hsi (1969) have contributed further to the theory of this subject.

10.4   Continuous responses

This book is chiefly concerned with dichotomous classification of subjects. At the other extreme are data for which responses are measured on a continuous scale. Statistical analysis of the relation between response and dose is then a standard regression problem (not necessarily linear regression), and methods are fully discussed in many text-books. The techniques of biological assay are often relevant when the dose-response relations of two or more materials are to be compared.

Many a statistician has wished that percentages had never been invented, because they have so often been abused. One not uncommon mistake is to confuse percentages of individuals that manifest a particular response with measurements expressed as percentages of a control. The first type comprises quantities that cannot fall outside the range 0 to 100, and that have an underlying binomial-type distribution (possibly distorted by heterogeneities). On the other hand, if animals receive various doses of a substance likely to inhibit growth, to express final weights as percentages of a mean for control animals on zero dose may be convenient, but animals on low doses may well include some that then exceed 100%. Moreover, conversion of a continuous variate such as response to a percentage scale does not alter the intrinsic character of its frequency distribution (possibly normal, but certainly not binomial), and any notion that each percentage value can be regarded as a quantal response based on $n = 100$ must be firmly rejected!

The standard statistical technique for relating responses measured on a continuous scale to dose is that of linear regression and its ramifications (polynomial regression, weighted regression, etc.). For many important problems, simple linear regression is adequate. Its use in biological assay is discussed elsewhere (Finney, 1964). Occaionally, a dose-response regression shows pronounced curvature at low doses or high doses or both. A sigmoidal curve is likely to be found, the rate of increase of response per unit increase in dose being low at very low doses, and low again at high doses as the response approaches a ceiling. The probit transformation, and an adaptation of the standard methods, can be useful (Bliss, 1941; Finney, 1943b).

If H is the mean response of subjects at very high doses (the true mean for the population, not merely the average of a sample), and u is the response of a single subject at a dose whose logarithm is x, the expected value of u for fixed x may be written

$$E(u|x) = HP, \qquad (10.24)$$

where P is a function of x increasing as x increases and having for all x
$$0 \leqslant P \leqslant 1.$$

The integral of the normal probability function, equation (3.2), is a possible form for P, as are corresponding functions introduced in § 3.8.

Methods of estimation analogous to the standard probit technique can be developed. For example, if P is of the normal type and H is known,
$$p = u/H \qquad (10.25)$$

can be written as an estimate of P for each subject. The probit calculations can then be conducted with a new weighting coefficient. If all u are believed to have the same variance about their expectations,
$$w = Z^2 \qquad (10.26)$$

should replace the usual weighting coefficient. If, also, H is unknown, the problem has affinities with that of Chapter 7, and an auxiliary variate is required in a rather similar analysis. Butler, Finney and Schiele (1943) gave a numerical example, discussed in earlier editions of this book. The problem is not very common, and probably to-day one would think of the logit transformation, equation (3.22), as more suitable than the probit.

10.5 Multiple classifications

Intermediate between dichotomous responses and the truly continuous are those that may be described as *polytomous*. As noted earlier (Example 1), Tattersfield, Gimingham and Morris (1925) classified their insects as dead, moribund, slightly affected, and unaffected, and thus recognized four levels of response instead of the two characteristics of quantal data. They reduced their results to a dichotomy, by assessing toxicity in terms of the percentage of insects which were either moribund or dead, and took no account of the subclassification of these insects or of the others into unaffected and slightly affected; standard probit methods are then applicable.

Better use might be made of the data if the numbers of insects at the four levels were combined into a single index of toxic effect. Fryer, Stenton, Tattersfield and Roach (1923) used the same four-fold classification but, in order to obtain a percentage response at each dose, scored moribund and slightly affected insects as one-half and one-quarter dead respectively. For example, if a batch of ten insects were classified as 4, 1, 3, 2, the proportionate response would be $(4 \times 1 + 1 \times 0.5 + 3 \times 0.25)/10$, or 52.5 %. This scoring did not give very much smoother response curves than that subsequently adopted by Tattersfield, and has never been widely used. Possibly discrimination between moribund and slightly affected is simpler than between other pairs of classifications, so that the information provided by the latter is relatively unreliable.

In recent years, probit methods have been developed for the systematic analysis of these kinds of data. Perhaps that most readily understood, and most in line with the ideas of earlier chapters, is due to Gurland, Lee and Dahm (1960). This can be adequately illustrated by reference to data with three classifications of subjects, say dead, moribund, unaffected. Suppose that n_i subjects received a dose of which the logarithm is x_i and are subsequently classified as r_{i1}, r_{i2}, r_{i3} in *descending* order of responsiveness, where

$$r_{i1} + r_{i2} + r_{i3} = n_i \qquad (10.27)$$

and $i = 1, 2, ..., k$. Then a reasonable model to adopt is a generalization of equation (3.2). Suppose that P_{i1} is the probability that a subject at dose x_i manifests the highest response and P_{i2} is the probability that a subject manifests the second response; the analogue of equation (3.2) is

$$\left.\begin{aligned}
P_{i1} &= \int_{-\infty}^{(x_i - \mu_1)/\sigma} \frac{1}{\sqrt{(2\pi)}} \exp\{-\tfrac{1}{2}u^2\}\, du, \\
P_{i2} &= \int_{(x_i - \mu_1)/\sigma}^{(x_i - \mu_2)/\sigma} \frac{1}{\sqrt{(2\pi)}} \exp\{-\tfrac{1}{2}u^2\}\, du.
\end{aligned}\right\} \qquad (10.28)$$

Corresponding to Q, one may conveniently define

$$P_{i3} = 1 - P_{i1} - P_{i2}. \qquad (10.29)$$

These correspond to statements that the thresholds between consecutive types of response are normally distributed with means μ_1, μ_2 and standard deviation σ. The inherent logic of the problem requires that

$$\mu_1 \geqslant \mu_2, \qquad (10.30)$$

and that the standard deviation shall be the same in both distributions, for otherwise there would be doses at which P_{i1} exceeded P_{i2}. Another way of expressing the model is to say that the two thresholds conform to the two probit regression equations

$$\left.\begin{array}{l} Y_1 = \alpha_1 + \beta x, \\ Y_2 = \alpha_2 + \beta x, \end{array}\right\} \qquad (10.31)$$

where $\beta = 1/\sigma$ and $\alpha_1 \leqslant \alpha_2$.

The probability of the particular set of responses is expressed by the *multinomial distribution*, a generalization of equation (2.7):

$$Pr(r_{i1}, r_{i2}, r_{i3} | n_i) = \frac{n_i!}{r_{i1}! \, r_{i2}! \, r_{i3}!} P_{i1}^{r_{i1}} P_{i2}^{r_{i2}} P_{i3}^{r_{i3}}. \qquad (10.32)$$

In place of equation (4.1), the log-likelihood of the observations is

$$L = S(r_{i1} \log P_{i1} + r_{i2} \log P_{i2} + r_{i3} \log P_{i3}), \qquad (10.33)$$

with S denoting summation over $i = 1, 2, ..., k$. The parameters can be estimated by maximizing L. Generalization of §4.2 gives an iterative procedure, but since it is rarely wanted it will not be presented here. The general computer program (Chapter 5) can be readily adapted to maximization of L, and to the production of the asymptotic variances of the estimates (Example 30).

The method for normal tolerance distributions requires little alteration to make it appropriate to the logistic or to any of the other functions in §3.8. Indeed Gurland *et al.* presented the logistic also. They chose to estimate by minimum χ^2 instead of maximum likelihood (§5.9), but when the computer program is used this no longer has any advantage in simplicity.

Aitchison and Silvey (1951) examined much the same problem by maximum likelihood, but did not introduce the constraint of having the same standard deviation for each tolerance distribution. This is a legitimate problem of estimation, but it does introduce the internal inconsistency of fitting a model for which the probit regression lines intersect. As an approximation, lines of different slope may fit particular data satisfactorily within the range of the observations, but usually the model constrained by parallelism will be preferable.

White and Graca (1958) developed similar methods in respect of experiments in which every subject was classified according to which of several successive time-intervals after treatment contained the first record of response. Ashford (1959), Ashford, Smith and Brown (1960), and Ashford and Smith (1965b) took these ideas further and, in particular, considered the consequences of the response variate being subject to errors of measurement. Claringbold (1958) described another form of analysis in which the separate classes of response were kept distinct, no transformation was used, and the classes were put through a canonical analysis; the objectives of this analysis are less clear, and it has not been widely adopted. A paper by Stevens (1948) is of interest in this context, although ostensibly it is concerned with an industrial problem, that of assessing conformity to specifications. Swan (1969a, b, c) described a related technique for estimating the mean and variance of a normal distribution from a single sample in which some members are merely classified as 'greater than A', 'less than B', or 'between C and D' instead of being measured exactly. He also listed a computer program (in Algol) for the analysis.

Ex. 30. *An assay of insecticide residues.* Gurland, Lee and Dahm (1960) reported an assay of the guthion content of an extract prepared from alfalfa that had been sprayed with this insecticide. In order to compensate for effects of other substances present in the extract, additional extract from unsprayed plants was included in each dose of both preparations to make up a total of 10 ml of extract. At each dose, 50 insects (species unspecified) were exposed and, after 17 hours, classified as dead, moribund, and alive. Table 10.3 shows the results.

Exactly as in §§ 6.1, 6.2, the validity of a biological assay requires that the regression coefficient is the same for both preparations. The only active constituent of the test preparation is guthion, and therefore a single factor must serve to convert the measure of any test dose into the quantity of guthion itself that

Table 10.3   Biological assay of guthion in alfalfa extract

Dose	No. of insects (n)	Alive (r_1)	Moribund (r_2)	Dead (r_3)
Standard preparation				
20 μg	50	44	1	5
35 μg	50	28	1	21
45 μg	50	8	7	35
Test preparation				
1.0 ml	50	37	1	12
1.5 ml	50	20	2	28
2.0 ml	50	8	6	36

is equally potent. This means that equally potent log doses will be a fixed distance apart, and corresponding regression lines must therefore be parallel: any real departure from parallelism indicates contamination of the test preparation or some other disturbance that invalidates the assay. The appropriate model for the test preparation will have P_{i1}, P_{i2}, P_{i3} defined by equations (10.28), (10.29). Then, for the test preparation, the same equations must hold, except that values of μ_1, μ_2 will in general be different. Moreover, the guthion content of the plant extract, and therefore the relative potency, is independent of whether the threshold of death or of moribundity is being examined, and therefore

$$\mu_{S1} - \mu_{T1} = \mu_{S2} - \mu_{T2}, \qquad (10.34)$$

where the first suffix denotes the preparation and the second the type of threshold.

Corresponding to equations (10.31) are the probit regression equations

$$\left.\begin{aligned}
Y_{S1} &= \alpha_{S1} + \beta x, \\
Y_{S2} &= \alpha_{S2} + \beta x, \\
Y_{T1} &= \alpha_{T1} + \beta x, \\
Y_{T2} &= \alpha_{T2} + \beta x,
\end{aligned}\right\} \qquad (10.35)$$

where equation (10.34) requires that

$$\alpha_{T2} = \alpha_{T1} + \alpha_{S2} - \alpha_{S1}. \qquad (10.36)$$

The general computer program for maximum likelihood estimation was applied to the data in Table 10.3, after first using equation (10.36) to express the definition of P_{i2} for the test preparation in terms of α_{S1}, α_{S2}, and α_{T1} (so leading to estimation constrained to conform to equation (10.34)). The estimates are

$$\left.\begin{aligned}
a_{S1} &= -3.0665, \\
a_{S2} &= -2.8768, \\
a_{T1} &= 4.1731, \\
b &= 5.1508.
\end{aligned}\right\} \qquad (10.37)$$

The matrix of variances and covariances of these is

$$V = \begin{pmatrix}
0.844\,045 & 0.842\,106 & 0.088\,341 & -0.543\,535 \\
0.842\,106 & 0.842\,024 & 0.087\,359 & -0.542\,856 \\
0.088\,341 & 0.087\,359 & 0.020\,859 & -0.057\,458 \\
-0.543\,535 & -0.542\,856 & -0.057\,458 & 0.355\,221
\end{pmatrix}.$$

$$(10.38)$$

Comparison of expected frequencies with the observations in Table 10.3 gives

$$\chi^2_{[8]} = 16.44, \qquad (10.39)$$

the original 12 degrees of freedom (2 within each of 6 doses) being reduced by the number of parameters estimated. This is just beyond the 5 % significance level (Table VII), but the discrepancies arise primarily from the small expectations for numbers of moribund flies (§ 4.6) and for the present illustrative purposes they are disregarded.

Calculation of relative potency can proceed from a_{S1}, a_{T1}, b alone, since the constraint (10.34) has ensured that the estimate takes account of the triple classification of flies. By equation (6.1),

$$\begin{aligned}
M &= (a_{T1} - a_{S1})/b \\
&= 1.4055.
\end{aligned} \qquad (10.40)$$

Fiducial limit calculations use the full form of Fieller's theorem, equation (4.35), because the constraint on the parameters has introduced a correlation between the numerator and denominator of M. The second row and column of (10.38) are unwanted. Subsidiary calculations are

$$\begin{aligned}
v_{11} &= V(a_{T1}) - 2Cov(a_{S1}, a_{T1}) + V(a_{S1}) \qquad (10.41) \\
&= 0.020\,859 - 2 \times 0.088\,341 + 0.844\,045 \\
&= 0.688\,222,
\end{aligned}$$

8

$$v_{12} = Cov\,(a_{T1}, b) - Cov\,(a_{S1}, b) \qquad (10.42)$$
$$= -\,0.057\,458 + 0.543\,535$$
$$= 0.486\,077,$$
$$v_{22} = V\,(b) \qquad (10.43)$$
$$= 0.355\,221.$$

Equation (4.35) gives 1.468, 1.348 as the 0.95 fiducial limits. Thus the potency of the extract is estimated as $25.44\,\mu g$ per ml, with fiducial limits at 29.35, $22.26\,\mu g$ per ml.

For comparison, one may note the consequences of combining the numbers of live and moribund flies and estimating from the dichotomous classification. The potency estimate is $25.72\,\mu g$ per ml, with limits at 30.30, 22.13, indicating that the triple classification of the flies improved the precision. Alternatively, if the moribund flies are grouped with the dead, the analysis leads to $25.22\,\mu g$ per ml as the estimate, with limits at 28.89, 22.16; somewhat surprisingly, this dichotomous analysis is almost as good as that using the triple classification. The reason seems to be that, for the particular data, the estimate of β happens to be greater when moribund and dead flies are grouped together, and this just compensates for the increased variance of the numerator of M arising from neglect of the sub-classification. With either dichotomy, the χ^2 (with 3 degrees of freedom) is small.

10.6 The Parker-Rhodes equation

O'Kane, Westgate, Glover and Lowry (1930) and O'Kane, Westgate and Glover (1934) found certain insecticidal studies to show evidence of the logarithm of the mortality probit being proportional to the log concentration of insecticide. An alternative statement is that the probit is proportional to a power of the concentration, but there seems no logic in a direct proportionality rather than a linear relation. Parker-Rhodes (1941, 1942 a, b, c, 1943 a, b) reported extensive tests of the toxicity of metallic salts and related compounds to spores of *Bacillus agri*, *Botrytis allii*, and *Macrosporium sarcinaeforme*; he proposed a probit regression of the form

$$Y = \alpha + \beta z^i, \qquad (10.44)$$

of which O'Kane's equation is a particular case (Bliss, 1935 c).

The equation implies that the dose metameter

$$x = z^i \qquad (10.45)$$

conforms to a normal tolerance distribution. Parker-Rhodes termed i the *index of variation*, and went on to define the *variability* of the spores relative to the toxic substance as

$$W_i(z) = \sigma^2/i^2\mu^2. \qquad (10.46)$$

Clearly $i = 1$ corresponds to a normal distribution of tolerance on the scale of z; less obviously, $i = 0$ corresponds to the limiting case of the log normal distribution (for which the variability takes the limiting value σ^2). If $i \neq 0$, equation (10.44) can only be an approximation: it suggests that (for $i > 0$) the response must approach a lower limit $Y = \alpha$ as z tends to zero.

Parker-Rhodes developed a theory of variability, and advanced chemical reasons for expecting i to be a simple rational fraction. He found many values close to $\frac{1}{4}$ and $\frac{1}{2}$, and even some close to $-\frac{1}{3}$ and $-\frac{2}{3}$ but his interpretation of the negative values involved inconsistencies. Discussion of the validity of his theory is outside the scope of this book. In previous editions, statistical techniques for simultaneous estimation of α, β, i were presented and some mistakes in Parker-Rhodes's analysis were corrected. To-day, the special iterative solution (dependent upon yet another auxiliary variate) is scarcely worth while, as a general program for maximizing the likelihood would be used unless the method were wanted often. The variability theory has attracted little attention in the past 20 years, and further discussion seems unnecessary.

10.7 Errors of dose

For some types of data to which probit methods seem relevant, the only available measure of dose is subject to uncertainty. When dose is a measurement such as weight of a chemical compound, errors should be only those of a physical measuring process and can usually be ignored. On occasion, however, dose may be a quantity such as age that has been recorded only in moderately broad groups. The Warsaw schoolgirls discussed in Example 12 had ages recorded in 3-month intervals, a grouping that is perhaps sufficiently fine to be neglected. Other similar data simply record age in years, so that a recorded age of 12 means $12.0 \leqslant$ age < 13.0; if the whole extent of the records is only 5–10 years, uncertainty within 1 year is not negligible.

Tocher (1949) discussed the problem of grouped data, making the assumption that subjects within a group can be taken as having their dose metameter uniformly distributed over the group. If equation (3.2) represents the probability of response for a subject for whom the true dose metameter is X, then the average probability for a subject whose dose is randomly situated in the interval $x_i \leqslant X < x_{i+1}$ is

$$\overline{P}_i = \int_{x_i}^{x_{i+1}} P \, dX / (x_{i+1} - x_i). \qquad (10.47)$$

As Tocher easily showed, this reduces to

$$\overline{P}_i = (Y_{i+1} P_{i+1} - Y_i P_i + Z_{i+1} - Z_i)/(Y_{i+1} - Y_i), \quad (10.48)$$

where

$$Y_i = (x_i - \mu)/\sigma,$$

$$\left. \begin{aligned} P_i &= \int_{-\infty}^{Y_i} \frac{1}{\sqrt{(2\pi)}} \exp\{-\tfrac{1}{2} u^2\} \, du, \\ \text{and} \qquad Z_i &= \frac{1}{\sqrt{(2\pi)}} \exp\{-\tfrac{1}{2} Y_i^2\}. \end{aligned} \right\} \qquad (10.49)$$

The log-likelihood, L, then takes the usual form (equation (4.1)), except that $\overline{P}$ replaces P.

Tocher's paper may be consulted for a detailed account of an iterative regression procedure for estimating the parameters, including a fully worked example. He followed the standard approach (as in §4.2), using differential coefficients of $\overline{P}$ instead of P, and found the computations rather heavy. He also constructed an approximation that first uses the mid-point of each group, $(x_i + x_{i+1})/2$, as though it were the true x in a simple analysis and then applied grouping corrections. Provided that each group is of the same width, so that for all i

$$x_{i+1} - x_i = d,$$

the correction amounts to multiplying the crude values of a, b by the factor $(1 + b^2 d^2/24)$. Once again, the need for a special computational technique has now lessened and a general maximum likelihood program can be applied. The effect of the grouping will always be to lose precision and to reduce the magnitude of b as obtained from the ordinary calculations. Tocher's example suggests only a very small loss when the grouping is into one-year intervals and the records extend over only eight years in all.

Patwary and Haley (1967) studied a similar problem, but instead of grouping they had doses that suffered probabilistic errors of administration. Suppose that an intended dose X has an associated probability density function $\phi(x|X)$ for the actual dose x; that is to say, conditional upon X being specified, the probability that the actual dose lies between x and $x+dx$ is $\phi(x|X)\,dx$. Then the probability of response for a subject recorded as receiving a dose x is

$$\bar{P} = \int_{-\infty}^{\infty} P\phi(x|X)\,dx. \qquad (10.50)$$

The likelihood function, written with $\bar{P}$ in place of P, will involve the usual parameters α and β as well as any new parameters used in specifying ϕ. Patwary and Haley discussed a particular case in which response depends upon numbers of bacterial particles and the normal tolerance distribution is simply an approximation to one based upon a Poisson distribution. For this, they presented a computational routine. Once more, the general computer analysis is available; the major difficulty may lie in the evaluation of equation (10.50), so as to have it in a form suitable for the likelihood calculations.

One example of this has already been given, in slightly different guise, in § 9.5. The derivation of equation (9.23) from equation (9.22) essentially involved

$$\phi(x|X) = \frac{1}{\gamma\sqrt{(2\pi)}} \exp\left\{ -\frac{(x-X)^2}{2\gamma^2} \right\}, \qquad (10.51)$$

but the symbol ξ was used in place of X; γ is an additional parameter for the variance of the conditional distribution of the actual dose. In this instance, γ cannot be estimated in addition to α, β, unless additional information of some kind is available.

11

Quantal responses to mixtures

11.1 Types of joint action

Full understanding of the toxic action of a group of insecticides or fungicides requires study of what happens when two or more are applied in mixture. Similarly in respect of other stimuli of types to which this book refers, the consequences of simultaneous application of two or more will often be important. The potency of a mixture may be greater than would be expected merely from knowledge of the potencies of its constituents applied separately, a situation likely to be of practical and economic importance. The alternative of a reduced potency for the mixture may also occur. Exact meaning can be given to these statements only after definition of the null state: if the dose-response relation be known for each of two stimuli used separately, how can the expected response for any mixture be calculated?

This is a general question for any type of response, and one that has not been as fully considered by biologists or statisticians as seems desirable. Its first systematic discussion in relation to quantal responses and probit analysis was by Bliss (1939a), who distinguished three types of joint toxic action:

'(1) *Independent joint action*. The poisons or drugs act independently and have different modes of toxic action. The susceptibility to one component may or may not be correlated with the susceptibility to the other. The toxicity of the mixture can be predicted from the dosage–mortality curve for each constituent alone and the correlation in susceptibility to the two poisons; the observed toxicity can be computed on this basis whatever the relative proportions of the components.

'(2) *Similar joint action*. The poisons or drugs produce similar but independent† effects, so that one component can

† This word is used with meaning entirely different from that in the previous paragraph.

[230]

be substituted at a constant proportion for the other; variations in individual susceptibility to the two components are completely correlated or parallel. The toxicity of a mixture is predictable directly from that of the constituents if their relative proportions are known.

'(3) *Synergistic action.* The effectiveness of the mixture cannot be assessed from that of the individual ingredients but depends upon a knowledge of their combined toxicity when used in different proportions. One component synergizes or antagonizes the other.'

Bliss explicitly excluded from his classification pairs of constituents which react chemically. In one sense, he covered all other types of joint action, independent and similar action being the simplest, and synergism including all forms of departure from these. *Antagonistic action,* in which the potency of a mixture is less than expected, had been described by Clark (1933, Chapter 17), who suggested various mathematical representations; his examples are so different in type as scarcely to fall within the scope of Bliss's paper or the present work. Antagonism will be regarded here simply as negative synergism.

The potency of a mixture whose constituents act similarly is generally greater than that of a mixture, in the same proportions, whose constituents are of the same individual potencies but act independently. Either type of action is specified by an exact law predicting the response produced by a mixture from the amounts of the constituents and their potencies. Hence a more exact definition of synergism is needed than Bliss's statement of its being 'characterized by a toxicity greater than that predicted from experiments with the isolated constituents'; at least it must be decided whether independent, or similar, or some other joint-action law is the norm to which any suspected case of synergism is to be referred. Bliss suggested alternative mathematical models for synergistic action, but neither of these nor any used by Clark has the more familiar concepts of independent or similar action as cases of zero synergism.

Finney (1942a) endeavoured to bring out the logical relation between different types of joint action. That of widest importance appears to be what Bliss termed similar action. In view of subsequent theoretical developments (§ 11.7), it will now be called *simple similar action* and discussed at considerable length.

11.2 Simple similar action

Two stimuli whose modes of action on the test subject are much alike (especially stimuli of closely related chemical constitutions) often show parallel regression lines of probits on log doses. The relative potency is then expressible by a single figure, the ratio of equally effective doses, which is constant at all levels of response (§ 6.1). Let the two regression lines be written

$$Y_1 = \alpha_1 + \beta \log z, \tag{11.1}$$

$$Y_2 = \alpha_2 + \beta \log z, \tag{11.2}$$

where z is the dose (as in § 2.1). The potency of the second material relative to the first is given by

$$\log \rho = (\alpha_2 - \alpha_1)/\beta. \tag{11.3}$$

Consequently, equation (11.2) can be written

$$Y_2 = \alpha_1 + \beta \log (\rho z). \tag{11.4}$$

Thus multiplication by ρ converts a measure of dose of the second material into an equivalent amount for the first.

A mixture containing amounts z_1, z_2 of the two preparations is said to show simple similar action if, within the limits of sampling variation, the response to it equals that which would be produced by a dose $(z_1 + \rho z_2)$ of the first alone. Hence simple similar action requires that any mixture shall have a response probit of the form

$$Y = \alpha_1 + \beta \log (z_1 + \rho z_2). \tag{11.5}$$

The mixture may alternatively be regarded as a total dose z in which the proportions of the two preparations are π_1, π_2. Equation (11.5) can be written

$$Y = \alpha_1 + \beta \log (\pi_1 + \rho \pi_2) + \beta \log z. \tag{11.6}$$

Here $\pi_1 = 1$, $\pi_2 = 0$ and $\pi_1 = 0$, $\pi_2 = 1$ correspond respectively to equations (11.1) and (11.4). Also

$$z_1 + \rho z_2 = z(\pi_1 + \rho \pi_2)$$

is the expression of a dose z of the mixture as an equivalent dose of the first preparation. If ζ_1, ζ_2, ζ are the ED 50's of the two preparations and a mixture in fixed proportions,

$$\zeta_2 = \zeta_1/\rho \qquad (11.7)$$

and $$\zeta = \zeta_1/(\pi_1 + \rho\pi_2), \qquad (11.8)$$

the denominator of (11.8) being the potency of the mixture relative to the first preparation. Bliss stated equivalent formulae.

The concept of simple similar action extends easily to three or more preparations, provided that all have the same probit regression coefficient. If a dose z of a mixture contains proportions π_1, π_2, π_3 of three preparations, the second and third having potencies ρ_2 and ρ_3 relative to the first, the regression for mixtures in constant proportions is

$$Y = \alpha_1 + \beta \log(\pi_1 + \rho_2\pi_2 + \rho_3\pi_3) + \beta \log z. \qquad (11.9)$$

Other relations follow in the obvious manner.

For experimental results, proportions such as π_1, π_2 are known from the conduct of the experiment, but α_1, α_2, β, ρ have to be replaced by estimates a_1, a_2, b, R, calculated as in Chapter 6 or similarly. Assessment of whether simple similar action is adequate to describe the interrelations of two preparations then depends upon whether the data accord satisfactorily with equations (11.1), (11.2), (11.6). Examples 33, 34 illustrate calculations and test procedures.

Ex. 31. *Simple similar action of rotenone and a deguelin concentrate.* The data used in Examples 1, 2, 3, 4, 6, 7, 10 were part of the results of an experiment fully reported by Martin (1942). In that experiment, batches of about 50 *Macrosiphoniella sanborni* were tested with doses of rotenone, a deguelin concentrate, or a 1:4 mixture of the two. Martin expressed doubts about the mortality at the lowest dose of the deguelin and the two lowest doses of the mixture. So far as fair interpretation of Martin's experiment is concerned, the rejection of anomalous observations may be questioned, but argument about this after so many years is unprofitable. Here the three low doses are omitted, solely for the purpose of illustrating numerical procedures.

Probit regression lines for the three preparations have been estimated, with the aid of the computer program and under a constraint of parallelism. The equations are:

$$\left.\begin{aligned} Y_1 &= 2.299 + 3.949x, \\ Y_2 &= 0.580 + 3.949x, \\ Y_3 &= 1.230 + 3.949x, \end{aligned}\right\} \qquad (11.10)$$

where $x = \log z$. From the first two, the potency of the deguelin concentrate relative to rotenone is estimated by

$$M = (0.580 - 2.299)/3.949$$

$$= -0.4353$$

whence $\qquad\qquad R = 0.367.$ $\qquad\qquad\qquad$ (11.11)

The potency of the $1:4$ mixture relative to rotenone is therefore estimated by

$$\pi_1 + R\pi_2 = 0.2 + 0.367 \times 0.8 = 0.494.$$

Hence, by equation (11.6), the probit regression line for the mixture can be estimated from the separate preparations as

$$Y = 2.299 + 3.949 \log (0.494) + 3.949x$$

$$= 1.090 + 3.949x. \qquad\qquad (11.12)$$

Comparison of equation (11.12) with the third of equations (11.10) shows that the mixture manifested slightly higher toxicity than was predicted by similar action. Example 33 shows the deviation not to be statistically significant.

In an experiment planned to throw light on the potencies of mixtures, the constituent materials will usually be tested separately as well as in mixture. This is very desirable but may not always be possible. If necessary, the internal consistency of results for three or more mixtures with the hypothesis of simple similar action may be judged without tests on the constituents separately, by expressing intermediate members of the series as though they were obtained by mixing, in appropriate proportions, the two most extreme mixtures.

Ex. 32. *Expression of one mixture in terms of two others.* Suppose tests have been made on three mixtures containing proportions

4 : 1, 7 : 3, and 2 : 3 of two materials. The first and the last are the most extreme in composition, and the second may be expressed in terms of them. The proportions of the two constituents in the mixtures are (0.8, 0.2), (0.7, 0.3), and (0.4, 0.6); the second mixture may therefore be regarded as composed of proportions π_1, $(1 - \pi_1)$ of the first and third, where

$$0.8\pi_1 + 0.4(1 - \pi_1) = 0.7,$$

whence $\pi_1 = 0.75$. If the 4 : 1 and 2 : 3 mixtures are themselves mixed in the ratio 3 : 1, a mixture of the original materials in the ratio 7 : 3 is obtained. The agreement of the results with predictions made from the hypothesis of simple similar action may then be examined in the standard manner.

11.3   Tests of simple similar action

If
$$Y_3 = \alpha_3 + \beta \log z \tag{11.13}$$

is the true probit regression equation for a mixture, a measure of deviation from simple similar action is the potency relative to the prediction in equation (11.6). The logarithm of this is

$$\frac{\alpha_3 - \alpha_1}{\beta} - \log (\pi_1 + \rho\pi_2), \tag{11.14}$$

which is estimated by

$$M_s = \frac{a_3 - a_1}{b} - \log (\pi_1 + R\pi_2). \tag{11.15}$$

No analogue of Fieller's theorem is available for assessing probabilities for M_s, and the intrinsic complexity of (11.15) suggests that none can be found. However, unless g is large, conclusions based on a variance of M_s are likely to be reasonably satisfactory; experiments concerned with mixtures will usually need more doses and more subjects than simple experiments, so that the precision of b will be increased and g correspondingly reduced. The variance of M_s can be neatly expressed as

$$V(M_s) = \left[\frac{\lambda^2}{_1Snw} + \frac{(1-\lambda)^2}{_2Snw} + \frac{1}{_3Snw} + \frac{\{\lambda\bar{y}_1 + (1 - \lambda)\,\bar{y}_2 - \bar{y}_3\}^2}{b^2 \Sigma S_{xx}} \right] \Big/ b^2. \tag{11.16}$$

In this formula, subscripts refer to the tests on the two constituent preparations and the mixture, Σ indicates summation over the three, and

$$\lambda = \frac{\pi_1}{\pi_1 + R\pi_2}. \qquad (11.17)$$

Computer output may give a_1, a_2, a_3 directly rather than values of $\bar{y}$, $\bar{x}$ for each preparation. If this is so, and if V is the 4×4 matrix of variances and covariances (with elements v_{ij}) of a_1, a_2, a_3, b, an alternative formula may be more convenient despite its greater length:

$$V(M_s) = \left[\lambda^2 v_{11} + 2\lambda(1-\lambda)v_{12} + (1-\lambda)^2 v_{22} - 2\lambda v_{13} - 2(1-\lambda)v_{23} \right.$$

$$+ v_{33} - 2\left\{ \frac{\lambda a_1 + (1-\lambda)a_2 - a_3}{b} \right\} \{\lambda v_{14} + (1-\lambda)v_{24} - v_{34}\}$$

$$\left. + \left\{ \frac{\lambda a_1 + (1-\lambda)a_2 - a_3}{b} \right\}^2 v_{44} \right] \bigg/ b^2. \qquad (11.18)$$

Ex. 33. *Test of simple similarity for rotenone and deguelin concentrate.* For the data analysed in Example 31, the variance–covariance matrix is

$$V = \begin{pmatrix} 0.056\,410 & 0.086\,967 & 0.079\,401 & -0.068\,180 \\ 0.086\,967 & 0.169\,900 & 0.143\,521 & -0.123\,238 \\ 0.079\,401 & 0.143\,521 & 0.144\,391 & -0.112\,516 \\ -0.068\,180 & -0.123\,238 & -0.112\,516 & 0.096\,615 \end{pmatrix}.$$

Also
$$\lambda = 0.2/(0.2 + 0.367 \times 0.8)$$
$$= 0.405.$$

Equation (11.18) gives

$$V(M_s) = [0.020\,603 - 2 \times 0.011\,698 \times 0.011\,576$$
$$+ 0.011\,698^2 \times 0.096\,615]/3.949^2$$
$$= 0.020\,345/15.5946$$
$$= 0.001\,305,$$

and therefore $M_s = 0.0355 \pm 0.0361.$

For these data, g at probability 0.95 was small, only 0.024. Even without any allowance for the fact that (11.16) and (11.18) will tend to underestimate the effective variance of M_s, the deviation of M_s from zero can be seen to be not large relative to the inherent variability of such experiments. Although the potency of the mixture appeared to be about 9 % greater than is predicted by the constituents, this could easily be due to chance.

If several mixtures with different proportionate constitutions have been tested, another method (Finney, 1942 a) may be used for examining the agreement of the median effective doses with predictions by simple similar action. This method explicitly employs only the median effective doses and their standard errors, but parallelism of the regressions is implicit. There are theoretical objections on account of assumptions of normality and independence; provided that the regression coefficient is estimated with reasonable precision, the tests of significance should be trustworthy.

Earlier paragraphs may seem to have implied that $\pi_1 + \pi_2 = 1$, but this is not essential. For consider a mixture of three materials, to which equation (11.9) applies. Now suppose that the third material is inert, so that $\rho_0 = 0$; equation (11.0) then reduces to (11.6), which therefore still applies when π_1, π_2 do not represent the totality of the mixture. Equation (11.8) can be written

$$\frac{1}{\zeta} = \left(\frac{1}{\zeta_1}\right)\pi_1 + \left(\frac{\rho}{\zeta_1}\right)\pi_2, \qquad (11.19)$$

which suggests a technique for estimating parameters and testing goodness of fit.

If values of ζ have been estimated for mixtures of two materials in several different proportions, $1/\zeta_1$ and ρ/ζ_1 may be estimated as the regression coefficients of $1/\zeta$ on π_1 and π_2. Write z for the estimate of ζ and m for the corresponding $\log \mathrm{ED} 50$ ($m = \log z$). Approximately,

$$V(1/z) = \frac{V(z)}{z^4} = \frac{V(m)}{z^2}, \qquad (11.20)$$

and consequently the weight to be attached to $1/z$ is

$$W = \frac{z^2}{V(m)}. \qquad (11.21)$$

The formula is appropriate to natural logarithms; if logarithms to base 10 are used, W should be multiplied by

$$(\log_{10} e)^2 = 0.18861.$$

The regression equation (11.19) contains no 'constant term'. It is constrained to make $1/\zeta$ zero when the content of both constituents is zero, and the sums of squares and products used in calculation must be totals, not adjusted to refer to deviations about means. Comparison of the values of ζ calculated for each mixture from equation (11.19) with the values from which the equation was estimated shows how well the observations are fitted by the similar action hypothesis and permits a test of significance of the agreement.

Tattersfield and Martin (1935; see Example 16 above) published results of toxicity tests with seven derris roots to *Aphis rumicis*. They expressed the ED 50 for each in terms of rotenone and a dehydro mixture; the latter (Martin, private communication) varied considerably from root to root in its proportions of different compounds. Though he recognized that any comparison of toxicities regarding the dehydro mixture as a single material was of doubtful validity, Bliss (1939 a) used the data as an illustration of synergistic action between two poisons. He found them to agree satisfactorily with a generalization of one of his formulae for synergism, except that one root (no. 2) behaved anomalously. His analysis was not strictly correct because the method of estimation of the ED 50's caused correlations, but the disturbance so caused can only be slight. Using the regression procedure, Finney (1942 a) showed equation (11.9) to fit well without recourse to synergism. Detailed calculations are not shown here, since the process is sufficiently illustrated by Example 34.

Ex. 34. *Similar action between the toxic constituents of derris root.* Toxicity tests on four derris roots (Martin, 1940) were discussed in Example 17. Comparative ED 50 values, adjusted for different levels of susceptibility of the aphids on the three days of testing, are given in Table 6.7. In calculating weights, $V(m)$ in equation (11.21) has been taken as the variance obtained in Example 17 without any adjustment for the heterogeneity suggested by the large χ^2 there found. The covariances between the parameters for the five preparations in Example 17 were relatively small (corresponding to correlation coefficients never exceeding 0.5 and usually much smaller); therefore the approximation implicit in a simplification which ignores them is likely to be satisfactory.

The probit regression coefficients for Martin's three days of experimentation are 5.72 ± 0.40, 5.38 ± 0.44, and 5.59 ± 0.55, evidently no indication of any true difference.

Martin divided the toxic constituents of the four roots into rotenone, a toxicarol fraction, and a deguelin fraction, and determined the proportions in each root. The second and third fractions were themselves mixtures, but chemical and toxicological evidence suggested that this further complexity could be ignored. An examination of the adequacy of similar action to explain the observations is therefore of interest. If equation (11.9) is adequate to describe the observed mortalities, the ED 50 for any root should be given by the obvious generalization of equation (11.19)

$$\frac{1}{\zeta} = \left(\frac{1}{\zeta_r}\right)\pi_r + \left(\frac{\rho_t}{\zeta_r}\right)\pi_t + \left(\frac{\rho_d}{\zeta_r}\right)\pi_d, \qquad (11.22)$$

in which π_r, π_t, π_d are the proportions of the three components, ζ_r is the ED 50 of rotenone, and ρ_t, ρ_d are the potencies of the toxicarol and deguelin fractions relative to rotenone.

Table 11.1 shows the first steps in fitting equation (11.22). The columns $1/V(m)$ and z are taken from Example 17 and the calculations that produced Table 6.7. From these, $u = 1/z$ and W from equation (11.21) are inserted. The proportions of rotenone, toxicarol, and deguelin are reproduced from Martin's Table 12. If c_1, c_2, c_3 are written for the estimates of $1/\zeta_r$, ρ_t/ζ_r, ρ_d/ζ_r respectively, the least squares equations for c_1, c_2, c_3 are

$$\left.\begin{array}{l} c_1\Sigma W\pi_r^2 + c_2\Sigma W\pi_r\pi_t + c_3\Sigma W\pi_r\pi_d = \Sigma W\pi_r u, \\[4pt] c_1\Sigma W\pi_r\pi_t + c_2\Sigma W\pi_t^2 + c_3\Sigma W\pi_t\pi_d = \Sigma W\pi_t u, \\[4pt] c_1\Sigma W\pi_r\pi_d + c_2\Sigma W\pi_t\pi_d + c_3\Sigma W\pi_d^2 = \Sigma W\pi_d u, \end{array}\right\} \quad (11.23)$$

where Σ denotes summation over rotenone and the four roots. Numerically these equations are

$$\left.\begin{array}{l} 116\,813c_1 + 182\,873c_2 + 239\,773c_3 = 17\,308.2, \\[4pt] 182\,873c_1 + 1\,430\,897c_2 + 913\,167c_3 = 46\,005.8, \\[4pt] 239\,773c_1 + 913\,167c_2 + 851\,264c_3 = 47\,928.7. \end{array}\right\}$$

The equations need to be solved to more decimal places than estimation of potencies itself needs, in order to give sufficient

Table 11.1 Test of simple similar action between the
constituents of derris roots

Material	$1/V(m)$	z	$u = 1/z$	$W/10^6$	π_r	π_t	π_d
Rotenone	1005	13.3	0.07519	0.0335	1.0000	0.000	0.000
W.211	5509	240.7	0.00415	60.20	0.0146	0.152	0.086
W.212	3769	138.4	0.00723	13.62	0.0414	0.043	0.124
W.213	3319	187.0	0.00535	21.89	0.0346	0.024	0.082
W.214	2299	87.5	0.01143	3.320	0.0794	0.026	0.122

accuracy in a test of significance of the departure from similar
action. The solutions are

$$
\left.
\begin{aligned}
c_1 &= 0.07878050, \\
c_2 &= 0.00099265, \\
c_3 &= 0.03304828.
\end{aligned}
\right\}
\tag{11.24}
$$

The residual sum of squares of deviations is

$$
\Sigma W u^2 - c_1 \Sigma W \pi_1 u - c_2 \Sigma W \pi_2 u - c_3 \Sigma W \pi_3 u = 2998.43 - 2993.18
$$

$$
= 5.25 \tag{11.25}
$$

with 2 degrees of freedom ($\Sigma W u^2$ has 5 degrees of freedom because
it is a sum of squares of five observations *not* adjusted for the
mean, and three parameters have been estimated). This residual
ought to be a χ^2, apart from the non-independence neglected in
the present approximation. However, Example 17 showed significant evidence of heterogeneity in the relative potencies
implicit in Table 6.6. Equation (6.18) suggests a heterogeneity
factor of about 6.1, but with only 2 degrees of freedom this is
untrustworthy. What is clear is that the deviation from simple
similarity, equation (11.19), is small relative to this amount of
heterogeneity. As a test of significance, the mean squares for the
two χ^2 values could be compared in a variance ratio test (Fisher
and Yates, 1964, Table V). To accept the hypothesis of similarity
on this evidence would be unwise, in that the issue of heterogeneity is not adequately settled by the evidence of the nine
ED50's. All that can be said is that, if the imperfect fit of the
regression lines in Example 17 can be attributed primarily to

heterogeneity of groups of insects, the results are not in conflict with simple similar action.

From the values of c_1, c_2, c_3, comparison with equation (11.19) leads to 12.69 mg/l as the estimate of the ED 50 for rotenone and 0.0126, 0.4195 as the estimated relative potencies of the toxicarol and deguelin fractions. These last compare with values of 0.067 and 0.200 used by Martin in computing rotenone equivalents for the four roots. The rotenone equivalent of any mixture is

$$\pi_r + \rho_t \pi_t + \rho_d \pi_d. \qquad (11.26)$$

Table 11.2 shows this quantity as estimated for each root, and the estimated ED 50 obtained from division of the rotenone ED 50 by the rotenone equivalent. Comparison with the earlier results from Table 6.7 shows very satisfying agreement; Martin's conclusions that some indications of synergism appeared depend on his use of relative potencies estimated from a previous investigation instead of the current experiments.

Table 11.2   Comparison of median lethal doses of derris roots with predictions from simple similar action

Preparation	ED50 in Table 6.7 (mg/l)	Predicted by first analysis		Predicted ED50 (mg/l)	
		% rotenone equivalent	ED50 (mg/l)	Second analysis	Third analysis
Rotenone	13.3	100.00	12.7	12.1	12.1
W.211	240.7	5.26	241.4	241.9	239.0
W.212	138.4	9.40	135.1	136.8	134.8
W.213	187.0	6.93	183.2	184.5	182.1
W.214	87.5	13.09	97.0	96.7	95.7

Although the analysis above, together with that in Example 17, shows simply how the similarity hypothesis can be examined in logical steps, it is not entirely satisfying. A better procedure would be to base the similarity calculations directly on Table 6.6. In the first part of Example 17, seven parameters were estimated (five for the materials and two for the three days). If simple similarity applies, five parameters should suffice, those for the roots being replaced by ζ_r (or $\mu_r = \log \zeta_r$), ρ_t, ρ_d. Equation (11.9) shows how estimates m_r, R_t, R_d can be used to express the entries in Table 6.6. Thus for W.211 on the second occasion, instead of $m_1 + d_2$, the estimated log ED 50 is expressed as

$$m_r - \log\,(0.0146 + 0.0152R_t + 0.086R_d) + d_2, \quad (11.27)$$

whereas that for rotenone on the third occasion is as before

$$m_r - d_1 - d_2.$$

The weighted sum of squares of deviations of the entries in Table 6.6 from these functions of m_r, R_t, R_d, d_1, d_2 can then be minimized with respect to the five quantities. The computations for doing so would have been intolerably laborious a few years ago because of the non-linear relations with R_t, R_d, but to-day a relatively simple computer program can deal with the problem.

The minimization leads to

$$\left.\begin{aligned} m_r &= 1.0841, \\ R_t &= 0.022\,62, \\ R_d &= 0.373\,76, \end{aligned}\right\} \qquad (11.28)$$

which may be compared with the values from the previous analysis. The residual mean square from the present calculations, that is to say the minimum achieved for the sum of squares of deviations, is 18.66 with 4 degrees of freedom (five parameters estimated for the nine values in Table 6.6). The residual from fitting seven parameters in Example 17 was 12.22, and the difference, 6.44 with 2 degrees of freedom, is therefore a measure of the extent to which simple similarity fails to describe the data adequately. Indeed, this corresponds with the 5.25 obtained in equation (11.25): both represent the same feature of the data but from alternative forms of analysis, and both could be approximately tested as χ^2 with 2 degrees of freedom if there were no residual heterogeneity.

Table 11.2 also shows rotenone equivalents and predicted ED 50's based upon the estimates from the second analysis. The two versions do not differ to any important extent. Indeed, inspection of the details of the calculations makes clear that the fit of the simple similarity model would have been almost as good if the toxicarol fraction had been assumed inactive and the constraint $\rho_t = 0$ introduced. Of the two analyses, the second is preferable. It makes the similarity calculations independently of those for unconstrained ED 50 estimation, and it avoids the approximation inherent in neglecting the fact that the estimates of ED 50's obtained in Example 17 were correlated; on the other hand, it is much the more laborious analysis without a computer.

Still better theoretically would be a comprehensive analysis of the type used in the second part of Example 17. A single new maximum likelihood estimation is undertaken, starting from the original records of doses and mortalities (sets of observations for three insecticidal preparations out of five, on each of three days), so as to estimate the probit regression equations with the constraints of simple similarity replacing the lesser constraints of constant relative potencies. Now seven parameters require estimation, in a mathematical model that would be very unpleasant without a computer. Nine probit regressions require simultaneous estimation, with special relations between them: the equation for W.211 on the second occasion is

$$Y = \alpha_r + \delta_2 + \beta \log (0.0146 + 0.0152 \rho_t + 0.086 \rho_d) + \beta x, \quad (11.29)$$

which may be compared with (11.27). As compared with the parametcrization in Example 17, the parameters are now α_r (for rotenone), δ_1 and δ_2 for the first two occasions, the regression coefficients β for all the lines, the relative potencies (ρ_t, ρ_d) for the toxicarol and deguelin fractions, and the natural mortality. The estimated natural mortality obtained from the computer analysis is 0.051 ± 0.013, small but not negligible. The parameter estimates of immediate importance are

$$\left.\begin{aligned} a_r &= -0.8038, \\ R_t &= 0.02024, \\ R_d &= 0.38517, \\ b &= 5.35340. \end{aligned}\right\} \quad (11.30)$$

Note that R_t, R_d are very similar to the values in equations (11.28). A third set of predicted ED 50's in Table 11.2 has been calculated from equations (11.30), as usual taking $m_r = (5 - a_r)/b$ for rotenone.

The three methods described in this example all give effectively the same estimates and predictions corresponding to the hypothesis of simple similarity. The third is undoubtedly the most laborious, although with a modern computer it requires only a few seconds. Its great merit is its internal self-consistency, leading to a full interpretation of the data. In fact this last analysis gave a residual measure of heterogeneity

$$\chi^2_{[46]} = 71.455, \tag{11.31}$$

which may be set alongside that for the nine-parameter analysis in equation (6.21). The difference, 5.252 with 2 degrees of freedom, has a mean square not very different from the heterogeneity factor in equation (6.22). Thus one may conclude that the data do not strongly indicate a need for the extra two parameters. If the similarity model is considered acceptable, a heterogeneity factor with 44 or 46 degrees of freedom can be used in assessments of variances and standard errors of estimates, in a manner that is certainly much more satisfactory than anything possible with the poorly determined heterogeneity of the two earlier analyses.

11.4 Planning tests of simple similarity

When experiments on the joint action of two materials have to be planned without information on the occurrence of synergism or antagonism, a working hypothesis of simple similar action may reasonably be adopted. Unless *a priori* requirements govern the mixtures to be tested, they should be so composed that their probit lines will be fairly evenly spaced between those for the constituents used separately. If the second constituent is ρ times as potent as the first, its probit line will be a distance $\log \rho$ to the left of the first. On the hypothesis of similar action, if a mixture in the proportion $\pi : (1 - \pi)$ yields a probit regression line at a distance $\theta \log \rho$ to the left of that for the first constituent, where θ is some fraction between 0 and 1, then

$$\log \{\pi + \rho(1 - \pi)\} = \theta \log \rho,$$

whence
$$\pi = \frac{\rho - \rho^\theta}{\rho - 1}. \tag{11.32}$$

In Table 11.3, π is tabulated, as a percentage, for a series of values of ρ and θ. If an approximation to ρ is available from earlier experiments, the table may be used in planning toxicity tests intended for investigation of similar action.

Table 11.3   The function $\pi = 100(\rho - \rho^{\theta})/(\rho - 1)$, for use in planning experiments on mixtures

ρ \ θ	0.1	0.2	0.3	0.4	0.5	0.6	0.7	0.8	0.9
1.1	90.4	81	71	61	51	41	31	21	10
1.5	91.7	83	74	65	55	45	34	23	12
2.0	92.8	85	77	68	59	48	38	26	13
3.0	94.2	88	80	72	63	53	42	30	16
4.0	95.0	89	83	75	67	57	45	32	17
5.0	95.6	90.5	84	77	69	59	48	34	19
6.0	96.1	91.4	86	79	71	61	50	36	20
7.0	96.4	92.1	87	80	73	63	52	38	21
8.0	96.7	92.6	88	81	74	65	53	39	21
9.0	96.9	93.1	88	82	75	66	54	40	22
10.0	97.1	93.5	89	83	76	67	55	41	23
15.0	97.8	94.9	91.0	86	79	71	60	45	25
20.0	98.2	95.7	92.3	88	82	74	62	47	27
25.0	98.4	96.2	93.2	89	83	75	65	49	29
30.0	98.6	96.6	93.9	90.0	85	77	66	51	30
40.0	98.9	97.2	04.8	91.4	86	79	69	54	32
50.0	99.0	97.6	95.4	92.3	88	81	70	55	33
60.0	99.1	97.9	95.9	93.0	89	82	72	57	34
80.0	99.3	98.2	96.6	94.0	90	84	74	59	36
100.0	99.4	98.5	97.0	94.6	90.9	85	76	61	37

Ex. 35. *The use of Table 11.3 in planning experiments.* Suppose that tests are to be planned for two materials and a mixture whose constitution should be such that, if simple similar action is operating, its probit regression line will be midway between the lines for the constituents. Suppose further that, from previous experience, the second material is believed to be about four times as potent as the first. From Table 11.3, with $\rho = 4$ in the column for $\theta = 0.5$, the required proportion of the first material is found to be 67 %, and therefore a 2 : 1 mixture should be used.

Again, suppose one wants three different mixtures of two similarly acting materials to have probit lines equally spaced between those for the constituents, the second constituent being known to be roughly eleven times as potent as the first. Interpolation in Table 11.3 for $\rho = 11$ and $\theta = 0.25$, 0.5, and 0.75 gives proportions of about 92 %, 77 %, and 50 %.

In order to obtain reliable evidence of the nature of the joint action of the constituents, at least four concentrations of each preparation should be tested. The doses should be roughly inversely proportional to their relative potencies. For the trials on two materials of relative potency 11 and their three mixtures, the relative potencies of the five preparations are, by equation (11.8), approximately $1:1.8:3.3:6.0:11$ and doses inversely proportional to these numbers should be used. Hence if the ED 50 for the first material were thought to be about eight units and experience had shown the suitability of a two-fold increase in dose between successive levels, the following sets of five doses would approximate reasonably to the stated requirements:

Percentage of first material

	100	92	77	50	0
	2	1.1	0.6	0.3	0.19
Total doses	4	2.2	1.2	0.6	0.38
	8	4.4	2.4	1.2	0.75
	16	8.8	4.8	2.4	1.50
	32	17.6	9.6	4.8	3.00

An experiment so planned can form a useful start for the study of a situation in which simple similar action seems likely to be at least a first approximation. The inclusion of mixtures in several different proportions should either confirm similarity or indicate clearly the type of deviation from this model.

11.5   Simple independent action

In rather different circumstances, a contrasting type of joint action may be expected. The two constituents may produce their effects in entirely distinct ways, although the observable response is the same for both. For example, two poisons may affect different systems within the subject, yet the only record of the outcome may be death or survival of the subject.

Suppose that a dose of a mixture contains amounts of two materials that, administered separately, would have response rates P_1, P_2. If each produces its effect without regard to that of the other, a proportion P_2 of the subjects that would not respond to the first material would respond to the second, so that the total response rate is

$$P = P_1 + P_2(1 - P_1).$$

This is symmetrical in P_1, P_2, for it can be written

$$P = 1 - (1 - P_1)(1 - P_2), \qquad (11.33)$$

or in the obvious notation

$$Q = Q_1 Q_2. \qquad (11.34)$$

Equation (11.34) makes apparent the extension to a greater number of constituents: for three,

$$Q = Q_1 Q_2 Q_3. \qquad (11.35)$$

Elementary though this concept is, the statistical consequences are more complicated than those of simple similar action. Even if the constituents of a mixture give probits linearly related to log dose and the two regression lines are parallel, the regression relation for the mixture will not be a straight line. A few illustrations of curves obtained from the independent action law, for mixtures whose constituents have normal distributions of log tolerances, will show the types of probit-dose relation that may be encountered.

Ex. 36. *Independent action between constituents with parallel probit regression lines.* Suppose two materials have the probit regression lines (with logarithms to base 10)

$$\left.\begin{array}{l} Y_1 = 6 + 2x, \\ Y_2 = 4 + 2x, \end{array}\right\} \qquad (11.36)$$

the first having ten times the potency of the second. In a 1:1 mixture, the dose of the first material in a total dose z will be $\frac{1}{2}z$, and the log concentration of the first will therefore be $(x - \log 2)$. For low values of x, the response produced by the second material will be negligible relative to that produced by the first; the total effect will be almost identical with that caused by the content of the first material alone, so that the response probit will be approximately

$$Y = 6 + 2(x - \log 2)$$

$$= 5.40 + 2x.$$

With increasing total concentration, the contribution from the second material begins to make the total response appreciably greater than that for the first alone. Points on the curve are easily calculated by tabulating Y_1, Y_2 for doses of the mixture, and using equation (11.33) to give P and Y from the corresponding P_1, P_2 (Fig. 11.1).

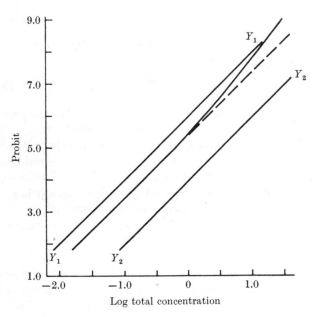

Fig. 11.1. Simple independent action for a 1:1 mixture of materials with probit regression lines $Y_1 = 6 + 2x$, $Y_2 = 4 + 2x$.
Curve: dose-response relation for mixture (Ex. 36).
Broken line: response to content of first material alone (cf. § 11.6).

The same type of curve will be found for other pairs of parallel lines corresponding to materials with relative potency 10.0. Mixtures in different proportions, or of materials with different relative potencies, will give qualitatively similar results.

Ex. 37. *Independent action between constituents with identical probit regression lines.* The contrast between simple similar and independent action is emphasized by the limiting case of materials with identical probit lines. Under simple similar action, a mixture of any number of these in any proportions would give the same line. Under independent action, for any number of constituents a curve somewhat like that of Fig. 11.1

is obtained, lying below the line for the constituents at low doses but above at high doses. If

$$Y_1 = Y_2 = 5 + 2x, \qquad (11.37)$$

the curve for a $1:1$ mixture is obtained from

$$Q = Q_1^2.$$

This is shown in Fig. 11.2, together with curves for mixtures of 4, 16, and 1000 components in equal proportions.

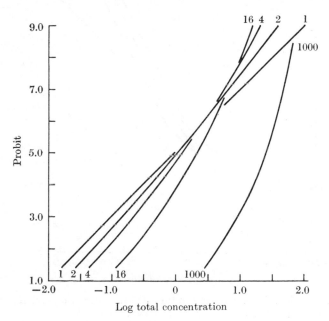

Fig. 11.2. Simple independent action for mixtures of several materials, each with probit regression line $Y = 5 + 2x$. Curves are drawn for 2, 4, 16, 1000 components in equal proportions (Ex. 37).

For a mixture of two independently acting materials with non-parallel probit lines, the situation is a little more complicated. At doses sufficiently low for the content of the material with the greater log-tolerance variance (smaller value of β) to be the more potent, a mixture produces almost the same effect as would this component alone. At higher doses, the curve for the mixture becomes steeper, eventually rising more steeply than the greater value of β.

Ex. 38. *Independent action between constituents with intersecting probit lines.* Fig. 11.3 shows the curve obtained for a 1:1 mixture of independently acting materials whose separate equations are

$$\left.\begin{array}{l} Y_1 = 5 + 2x, \\ Y_2 = 5 + 4x. \end{array}\right\} \qquad (11.38)$$

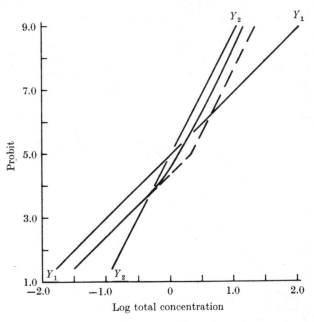

Fig. 11.3. Simple independent action for a 1:1 mixture of materials with probit regression lines $Y_1 = 5 + 2x$, $Y_2 = 5 + 4x$.
Curve: dose-response relation for mixture (Ex. 38).
Broken line: response to content of more potent material alone
(cf. § 11.6).

Presumably the curve must eventually intersect the Y_2 line, since it is rising more steeply, but usually this intersection will not occur until a very high response rate is reached.

Ex. 39. *Analysis of independent action of digitalis and quinidine.* Professor J. F. Crow, of the University of Wisconsin, kindly made available the data in Table 11.4, from an experiment on the toxicity to frogs of digitalis, quinidine, and a mixture. The five doses of the mixture were chosen, on the results of a preliminary analysis, to contain amounts of each preparation that would separately kill about 10 %, 30 %, 50 %, 70 %, 90 % respectively.

Table 11.4   Toxicity of digitalis, quinidine, and a mixture to frogs

Digitalis				Quinidine			
Dose†	n	r	p	Dose‡	n	r	p
8.0	17	1	6	10.0	26	0	0
9.4	19	1	5	15.0	24	2	8
12.5	17	7	41	22.5	23	9	39
15.6	17	13	76	33.7	24	17	71
19.5	17	16	94	50.5	26	20	77

Mixture				
Doses				
Digitalis†	Quinidine‡	n	r	p
9.8	13.8	17	4	24
11.7	21.2	19	11	58
13.2	28.6	19	15	79
15.0	38.5	19	17	89
17.9	59.0	19	18	95

† Measured in units of 10^{-3} ml of a standard laboratory solution per g body weight.
‡ Measured in units of 10^{-2} mg per g body weight.

The data for the separate poisons are well fitted by the standard probit model, although the subjects were so few that only gross deviations could be detected by χ^2 or any other tests. The manner of determination of the mixed doses was such that the proportions of the two constituents were not constant. Consequently, diagrammatic representation is unhelpful. The model expressed by equation (11.34) could be fitted to the totality of data by maximizing the likelihood. Once again the log-likelihood is as in equation (5.4); this is formed in the usual way for digitalis and quinidine separately, and for the mixture P must be defined by equation (11.33) with P_1, P_2 being the quantities appropriate to the separate constituents. Iterative calculations analogous to those of Chapter 4 could be developed, but the simple pattern of weighted regression disappears and the arithmetic becomes very tedious on desk calculators. A computer programmed to determine the maximum of an arbitrarily defined function can tackle the problem very easily, so that to-day analyses are quickly completed which a few years ago were regarded as so complicated that data rarely deserved such elaborate attention. In this instance, the computations lead to

$$Y_d = -4.474 + 8.478x \qquad (11.39)$$

for digitalis and $\qquad Y_q = -0.943 + 4.063x \qquad (11.40)$

for quinidine, x being the logarithm of dose in the units of Table 11.4. The expectations can be found by determining P corresponding to each Y, or by combining two values of P for the mixture. The combined statistic for heterogeneity of deviations is

$$\chi^2_{[11]} = 7.128, \qquad (11.41)$$

four parameters having been estimated. Evidently there is no reason to suspect inadequacy of the model, even the fact that some expected frequencies are very small having failed to give any large contributions to χ^2 (cf. Example 26). Indeed, agreement between observations and expectations is so good that nothing is gained by displaying a tabulation, and no purpose would be served by seeking to explore the more complicated models discussed below.

11.6   Correlated independent action

Plackett and Hewlett (1948) proposed a more fundamental approach to independent action, which in turn leads to an important generalization. Bliss (1939a) had earlier suggested an equation that took account of a correlation between the susceptibilities of the subjects in respect of the two stimuli. For various reasons this equation is unsatisfactory. The Plackett and Hewlett approach is to represent the distribution of log tolerances for the two components of a mixture by a bivariate frequency function. The obvious analogue of equation (2.6) is the bivariate normal function

$$dP = \frac{1}{2\pi\sigma_1\sigma_2(1-\phi^2)^{\frac{1}{2}}} \exp\left[-\frac{1}{2(1-\phi^2)}\left\{\frac{(x_1-\mu)^2}{\sigma_1^2}\right.\right.$$
$$\left.\left.-\frac{2\phi(x_1-\mu_1)(x_2-\mu_2)}{\sigma_1\sigma_2} + \frac{(x_2-\mu_2)^2}{\sigma_2^2}\right\}\right]dx_1\,dx_2, \quad (11.42)$$

in which x_1, x_2 are the two log doses, μ_1, σ_1 and μ_2, σ_2 are the means and standard deviations of the tolerance distributions, and ϕ is the correlation coefficient between the two tolerances of a subject. When a subject receives a dose of a mixture, it will respond if *either* constituent is present in excess of its tolerance.

A large positive ϕ means that most subjects with high tolerances for one material will have high tolerances for the other. Consequently, the response rate to a mixture is only a little greater than that for the amount of the stronger constituent alone. The extreme of $\phi = +1$ makes the response rate exactly equal to that for the stronger constituent, and in this is identical with the extreme version of Bliss's equation; in Fig. 11.1, the equation $Y_1 = 6 + 10x$ also represents the mixture with $\phi = +1$, but in Fig. 11.3 this value of ϕ would give segments of two straight lines ($Y = 5 + 2x$ if $x < 0$, $Y = 5 + 4x$ if $x > 0$). A negative ϕ means that subjects with high tolerances for one constituent tend to have low tolerances for the other, so that a mixture is much more potent than its content of either material alone. The extreme of $\phi = -1$, perfect negative correlation, replaces equation (11.33) by

$$P = \text{the lesser of } (P_1 + P_2) \text{ and } 1; \quad (11.43)$$

at low doses, the response to the mixture is equal to the sum of the responses appropriate to the constituents applied separately, but at some dose this sum is unity and for all higher doses the mixture will cause every subject to respond. Obviously, if constituents with $\phi = -1$ could be found, they would have immense practical importance in some fields.

Plackett and Hewlett examined in greater detail the consequences of $\phi = 0$, which corresponds exactly with equation (11.33), and confirmed the generality of statements earlier in this section (cf. Figs. 11.1, 11.3). For a 1:1 mixture in which the doses of the constituents have

$$Y_1 = \alpha_1 + \beta_1 x, \qquad (11.44)$$

$$Y_2 = \alpha_2 + \beta_2 x, \qquad (11.45)$$

at low doses the relation of probit to dose will be well approximated by (asymptotic to, as x becomes small) equation (11.44) if $\beta_1 < \beta_2$, or if $\beta_1 = \beta_2$ and $\alpha_1 > \alpha_2$, and otherwise by equation (11.45). At high doses, the relation approaches a different and steeper line, the equation† of which is

$$Y = (\alpha_1\beta_1 + \alpha_2\beta_2)/\beta + \beta x, \qquad (11.46)$$

where $\qquad\qquad \beta^2 = \beta_1^2 + \beta_2^2, \qquad (11.47)$

but the approach to linearity may be slow and appreciable curvature may persist up to very high values of Y. If β_1, β_2 are not very different, results of an experiment are likely to be well fitted by a straight line of slope $(\beta + \beta_1)/2$ (for $\beta_1 < \beta_2$), or $1.207\beta_1$ if $\beta_1 = \beta_2$. Generalizations to mixtures of three or more constituents are easy in principle.

If two preparations have parallel probit regression lines, the response rate to a mixture calculated on the basis of simple similar action will commonly be greater than that calculated on the basis of independent action, whether with or without a correlation of tolerances. Though this tends to be so at moderate levels of dose, it is not universally true (as Bliss (1939a) and Finney (1942a) implied), because at high doses the increased slope of the curve for independent action must be dominant.

11.7 Tests for independent action

To test whether observations on a mixture are adequately fitted by an hypothesis of independent action would be relatively easy if the regression lines for the two constituents were known exactly. The values of P for a mixture containing stated doses of the constituents can be obtained by integrating equation (11.42)

---

† Note particularly that equation (11.46) is expressed in theoretical terms and relates to normal equivalent deviates (§ 3.2); if equations (11.44), (11.45) are for probits, then the right-hand side of equation (11.46) must be decreased by $5(\beta_1 + \beta_2 - \beta)/\beta$.

from $-\infty$ up to the appropriate values of x_1, x_2; this can be done for a known ϕ or for a set of trial values of ϕ. Tables exist that will aid this integration, or it can be done by computer. From P can be calculated

$$\frac{(r - nP)^2}{nP(1 - P)}, \tag{11.48}$$

where n is the number of subjects observed, among which r responded. Summation over k doses of the mixture gives a χ^2 with k *degrees of freedom*, because no information from these doses has been used in estimating parameters (cf. Example 39). If ϕ is unknown, as is often the case, it may be estimated by maximizing the usual log-likelihood expression in equation (5.4) with respect to ϕ and reducing by one the degrees of freedom for χ^2. A preliminary scanning of the data is advisable, because noticing that r/n in general exceeds $(P_1 + P_2)$ or in general is less than the greater of P_1, P_2 will suggest that no value of ϕ in the permissible range $(-1 \leqslant \phi \leqslant 1)$ can be satisfactory.

Hewlett and Plackett (1950) outlined methods that are extensions of these. They suggested making allowance for the fact that the separate regression lines are themselves estimated from other data by adding a term for the variance of P to the denominator of (11.48). They also modified this denominator so as to take account of a natural response rate as in Chapter 7. They comment that estimation of α_1, α_2, β_1, β_2, ϕ from all the data simultaneously would be preferable, but that the computations 'would be formidable'. With modern computers, this estimation could be undertaken, but as far as is known it has never been tried.

Hewlett and Plackett also presented data on the mortalities of *Tribolium castaneum* exposed to mixtures of pairs of the insecticides pyrethrins, DDT, BHC. They found that, under some conditions, independent action adequately represented the data, and they fully described the computations. Their examples produced estimates of ϕ ranging from 0.7 to -0.7. In view of the little interest subsequently shown in independent action, and the absence of publications of data that appear to conform to this model, detailed account of statistical techniques seems unnecessary; Hewlett and Plackett should be consulted for guidance.

11.8 A general theory

After a preliminary short report (Hewlett and Plackett, 1952), a major advance came with publication of a new general theory (Plackett and Hewlett, 1952) which enabled the concepts of similar and independent action to be combined. At the heart was the authors' realization that independent action (in its original specification by Bliss and in their own more systematic development) involved consideration of individual thresholds, whereas current ideas on similar action involved only statements about the frequency distributions of threshold values for two drugs without reference to individual subjects. Their deeper theory removed the requirement that regression lines be parallel as a basis for similar action. The outline that follows is based upon later papers (Hewlett and Plackett, 1959, 1961) which clarified and further generalized the argument. The presentation here is intended to preserve the essentials of the theory but may be less general; the notation is totally different, conforming more closely to the pattern of this book but inevitably using some symbols with changed meanings. The theory is complicated, and the mathematics necessarily more difficult than that for most of the earlier chapters.

Hewlett and Plackett distinguish between the dose of a preparation administered to a subject and the amount that reaches the site of action, the difference being lost at some stage (stored in tissues, metabolized to a less active substance, or excreted). Suppose that administration of a dose z causes an amount w to reach the site of action, where

$$w = fz^\eta \quad (f > 0, \eta > 0). \tag{11.49}$$

This cannot be exact, because w cannot exceed z, but it may be adequate over a range of values of z. The parameter η is assumed constant for all subjects (and indeed $\eta = 1$ is an important possibility), though it need not be the same for different drugs or other preparations under test; on the other hand, f may vary from subject to subject. Suppose further that the tolerance of the subject is represented by a threshold value w^*, of w at the site of action. Define z^* by

$$w^* = f(z^*)^\eta, \tag{11.50}$$

and write

$$u = w/w^* = (z/z^*)^\eta. \tag{11.51}$$

Thus the subject will respond if and only if its dose has

$$w > w^* \Big\}$$

or
$$u > 1. \Big\}$$
(11.52)

Over the population of subjects, the individual tolerances will have a frequency distribution. If this distribution is normal in $\log w$, the probability that a dose w will cause a randomly selected subject to respond can be represented by

$$Y = A + B \log w. \qquad (11.53)$$

Here Y is the probit (in theoretical discussion, the normal equivalent deviate) of the proportion of subjects whose tolerances are less than w. Now

$$\log w = \log f + \eta \log z. \qquad (11.54)$$

If f were constant, this would permit immediate expression of (11.53) in terms of z. Variability in f establishes essentially the situation described in § 9 5· the effective dose received deviates from that measured, in a random manner dependent upon the frequency distribution of f. Suppose that $\log f$ is normally distributed with variance γ^2, and is uncorrelated with the tolerance, w^*. Then equations (11.53), (11.54), (9.27) lead to

$$Y = \alpha + \beta \log z, \qquad (11.55)$$

where
$$\beta = \frac{B\eta}{(1 + B^2\gamma^2)^{\frac{1}{2}}} \qquad (11.56)$$

and α is a function of A, B, γ, and the mean of $\log f$. Thus the probit regression of response on z remains linear, but the regression coefficient is reduced by the non-zero variance of $\log f$. Equations (11.53) and (11.55) do not represent exactly the same Y; one refers to a randomly selected subject in which an amount w of the active material reaches the site of action, whereas the other refers to administration of a dose z averaged over all subjects so that f and w are no longer fixed.

Now suppose that two preparations identified by subscripts 1, 2, separately conform to this model. The probability that a subject does not respond to a dose z_1 is

$$Q_1 = Pr(u_1 \leqslant 1), \qquad (11.57)$$

with a similar equation for the second preparation. Any equation for joint action must reduce to these forms when $u_1 = 0$ or $u_2 = 0$. If responses are produced independently, the probability that a subject does not respond to a dose containing z_1, z_2 simultaneously is the probability that neither dose exceeds its own tolerance:

$$Q = Pr(u_1 \leqslant 1, u_2 \leqslant 1),   (11.58)$$

exactly as in § 11.6 (Plackett and Hewlett, 1948) where the further supposition of a bivariate normal distribution for $\log z_1^*$, $\log z_2^*$ was expressed by equation (11.42).

Alternatively, if the preparations produce their effects on the same physiological system at the same site of action, similar action can be regarded as determination of the response by the sum of the fractions of the tolerances at that site:

$$Q = Pr(u_1 + u_2 \leqslant 1).   (11.59)$$

Equation (11.59) can be written

$$Q = Pr\left(w_1 + \frac{w_1^* w_2}{w_2^*} \leqslant w_1^*\right),   (11.60)$$

which shows the analogy with the ideas of § 11.2; w_1^*/w_2^* is analogous to the relative potency, ρ, in equation (11.5) but it need not be a constant for all subjects. Equation (11.59) implies no restrictions on the two examples of equations (11.55).

The first achievement of Hewlett and Plackett was to reconcile equations (11.58) and (11.59). They suggested

$$Q = Pr(u_1^{1/\lambda} + u_2^{1/\lambda} \leqslant 1)   \text{for}   0 < \lambda \leqslant 1,   (11.61)$$

which approaches (11.58) as a limit when $\lambda \to 0$ and reduces to (11.59) when $\lambda = 1$. An alternative is

$$Q = Pr(\lambda u_1 + u_2 \leqslant 1, u_1 + \lambda u_2 \leqslant 1)   \text{for}   0 \leqslant \lambda \leqslant 1,   (11.62)$$

which takes the required forms when $\lambda = 0$, $\lambda = 1$ respectively. Hewlett and Plackett preferred equation (11.61) to (11.62), though without stating any compelling reasons. They considered the possibility of having distinct parameters λ_1, λ_2 (in either equation), but concluded this to be an unnecessary complication on present evidence.

Equation (11.55) is the regression equation of direct concern for a single preparation, since it relates to what may be studied experimentally; equation (11.53) remains in the background, but cannot be observed. Consider now z_1 as the dose administered to a particular subject for which the tolerance is z_1^*, and write

$$Y_1 = \alpha_1 + \beta_1 \log z_1, \tag{11.63}$$

$$Y_1^* = \alpha_1 + \beta_1 \log z_1^*. \tag{11.64}$$

From equation (11.51), for the subject

$$\log u_1 = \eta_1(\log z_1 - \log z_1^*)$$

$$= \eta_1(Y_1 - Y_1^*)/\beta. \tag{11.65}$$

Equation (11.61) becomes

$$Q = Pr\left[\exp\left\{\frac{\eta_1(Y_1 - Y_1^*)}{\beta_1\lambda}\right\} + \exp\left\{\frac{\eta_2(Y_2 - Y_2^*)}{\beta_2\lambda}\right\} \leqslant 1\right]. \tag{11.66}$$

This equation represents the Hewlett–Plackett generalization of similar action when $\lambda = 1$, and independent action when $\lambda \to 0$.

Suppose the joint distribution of log tolerances of administered doses to be bivariate normal, with correlation coefficient ϕ. This is essentially equation (11.42), with $\log z$ for x, $1/\beta$ for σ, and $-\alpha/\beta$ for μ. The frequency function may be regarded as one for Y_1^*, Y_2^* for which

$$dP = \frac{1}{2\pi(1-\phi^2)^{\frac{1}{2}}} \exp\left[-\frac{y_1^2 - 2\phi y_1 y_2 + y_2^2}{2(1-\phi^2)}\right]. \tag{11.67}$$

Note that this quite rightly gives normal distributions of mean zero and variance 1 for either of Y_1^*, Y_2^* averaged over the other. If all parameters were known, the probability of response to a joint dose z_1, z_2 could be obtained by using equation (11.63) to calculate Y_1 and similarly Y_2, and then integrating the function dP in (11.67) over that region for which

$$\exp\left\{\frac{\eta_1(Y_1 - y_1)}{\beta_1\lambda}\right\} + \exp\left\{\frac{\eta_2(Y_2 - y_2)}{\beta_2\lambda}\right\} \leqslant 1. \tag{11.68}$$

No explicit general evaluation seems possible, but numerical integration procedures could be used for particular instances. Hewlett and Plackett prepared diagrams to show the probit response curves for mixtures with $\lambda = 0, \frac{1}{2}, 1$ and $\phi = -1, 0, 1$, first for two preparations with $\beta_1 = \beta_2$ and secondly for two with $2\beta_1 = \beta_2$, but based upon equation (11.62) instead of (11.61).

If $\phi = 1$, the bivariate frequency function takes the limiting form of a univariate normal distribution of mean zero, variance 1 on the line $y_1 = y_2$. The probit, Y, for a joint dose z_1, z_2 therefore satisfies

$$\exp\left\{\frac{\eta_1(Y_1 - Y)}{\beta_1\lambda}\right\} + \exp\left\{\frac{\eta_2(Y_2 - Y)}{\beta_2\lambda}\right\} = 1, \quad (11.69)$$

an equation which uniquely determines Y. Especially interesting is the case

$$\frac{\beta_1\lambda}{\eta_1} = \frac{\beta_2\lambda}{\eta_2} = \theta, \quad \text{say.} \quad (11.70)$$

This reduces equation (11.69) to

$$\exp(Y/\theta) = \exp(Y_1/\theta) + \exp(Y_2/\theta)$$

or

$$Y = \theta \log(e^{Y_1/\theta} + e^{Y_2/\theta}). \quad (11.71)$$

If also

$$\beta_1 = \beta_2 = \beta \quad \text{and} \quad \eta_1 = \eta_2 = \eta, \quad (11.72)$$

a simpler form can be derived for mixtures in the fixed proportion $\pi_1 : \pi_2$. Writing

$$z_1 = \pi_1 z, \quad z_2 = \pi_2 z \quad (11.73)$$

in equations (11.63), (11.71), it follows that

$$Y = \theta \log(e^{\alpha_1/\theta}\pi_1^{\beta/\theta} + e^{\alpha_2/\theta}\pi_2^{\beta/\theta}) + \beta \log z, \quad (11.74)$$

the equation of a line parallel to the lines for the separate preparations. If, as in equation (11.13), this is written

$$Y = \alpha_3 + \beta \log z, \quad (11.75)$$

the potencies of the mixture relative to each of the preparations, ρ_{31} and ρ_{32}, are given by

$$\log\rho_{31} = (\alpha_3 - \alpha_1)/\beta, \quad (11.76)$$

$$\log\rho_{32} = (\alpha_3 - \alpha_2)/\beta. \quad (11.77)$$

Equations (11.74), (11.75) lead to

$$\left(\frac{\pi_1}{\rho_{31}}\right)^{\beta/\theta} + \left(\frac{\pi_2}{\rho_{32}}\right)^{\beta/\theta} = 1. \quad (11.78)$$

Further simplification occurs if also

$$\lambda = \eta, \quad (11.79)$$

which implies $\beta/\theta = 1$. Since ρ_{21}, the potency of the second preparation relative to the first, is given by

$$\rho_{21} = \rho_{31}/\rho_{32}, \qquad (11.80)$$

equation (11.78) can now be written

$$\rho_{31} = \pi_1 + \rho_{21}\pi_2, \qquad (11.81)$$

effectively the same as equation (11.6). This includes the simple similar action discussed in § 11.2, but is more general in that it does not assume the stimuli at the site of action to be equal to or even proportional to ($\eta = 1$) the doses administered.

Plackett and Hewlett (1963) presented this group of special cases and also that with equation (11.79) satisfied but $\beta_1 \neq \beta_2$. The latter shows the possibility of an easily expressed form of joint action:

$$\exp\left(\frac{Y_1 - Y}{\beta_1}\right) + \exp\left(\frac{Y_2 - Y}{\beta_2}\right) = 1, \qquad (11.82)$$

in which the original lines need not be parallel. The authors illustrated iterative procedures for maximizing the likelihood for various of these cases. Computation of the iterative cycles is complicated and undoubtedly programming for a high-speed computer is to be preferred. They produced evidence of experimental results more satisfactorily fitted by equation (11.66) or one of its special cases than by the restricted forms of similar and independent action presented in §§ 11.2, 11.5. Hewlett (1963 a, b, c) reported numerous experiments in which insecticidal action appeared to be in accordance with the models described above. Sawicki, Elliott, Gower, Snarey and Thain (1962) found the obvious extensions of equations (11.71) or (11.82) to mixtures of four components very satisfactory for experiments on the insecticidal constituents of pyrethrum extract. Ashford (1958) discussed much the same model for the logistic transformation, with similar action ($\lambda = 1$), but was more concerned with computational issues than with underlying theory.

11.9 Interactive joint action

The work of Hewlett and Plackett has necessitated reassessment of the ideas of synergism and antagonism in § 11.1. Experiments such as those of Turner and Bliss (1953) are scarcely interpretable as evidence for or against synergism unless examined relative to exactly specified theory of non-synergistic action. Bliss himself (1939a) proposed two equations for representing synergistic action. These were not related to his equations for similar and independent action, and were also unsatisfactory because of discontinuities at low doses. Finney (1942a) suggested an alternative related to similar action. In place of equation (11.6), he wrote

$$Y = \alpha + \beta \log [\pi_1 + \rho \pi_2 + \kappa(\rho \pi_1 \pi_2)^{\frac{1}{2}}] + \beta x, \quad (11.83)$$

so that $\kappa = 0$ corresponds to simple similar action and positive and negative values of κ correspond to synergistic and antagonistic deviations respectively. There was no expectation that this would provide anything more than a point of reference for further investigation.

Plackett and Hewlett (1952) termed the types of joint action discussed in earlier sections *non-interactive*; neither preparation affects the amount of the other reaching the site of action or the changes induced by the other at the site of action. This includes independent and simple similar action (they would call all types of similarity described above 'simple'). In two subsequent papers (Hewlett and Plackett, 1964; Plackett and Hewlett, 1967), they went on to consider *interactive* action, in which the presence of one preparation affects either the amount or the effect of the other at the site of action. If the interaction is in the sense of making the amount of the second preparation at the site of action behave as though it were greater than when the first was absent, the first can be said to *synergize* the second; if otherwise, the first *antagonizes* the second. Synergism of the second by the first is not inconsistent with antagonism of the first by the second, since the concepts are based upon effective amounts at the two sites of action.

The authors discussed biological mechanisms that might produce interactive action. They outlined the construction and properties of mathematical models for *complex similar action* and for *dependent action*, the latter being contrasted with independent action as alternative forms of dissimilarity. Their approach becomes increasingly based upon molecular pharmacology and receptor theory, rather than being primarily statistical. This remark is intended as commendation, not condemnation, but it does mean that further discussion here would be out of place.

So much special thought has been given to this range of problems (see the papers quoted for further references) that a monograph on the pharmacological and toxicological effects of mixtures of chemical substances is desirable. Ashford and Smith (1964, 1965a) have criticized the Hewlett–Plackett theory, but appear to have been answered (Plackett and Hewlett, 1967).

Hewlett (1969) reported data that appear to conform to a model suggested earlier (Plackett and Hewlett, 1952) for inter-active action between two preparations of which one is inactive in the absence of the other. The equation suggested is

$$Y = \alpha + \beta_1 \log z_1 + \frac{\beta_2 z_2}{z_2 + \gamma}. \qquad (11.84)$$

If $z_1 = 0$, Y is infinitely small and the probability of response therefore zero irrespective of the value of z_2. On the other hand, for non-zero z_1, Y will increase as z_2 increases (assuming $\beta_2 > 0$) though the increment in Y will approach the limit β_2 as z_2 becomes large. Example 40 presents a re-analysis of Hewlett's data.

Ex. 40. *The influence of piperonyl butoxide on the toxicity of pyrethrins to* Tribolium confusum. Batches of about 50 *T. casta-neum* were sprayed with concentrations of an oil solution of pyrethrins, to some of which had been added one of three different concentrations of piperonyl butoxide. At each dose, about 150 insects in all were tested, and 200 receiving zero dose were also recorded. Previous work had shown piperonyl butoxide alone to be non-toxic, and nothing in the new experiment conflicted with this. Table 11.4 contains Hewlett's results.

Hewlett accepted the estimate $c = 0.005$ for the control mortality, since this does not markedly conflict with the evidence of any of the non-zero doses. He then guessed 0.95 for γ, after which he could use the standard iterative probit calculations for estimating α, β_1, β_2. He obtained

$$Y = -2.26 + 7.20 \log z_1 + 3.48 z_2 / (z_2 + 0.95). \quad (11.85)$$

In fact, with a computer, estimating the four parameters of equation (11.84) by maximizing the likelihood is not difficult. The likelihood function has the usual form, but P must now be expressed in terms of the current version of Y. Indeed, to estimate a fifth parameter for the natural mortality rate is no

Table 11.5   Mortality of *Tribolium castaneum* subjected to doses
or pyrethrins and piperonyl butoxide (PB)

Pyrethrins ($\%$ w/v)	PB ($\%$ v/v)	No. of beetles (n)	No. dead (r)	Mortality ($\%$)	Expected no. dead from eqn. (11.86)
1.50	0	150	138	92.0	134.8
1.06	0	149	75	50.3	86.1
0.75	0	150	32	21.3	28.9
1.10	0.25	151	129	85.4	124.0
0.78	0.25	151	65	43.0	66.7
0.55	0.25	150	19	12.7	16.8
0.80	2.5	149	143	96.0	142.2
0.57	2.5	150	112	74.7	110.5
0.40	2.5	140	37	26.4	45.1
0.65	10.0	150	141	94.0	144.5
0.46	10.0	150	117	78.0	114.4
0.32	10.0	149	56	37.6	50.9
0	0	200	1	0.5	1.1

more difficult, requiring only a further modification as in Chapter 7. The agreement of the data with the equation is assessed by

$$\chi^2 = 11.72, \text{ with 8 degrees of freedom.}$$

Certainly this is not large enough to cause concern; the expected frequencies, inserted as a final column in Table 11.5, confirm the good fit. The natural response rate is estimated to be very close to that determined empirically for control insects, but with a relatively large standard error: 0.0054 ± 0.0053. Almost no difference is made if $C = 0.005$ is assumed. The equation as finally estimated is

$$Y = -2.150 + 7.159 \log z_1 + 3.484 z_2/(z_2 + 1.179). \quad (11.86)$$

Agreement with Hewlett's equation (11.85) is close, except that γ has been appreciably increased. The computer program estimated the standard errors of the four parameters in equation (11.86) to be ± 0.326, ± 0.317, ± 0.175, and ± 0.184, so that the change in γ is the only one exceeding a small fraction of the standard error.

11.10   Compound response curves

In some circumstances, the probits of quantal responses may show a relation to dose much more complicated than any mentioned above. A great deal of evidence would be needed before adequate theories and methods of statistical analysis could be devised, but a little speculation about one particular situation may stimulate investigation.

Dimond, Horsfall, Heuberger and Stoddard (1941) reported tests of the toxicity of tetramethylthiuram disulphide to spores of *Macrosporium sarcinaeforme*. Dr Dimond kindly made available additional information on one major experiment. Five concentrations, ranging from 0.000 02 % to 0.2 %, were sprayed on to glass slides. By using spray times from 5 to 50 seconds, deposits from 0.001 05 to 35 μg/cm^2 were obtained, three being duplicated by different combinations of concentration and time. Spore suspensions were added to the spray residues, the concentration of the toxicant in the drop of suspension being proportional to the density of the dried deposit on the slide. The percentage inhibition of spore germination was measured. Fig. 11.4 (from Dimond *et al.* 1941) shows the relation between inhibition probits and log deposits. The agreement between pairs of results for equal deposits obtained by different combinations of concentration and time suggests that total deposit was the chief determinant of response. The curve (drawn by eye) indicates that tetramethylthiuram disulphide had a maximum potency at about 0.06 μg/cm^2. Increase in dose above this level decreased the inhibition of germination, a minimum occurring at about 0.3 μg/cm^2. At still higher doses, the potency again increased, rising beyond the previous maximum and achieving complete inhibition with the six largest deposits tested.

The authors commented that 'A possible explanation of this type of behaviour is that dissociation or association of the toxicant occurs'. They went on to suggest that, in dilute solution, the tetramethylthiuram disulphide molecule might dissociate into a complex with a greater toxicity. Thus the high concentrations arising from the large deposits would consist mostly of undissociated molecules; the low concentrations from the small deposits would be largely dissociated, and so could still manifest considerable toxicity despite the lesser quantities of active material. Between the extremes, a smooth trend in the proportionate dissociation would occur, so exposing the spores to the toxic action of a mixture. The law determining the extent of the dissociation might be such as to produce a curve like that of Fig. 11.4. Montgomery and Shaw (1943) found similar curves when they tested the toxicity of thiuram sulphides to spores of *Venturia inaequalis*.

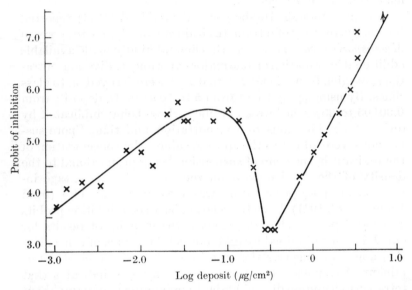

Fig. 11.4. Toxicity of tetramethylthiuram disulphide to spores of *M. sarcinaeforme*. Each point is based on a count of 100 spores. Complete inhibition was recorded for log deposits of 0.85, 1.02, 1.15, 1.32, 1.45, and 1.54.

The form of curve thus produced depends on the mode of joint action of mixtures of the dissociated and undissociated molecules and also on the law governing the dissociation. Although the percentage of dissociated molecules would increase as concentration decreases, one would not ordinarily expect the total concentration of dissociated material to increase. A curve like Fig. 11.4 would then require some antagonism between the activities of the two constituents or some correlation of susceptibilities, in order to produce the decline in response rate at intermediate degrees of dissociation.

Ex. 41. *Illustration of a compound response curve.* Purely to illustrate what could occur when the constitution of a mixture depends upon the total concentration, an artificial model based upon simple similar action has been examined. Quite apart from the need for a different formulation of dissociation, this cannot explain the Dimond results for which the two linear phases are not parallel. Yet it can produce a curve surprisingly like that in Fig. 11.4.

Suppose that when z, the total dose, is high

$$Y = \alpha + \beta \log z, \qquad (11.87)$$

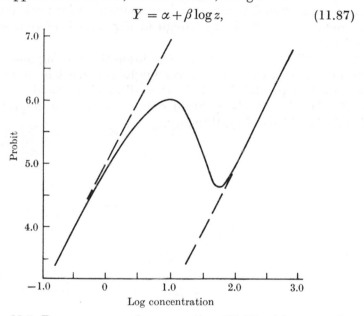

Fig. 11.5. Response curve from equation (11.89) with $\alpha = 5$, $\beta = 2$, $\rho = 100$, $c = 0.05$ (Ex. 41).

but that at low values of z the material 'dissociates' into an alternative form that has a potency ρ relative to the 'undissociated'. Suppose further that the fraction of the second form present in a total dose z is

$$\pi = 10^{-cz}, \qquad (11.88)$$

where c is a positive constant. If simple similar action obtains, equation (11.6) shows that the equation appropriate to all values of z is

$$Y = \alpha + \beta \log (1 - 10^{-cz} + \rho 10^{-cz}) + \beta \log z. \quad (11.89)$$

This expression for Y has both a maximum and a minimum if $\rho > e^2 + 1$, or approximately $\rho > 8.4$; for any smaller ρ, Y will always increase as z increases, but will of course display a marked change of curvature as it moves from the one linear phase to the other. Fig. 11.5 shows the curve obtained by starting with

$$Y = 5 + 2 \log z$$

and taking $\rho = 100$, $c = 0.05$. These values were chosen to have a rough correspondence with Fig. 11.4; they succeed in producing a recognizably similar compound curve with a wide peak and a narrow trough, though the trough in Fig. 11.4 is more accentuated.

For smaller ρ, the curves are less interesting, being much of the form that a railway engineer might construct in order to connect two parallel lines! Further discussion would be unprofitable without either more extensive data or improved chemical and toxicological theory.

References

Abbott, W. S. (1925). A method of computing the effectiveness of an insecticide. *J. Econ. Ent.*, **18**, 265–7.

Aitchison, J. and Silvey, S. D. (1951). The generalization of probit analysis to the case of multiple responses. *Biometrika*, **44**, 131–40.

Anscombe, F. J. (1948). The validity of comparative experiments. *J. R. Statist. Soc.*, **A111**, 181–211.

Anscombe, F. J. (1949). Note on a problem in probit analysis. *Ann. Appl. Biol.*, **36**, 203–5.

Armitage, P. and Allen, I. (1950). Methods of estimating the LD50 in quantal response data. *J. Hyg., Camb.*, **48**, 298–322.

Ashford, J. R. (1958). Quantal responses to mixtures of poisons under conditions of simple similar action – the analysis of uncontrolled data. *Biometrika*, **45**, 74–88.

Ashford, J. R. (1959). An approach to the analysis of data for semi-quantal responses in biological assay. *Biometrics*, **15**, 573–81.

Ashford, J. R. and Smith, C. S. (1964). General models for quantal response to the joint action of a mixture of drugs. *Biometrika*, **51**, 413–28.

Ashford, J. R. and Smith, C. S. (1965a). An alternative system for the classification of mathematical models for quantal responses to mixtures of drugs in biological assay. *Biometrics*, **21**, 181–88.

Ashford, J. R. and Smith, C. S. (1965b). An analysis of quantal response data in which the measurement of response is subject to error. *Biometrics*, **21**, 811–25.

Ashford, J. R., Smith, C. S. and Brown, S. (1960). The quantal response analysis of a series of biological assays on the same subjects. *Biometrika*, **47**, 23–32.

Bacharach, A. L., Coates, M. E. and Middleton, T. R. (1942). A biological test for vitamin P activity. *Biochem. J.*, **36**, 407–12.

Barr, M. and Nelson, J. W. (1949). An accurate and economical method for the biological assay of aconite tincture. *J. Amer. Pharm. Ass.*, **38**, 518–21.

Bartlett, M. S. (1946). A modified probit technique for small probabilities. *J. R. Statist. Soc. Suppl.*, **8**, 113–17.

Behrens, B. (1929). Zur Auswertung der Digitalisblätter im Froschversuch. *Arch. Exp. Path. Pharmak.*, **140**, 237–56.

Berkson, J. (1944). Application of the logistic function to bioassay. *J. Amer. Statist. Ass.*, **39**, 357–65.

Berkson, J. (1946). Approximation of chi-square by 'probits' and by 'logits'. *J. Amer. Statist. Ass.*, **41**, 70–4.

Berkson, J. (1949). Minimum χ^2 and maximum likelihood solution in terms of a linear transform, with particular reference to bioassay. *J. Amer. Statist. Ass.*, **44**, 273–8.

Berkson, J. (1950). Some observations with respect to the error of bio-assay. *Biometrics*, **6**, 432–34.

Berkson, J. (1951). Why I prefer logits to probits. *Biometrics*, **7**, 327–39.

Berkson, J. (1953). A statistically precise and relatively simple method of estimating the bio-assay with quantal response based on the logistic function. *J. Amer. Statist. Ass.*, **48**, 565–99.

Berkson, J. (1955a). Maximum likelihood and minimum χ^2 estimates of the logistic function. *J. Amer. Statist. Ass.*, **50**, 130–62.

Berkson, J. (1955b). Estimate of the integrated normal curve by minimum normit chi-square with particular reference to bioassay. *J. Amer. Statist. Ass.*, **50**, 529–49.

Berkson, J. (1955c). Estimation by least squares and by maximum likelihood. *Proceedings of the Third Berkeley Symposium on Mathematical Statistics and Probability*, **1**, 1–11.

Berkson, J. (1957). Tables for use in estimating the normal distribution function by normit analysis. *Biometrika*, **44**, 411–53.

Berkson, J. (1960). Nomograms for fitting the logistic function by maximum likelihood. *Biometrika*, **47**, 121–41.

Biggers, J. D. (1952). The calculation of the dose-response line in quantal assays with special reference to oestrogen assays by the Allen–Doisy technique. *J. Endoc.*, **8**, 169–78.

Black, A. (1950). Weighted probits and their use. *Biometrika*, **37**, 158–67.

Bliss, C. I. (1934a). The method of probits. *Science*, **79**, 38–9.

Bliss, C. I. (1934b). The method of probits – a correction. *Science*, **79**, 409–10.

Bliss, C. I. (1935a). The calculation of the dosage–mortality curve. *Ann. Appl. Biol.*, **22**, 134–67.

Bliss, C. I. (1935b). The comparison of dosage–mortality data. *Ann. Appl. Biol.*, **22**, 307–33.

Bliss, C. I. (1935c). Estimating the dosage–mortality curve. *J. Econ. Ent.*, **28**, 646–7.

Bliss, C. I. (1938). The determination of dosage–mortality curves from small numbers. *Quart. J. Pharm.*, **11**, 192–216.

Bliss, C. I. (1939a). The toxicity of poisons applied jointly. *Ann. Appl. Biol.*, **26**, 585–615.

Bliss, C. I. (1939b). Fly spray testing: A discussion on the theory of evaluating liquid household insecticides by the Peet–Grady method. *Soap*, **15**, no. 4, 103–11.

Bliss, C. I. (1940). The relation between exposure time, concentration and toxicity in experiments on insecticides. *Ann. Ent. Soc. Amer.*, **33**, 721–66.

Bliss, C. I. (1941). Biometry in the service of biological assay. *Industr. Engng Chem.* (Anal. ed.), **13**, 84–8.

Bliss, C. I. (1952). *The Statistics of Bioassay*. New York: Academic Press Inc.

Bliss, C. I. and Cattell, McK. (1943). Biological assay. *Ann. Rev. Physiol.*, **5**, 479–539.

Bliss, C. I. and Hanson, J. C. (1939). Quantitative estimation of the potency of digitalis by the cat method in relation to secular variation. *J. Amer. Pharm. Ass.*, **28**, 521–30.

Bliss, C. I. and Pabst, M. L. (1955). Assays for standardizing adrenal cortex extract in production. *Bull. Int. Statist. Inst.*, **34** (4), 317–38.

Bliss, C. I. and Packard, C. (1941). Stability of the standard dosage-effect curve for radiation. *Amer. J. Roentgenol.*, **46**, 400–4.

Brown, W. and Thomson, G. H. (1940). *The Essentials of Mental Measurement* (4th ed.). London: Cambridge University Press.

Brownlee, K. A., Hodges, J. L. and Rosenblatt, M. (1953). The up-and-down method with small samples. *J. Amer. Statist. Ass.*, **48**, 262–77.

Burn, J. H., Finney, D. J. and Goodwin, L. G. (1950). *Biological Standardization* (2nd ed.). London: Oxford University Press.

Burrell, R. J. W., Healy, M. J. R. and Tanner, J. M. (1961). Age at menarche of South African Bantu school girls living in the Transkei reserve. *Human Biology*, **33**, 250–61.

Busvine, J. R. (1938). The toxicity of ethylene oxide to *Calandra oryzae*, *C. granaria*, *Tribolium castaneum*, and *Cimex lectularius*. *Ann. Appl. Biol.*, **25**, 605–32.

Butler, C. G., Finney, D. J. and Schiele, P. (1943). Experiments on the poisoning of honeybees by insecticidal and fungicidal sprays used in orchards. *Ann. Appl. Biol.*, **30**, 143–50.

Campbell, F. L. (1930). A comparison of four methods for estimating the relative toxicity of stomach poison insecticides. *J. Econ. Ent.*, **23**, 357–70.

Campbell, F. L. and Filmer, R. S. (1929). A quantitative method of estimating the relative toxicity of stomach-poison insecticides. *Int. Congr. Ent.*, **4**, 523–33.

Campbell, F. L. and Moulton, F. R. (editors) (1943). *Laboratory*

Procedures in Studies of the Chemical Control of Insects. Washington, D.C.; Amer. Ass. Adv. Sci.

Chambers, E. A. and Cox, D. R. (1967). Discrimination between alternative binary response models. *Biometrika*, **54**, 573–78.

Claringbold, P. J. (1955). Matrices in quantal analysis. *Biometrics*, **11**, 481–501.

Claringbold, P. J. (1956). The within-animal bioassay with quantal responses. *J. R. Statist. Soc.*, B **18**, 133–37.

Claringbold, P. J. (1958). Multivariate quantal analysis. *J. R. Statist. Soc.*, B **20**, 398–405.

Claringbold, P. J., Biggers, J. D. and Emmens, C. W. (1953). The angular transformation in quantal analysis. *Biometrics*, **9**, 467–84.

Clark, A. J. (1933). *The Mode of Action of Drugs upon Cells.* London: Edward Arnold and Co.

Cochran, W. G. (1938). Appendix to a paper by Tattersfield, F. and Martin, J. T. *Ann. Appl. Biol.*, **25**, 426–9.

Cochran, W. G. (1942). The χ^2 correction for continuity. *Iowa St. Col. J. Sci.*, **16**, 421–36.

Cochran, W. G. and Cox, G. M. (1957). *Experimental Designs* (2nd ed.). New York: John Wiley and Sons.

Cochran, W. G. and Davis, M. (1963). Sequential experiments for estimating the median lethal dose. *Le Plan d'Expériences*, 181–94. Paris: Centre National de la Recherche Scientifique.

Cochran, W. G. and Davis, M. (1964). Stochastic approximation to the median effective dose in bioassay. *Stochastic Models in Medicine and Biology*, 281–97. Madison: University of Wisconsin Press.

Cochran, W. G. and Davis, M. (1965). The Robbins–Monro method for estimating the median lethal dose. *J. R. Statist. Soc.*, B **27**, 28–44.

Cohen, H. H. and Leppink, G. J. (1956). Selection of *Hemophilus pertussis* strains for vaccine production in the mouse protection test in a balanced design. *J. Immun.*, **77**, 299–304.

Cohen, H., van Ramshorst, J. D. and Tasman, A. (1959). Consistency in potency assay of tetanus toxoid in mice. *Bull. Wld Hlth Org.*, **20**, 1133–50.

Cohen, L. (1951). Estimation of biological dosage factors in clinical radiotherapy. *Br. J. Cancer*, **5**, 180–94.

Cornfield, J. (1964). Comparative bioassay and the rôle of parallelism. *J. Pharmacol.*, **144**, 143–49.

Cornfield, J. and Mantel, N. (1950). Some new aspects of the application of maximum likelihood to the calculation of the dosage response curve. *J. Amer. Statist. Ass.*, **45**, 181–210.

Cox, D. R. (1958). *Planning of Experiments*. New York: John Wiley and Sons.

Cramer, E. M. (1962). A comparison of three methods of fitting the normal ogive. *Psychometrika*, **27**, 183–92.

Cramer, E. M. (1964). Some comparisons of methods of fitting the dosage response curve for small samples. *J. Amer. Statist. Ass.*, **59**, 779–93.

Cramér, H. (1946). *Mathematical Methods of Statistics*. Princeton: University Press.

De Beer, E. J. (1941). A scale for graphically determining the slopes of dose-response curves. *Science*, **94**, 521–2.

De Beer, E. J. (1945). The calculation of biological assay results by graphic methods. The all-or-none type of response. *J. Pharmacol.*, **85**, 1–13.

Dimond, A. E., Horsfall, J. G., Heuberger, J. W. and Stoddard, E. M. (1941). Role of the dosage-response curve in the evaluation of fungicides. *Bull. Conn. Agric. Exp. Sta.*, no. 451.

Dixon, W. J. (1965). The up-and-down method for small samples. *J. Amer. Statist. Assoc.*, **60**, 967–78.

Dixon, W. J. and Mood, A. M. (1948). A method for obtaining and analyzing sensitivity data. *J. Amer. Statist. Ass.*, **43**, 109–26.

Dragstedt, C. A. and Lang, V. F. (1928). Respiratory stimulants in acute cocaine poisoning in rabbits. *J. Pharmacol.*, **32**, 215–22.

Fechner, G. T. (1860). *Elemente der Psychophysik*. Leipzig: Breitkopf und Härtel.

Ferguson, G. A. (1942). Item selection by the constant process. *Psychometrika*, **7**, 19–29.

Fieller, E. C. (1944). A fundamental formula in the statistics of biological assay, and some applications. *Quart. J. Pharm.*, **17**, 117–23.

Fieller, E. C. (1954). Some problems in interval estimation. *J. R. Statist. Soc.*, B**16**, 175–85.

Finney, D. J. (1942a). The analysis of toxicity tests on mixtures of poisons. *Ann. Appl. Biol.*, **29**, 82–94.

Finney, D. J. (1942b). Examples of the planning and interpretation of toxicity tests involving more than one factor. *Ann. Appl. Biol.*, **29**, 330–2.

Finney, D. J. (1943a). The statistical treatment of toxicological data relating to more than one dosage factor. *Ann. Appl. Biol.*, **30**, 71–9.

Finney, D. J. (1943b). The design and interpretation of bee experiments. *Ann. Appl. Biol.*, **30**, 197.

Finney, D. J. (1944a). The application of the probit method to toxicity test data adjusted for mortality in the controls. *Ann. Appl. Biol.*, **31**, 68–74.

Finney, D. J. (1944 b). The application of probit analysis to the results of mental tests. *Psychometrika*, **9**, 31–9.

Finney, D. J. (1946). The analysis of a factorial series of insecticide tests. *Ann. Appl. Biol.*, **33**, 160–5.

Finney, D. J. (1947 a). The principles of biological assay. *J. R. Statist. Soc. Suppl.*, **9**, 46–91.

Finney, D. J. (1947 b). The estimation from individual records of the relationship between dose and quantal response. *Biometrika*, **34**, 320–34.

Finney, D. J. (1949 a). On a method of estimating frequencies. *Biometrika*, **36**, 233–4.

Finney, D. J. (1949 b). The adjustment for a natural response rate in probit analysis. *Ann. Appl. Biol.*, **36**, 187–95.

Finney, D. J. (1949 c). The estimation of the frequency of recombinations. I. Matings of known phase. *J. Genet.*, **49**, 159–76.

Finney, D. J. (1949 d). The estimation of the parameters of tolerance distributions. *Biometrika*, **36**, 239–56.

Finney, D. J. (1950). The estimation of the mean of a normal tolerance distribution. *Sankhyā*, **10**, 341–60.

Finney, D. J. (1951). Subjective judgement in statistical analysis – An experimental study. *J. R. Statist. Soc.*, B **13**, 284–97.

Finney, D. J. (1953). The estimation of the ED50 for a logistic response curve. *Sankhyā*, **12**, 121–36.

Finney, D. J. (1960). *An Introduction to the Theory of Experimental Design*. Chicago: University of Chicago Press.

Finney, D. J. (1964). *Statistical Methods in Biological Assay* (2nd ed.). London: Charles Griffin and Co. Ltd.

Finney, D. J. (1965). The meaning of bioassay. *Biometrics*, **21**, 785–98.

Finney, D. J. and Stevens, W. L. (1948). A table for the calculation of working probits and weights in probit analysis. *Biometrika*, **35**, 191–201.

Fisher, R. A. (1922). On the mathematical foundations of theoretical statistics. *Philos. Trans.* A, **222**, 309–68.

Fisher, R. A. (1935). Appendix to Bliss, C. I.: The case of zero survivors. *Ann. Appl. Biol.*, **22**, 164–5.

Fisher, R. A. (1954). The analysis of variance with various binomial transformations. *Biometrics*, **10**, 130–39.

Fisher, R. A. (1966). *The Design of Experiments* (8th ed.). Edinburgh: Oliver and Boyd.

Fisher, R. A. (1969). *Statistical Methods for Research Workers* (14th ed.). Edinburgh: Oliver and Boyd.

Fisher, R. A. and Yates, F. (1964). *Statistical Tables for Biological*,

Agricultural and Medical Research (6th ed.). Edinburgh: Oliver and Boyd.

Freeman, H. (1963). *Introduction to Statistical Inference*. Reading, Massachusetts: Addison–Wesley Publishing Company, Inc.

Fryer, J. C. F., Stenton, R., Tattersfield, F. and Roach, W. A. (1923). A quantitative study of the insecticidal properties of *Derris elliptica* (tuba root). *Ann. Appl. Biol.*, **10**, 18–34.

Gaddum, J. H. (1933). Reports on biological standards. III. Methods of biological assay depending on a quantal response. *Spec. Rep. Ser. Med. Res. Coun., Lond.*, no. 183.

Galton, F. (1879). The geometric mean in vital and social statistics. *Proc. Roy. Soc.*, **29**, 365–7.

Garwood, F. (1941). The application of maximum likelihood to dosage–mortality curves. *Biometrika*, **32**, 46–58.

Gilliatt, R. W. (1947). Vaso-constriction in the finger following deep inspiration. *J. Physiol.*, **107**, 76–88.

Greenberg, L. (1953). International standardization of diphtheria toxoids. *Bull. Wld Hlth Org.*, **9**, 829–36.

Greenwood, M. and Yule, G. U. (1914). The statistics of anti-typhoid and anti-cholera inoculations and the interpretation of such statistics in general. *Proc. R. Soc. Med.*, **8**, 113–94.

Grewal, R. S. (1952). A method for testing analgesics in mice. *British J. Pharmac. Chemother.*, **7**, 433–37.

Gurland, J., Lee, I. and Dahm, P. A. (1960). Polychotomous quantal response in biological assay. *Biometrics*, **16**, 382–97.

Haag, J. (1926). Vérification expérimentale de la loi de Gauss en artillerie. *Mémorial de l'Artillerie française*, **5**, 449–76.

Harvard Computation Laboratory (1955). *Tables of the Cumulative Binomial Probability Distribution*. Cambridge, Massachusetts: Harvard University Press.

Hatcher, R. A. and Brody, J. G. (1910). The biological standardization of drugs. *Amer. J. Pharm.*, **82**, 360–72.

Hazen, A. (1914). Storage to be provided in impounding reservoirs for municipal water supply. *Trans. Amer. Soc. Civ. Engrs*, **77**, 1539–669.

Healy, M. J. R. (1950). The planning of probit assays. *Biometrics*, **6**, 424–31.

Healy, M. J. R. (1952). A table of Abbott's correction for natural mortality. *Ann. Appl. Biol.*, **39**, 211–12.

Hemmingsen, A. M. (1933). The accuracy of insulin assay on white mice. *Quart. J. Pharm.*, **6**, 39–80 and 187–217.

Henry, P. (1894). *Probabilités du Tir*. (Cours professé a l'École d'application d'Artillerie et du Génie, Fontainebleau.) Reprinted in 1926 in *Mémorial de l'Artillerie française*, **5**, 297–447.

Hewlett, P. S. (1963a). Toxicological studies on a beetle, *Alphitobius laevigatus* (F.). III. The joint action of doses of each of four toxicants put on two parts of the body. *Ann. Appl. Biol.*, **52**, 305–11.

Hewlett, P. S. (1963b). Toxicological studies on a beetle, *Alphitobius laevigatus* (F.). IV. Joint action experiments with two alkyl dinitrophenols. *Ann. Appl. Biol.*, **52**, 313–19.

Hewlett, P. S. (1963c). Toxicological studies on a beetle, *Alphitobius laevigatus* (F.). V. The joint actions of some pairs of like and unlike toxicants. *Ann. Appl. Biol.*, **52**, 351–9.

Hewlett, P. S. (1969). The toxicity to *Tribolium castaneum* (Herbst) (Coleoptera, Tenebrionidae) of mixtures of pyrethrins and piperonyl butoxide: Fitting a mathematical model. *J. Stor. Prod. Res.*, **5**, 1–9.

Hewlett, P. S. and Plackett, R. L. (1950). Statistical aspects of the independent joint action of poisons, particularly insecticides. II. Examination of data for agreement with the hypothesis. *Ann. Appl. Biol.*, **37**, 527–52.

Hewlett, P. S. and Plackett, R. L. (1952). Similar joint action of insecticides. *Nature, Lond.*, **169**, 198–99.

Hewlett, P. S. and Plackett, R. L. (1959). A unified theory for quantal responses to mixtures of drugs: non-interactive action. *Biometrics*, **15**, 591–610.

Hewlett, P. S. and Plackett, R. L. (1961). Models for quantal responses to mixtures of two drugs. *Symposium on Quantitative Methods in Pharmacology, Leyden, 1960*, 328–36. Amsterdam: North-Holland Publishing Company.

Hewlett, P. S. and Plackett, R. L. (1964). A unified theory for quantal responses to mixtures of drugs: competitive action. *Biometrics*, **20**, 556–75.

Hsi, B. P. (1969). The multiple sample up-and-down method in bioassay. *J. Amer. Statist. Ass.*, **64**, 147–62.

Irwin, J. O. (1937). Statistical method applied to biological assay. *J. R. Statist. Soc. Suppl.*, **4**, 1–60.

Irwin, J. O. and Cheeseman, E. A. (1939a). On an approximate method of determining the median effective dose and its error in the case of a quantal response. *J. Hyg., Camb.*, **39**, 574–80.

Irwin, J. O. and Cheeseman, E. A. (1939b). On the maximum likelihood method of determining dosage-response curves and approximations to the median effective dose, in cases of a quantal response. *J. R. Statist. Soc. Suppl.*, **6**, 174–85.

Jerne, N. K. and Wood, E. C. (1949). The validity and meaning of the results of biological assays. *Biometrics*, **5**, 273–99.

Kapteyn, J. C. (1903). *Skew Frequency Curves in Biology and Statistics*. Groningen: P. Noordhoff.

Kapteyn, J. C. and Van Uven, M. J. (1916). *Skew Frequency Curves in Biology and Statistics*. Groningen: Hoitsema Brothers.

Kärber, G. (1931). Beitrag zur kollektiven Behandlung pharmakologischer Reihenversuche. *Arch. Exp. Path. Pharmak.*, **162**, 480–7.

Kendall, M. G. and Stuart, A. (1967). *The Advanced Theory of Statistics*, vol. 2 (2nd ed.). London: Charles Griffin and Co. Ltd.

Knudsen, L. F. and Curtis, J. M. (1947). The use of the angular transformation in biological assays. *J. Amer. Statist. Ass.*, **42**, 282–96.

Lawley, D. N. (1943). On problems connected with item selection and test construction. *Proc. Roy. Soc. Edinb.* A, **61**, 273–87.

Lawley, D. N. (1944). The factorial analysis of multiple item tests. *Proc. Roy. Soc. Edinb.* A, **62**, 74–82.

Leslie, P. H., Perry, J. S. and Watson, J. S. (1945). The determination of the median body-weight at which female rats reach maturity. *Proc. Zool. Soc.*, **115**, 473–88.

Litchfield, J. T. and Fertig, J. W. (1941). On a graphic solution of the dosage-effect curve. *Johns Hopk. Hosp. Bull.*, **69**, 276–86.

Litchfield, J. T. and Wilcoxon, F. (1949). A simplified method of evaluating dose-effect experiments. *J. Pharmacol.*, **95**, 99–113.

McLeod, W. S. (1944). Further refinements of a technique for testing contact insecticides. *Canad. J. Res.* D, **22**, 87–104.

Martin, J. T. (1940). The problem of the evaluation of rotenone-containing plants. v. The relative toxicities of different species of derris. *Ann. Appl. Biol.*, **27**, 274–94.

Martin, J. T. (1942). The problem of the evaluation of rotenone-containing plants. vi. The toxicity of *l*-elliptone and of poisons applied jointly, with further observations on the rotenone equivalent method of assessing the toxicity of derris root. *Ann. Appl. Biol.*, **29**, 69–81.

Martin, J. T. (1943). The preparation of a standard pyrethrum extract in heavy mineral oil, with observations on the relative toxicities of the pyrethrins in oil and aqueous media. *Ann. Appl. Biol.*, **30**, 293–300.

Martin, L. (1951). Probits normaux classiques et probits relatifs à la répartition triangulaire. *Bulletin de l'Institut Agronomique et des Stations de Recherches de Gembloux*, **19**, 364–86.

Milicer, H. (1968). Age at menarche of girls in Wrocław, Poland, in 1966. *Human Biology*, **40**, 249–59.

Milicer, H. and Szczotka, F. (1966). Age at menarche in Warsaw girls in 1965. *Human Biology*, **38**, 199–203.

Miller, L. C., Bliss, C. I. and Braun, H. A. (1939). The assay of digitalis. I. Criteria for evaluating various methods using frogs. *J. Amer. Pharm. Ass.*, **28**, 644–57.

Miller, L. C. and Tainter, M. L. (1944). Estimation of the ED50 and its error by means of logarithmic-probit graph paper. *Proc. Soc. Exp. Biol.*, *N.Y.*, **57**, 261–4.

Mollison, P. L. and Armitage, P. (1953). Further analysis of controlled trials of treatment of haemolytic disease of the newborn. *J. Obstet. Gynaec. Br. Commonw.*, **60**, 605–20.

Montgomery, H. B. S. and Shaw, H. (1943). Behaviour of thiuram sulphides, etc., in spore germination tests. *Nature, Lond.*, **151**, 333.

Mood, A. M. and Graybill, F. A. (1963). *Introduction to the Theory of Statistics* (2nd ed.). New York: McGraw-Hill Book Company, Inc.

Moore, W. and Bliss, C. I. (1942). A method for determining insecticidal effectiveness using *Aphis rumicis* and certain organic compounds. *J. Econ. Ent.*, **35**, 544–53,

Müller, G. E. (1879). Ueber die Maassbestimmungen des Ortsinnes der Haut mittels der Methode der richtigen und falschen Fälle. *Pflüg. Arch. ges. Physiol.*, **19**, 191–235.

Murray, C. A. (1937). A statistical analysis of fly mortality data. *Soap*, **13**, no. 8, 89–105.

Nelder, J. A. and Mead, R. (1965). A simplex method for function minimization. *Comput. J.*, **7**, 308–13.

O'Kane, W. C., Westgate, W. A., Glover, L. C. and Lowry, P. R. (1930). Studies of contact insecticides. I. *Tech. Bull. N. H. Agric. Exp. Sta.*, no. 39.

O'Kane, W. C., Westgate, W. A. and Glover, L. C. (1934). Studies of contact insecticides. VII. *Tech. Bull. N. H. Agric. Exp. Sta.*, no. 58.

Ostwald, W. and Dernoschek, A. (1910). Über die Beziehungen zwischen Adsorption und Giftigkeit. *Kolloidzschr.*, **6**, 297–307.

Parker-Rhodes, A. F. (1941). Studies on the mechanism of fungicidal action. I. Preliminary investigation of nickel, copper, silver, and mercury. *Ann. Appl. Biol.*, **28**, 389–405.

Parker-Rhodes, A. F. (1942a). Studies on the mechanism of fungicidal action. II. Elements of the theory of variability. *Ann. Appl. Biol.*, **29**, 126–35.

Parker-Rhodes, A. F. (1942b). Studies on the mechanism of fungicidal action. III. Sulphur. *Ann. Appl. Biol.*, **29**, 136–43.

Parker-Rhodes, A. F. (1942c). Studies on the mechanism of fungicidal action. IV. Mercury. *Ann. Appl. Biol.*, **29**, 404–11.

Parker-Rhodes, A. F. (1943a). Studies in the mechanism of fungicidal action. V. Non-metallic and sodium dithiocarbamic acid derivatives. *Ann. Appl. Biol.*, **30**, 170–8.

Parker-Rhodes, A. F. (1943b). Studies in the mechanism of fungicidal action. VI. Water. *Ann. Appl. Biol.*, 30, 372–9.

Parsons, S. D., Hunter, G. L. and Rayner, A. A. (1967). Use of probit analysis in a study of the effect of the ram on time of ovulation in the ewe. *J. Reprod. Fert.*, 14, 71–80.

Patwary, K. M. and Haley, K. D. C. (1967). Analysis of quantal response assays with dosage errors. *Biometrics*, 23, 747–60.

Plackett, R. L. and Hewlett, P. S. (1948). Statistical aspects of the independent joint action of poisons, particularly insecticides. I. The toxicity of a mixture of poisons. *Ann. Appl. Biol.*, 35, 347–58.

Plackett, R. L. and Hewlett, P. S. (1952). Quantal responses to mixtures of poisons. *J. R. Statist. Soc.*, B14, 141–63.

Plackett, R. L. and Hewlett, P. S. (1963). A unified theory for quantal responses to mixtures of drugs: the fitting to data of certain models for two non-interactive drugs with complete positive correlation of tolerances. *Biometrics*, 19, 517–31.

Plackett, R. L. and Hewlett, P. S. (1967). A comparison of two approaches to the construction of models for quantal responses to mixtures of drugs. *Biometrics*, 23, 27–44.

Potter, C. and Gillham, E. M. (1946). Effect of atmospheric environment, before and after treatment, on the toxicity to insects of contact poisons. *Ann. Appl. Biol.*, 33, 142–59.

Rao, C. R. (1961). Asymptotic efficiency and limiting information. *Proceedings of the Fourth Berkeley Symposium on Mathematical Statistics and Probability*, 1, 531–46.

Rao, C. R. (1962). Efficient estimates and optimum inference procedures in large samples. *J. R. Statist. Soc.*, B24, 46–72.

Reed, L. J. and Muench, H. (1938). A simple method of estimating fifty per cent endpoints. *Amer. J. Hyg.*, 27, 493–7.

Richards, F. J. (1941). The diagrammatic representation of the results of physiological and other experiments designed factorially. *Ann. Bot., Lond.*, N.S. 5, 249–62.

Robbins, H. and Monro, S. (1951). A stochastic approximation method. *Ann. Math. Statist.*, 29, 400–7.

Sawicki, R. M., Elliott, M., Gower, J. C., Snarey, M. and Thain, E. M. (1962). Insecticidal activity of pyrethrum extract and its four insecticidal constituents against house flies. I. Preparation and relative toxicity of the pure constituents; statistical analysis of the action of mixtures of these components. *J. Sci. Fd Agric.*, 13, 172–85.

Shackell, L. F. (1923). Studies in protoplasm poisoning. I. Phenols. *J. Gen. Physiol.*, 5, 783–805.

Snedecor, G. W. and Cochran, W. G. (1968). *Statistical Methods* (6th ed.). Ames, Iowa: Iowa State University Press.

Spearman, C. (1908). The method of 'right and wrong cases' ('constant stimuli') without Gauss's formulae. *Brit. J. Psychol.*, **2**, 227–42.

Stevens, W. L. (1948). Control by gauging. *J. R. Statist. Soc.*, B **10**, 54–108.

Strand, A. L. (1930). Measuring the toxicity of insect fumigants. *Industr. Engng Chem.* (Anal. ed.), **2**, 4–8.

Swan, A. V. (1969*a*). Computing maximum-likelihood estimates for parameters of the normal distribution from grouped and censored data. *Appl. Statist.*, **18**, 65–9.

Swan, A. V. (1969*b*). Maximum likelihood estimation from grouped and censored normal data. *Appl. Statist.*, **18**, 110–14.

Swan, A. V. (1969*c*). The reciprocal of Mills's ratio. *Appl. Statist.*, **18**, 115–16.

Tattersfield, F., Gimingham, C. T. and Morris, H. M. (1925). Studies on contact insecticides. I. Introduction and methods. *Ann. Appl. Biol.*, **12**, 60–5.

Tattersfield, F. and Martin, J. T. (1935). The problem of the evaluation of rotenone-containing plants. I. *Derris elliptica* and *Derris malaccensis*. *Ann. Appl. Biol.*, **22**, 578–605.

Tattersfield, F. and Martin J. T. (1938). The problem of the evaluation of rotenone-containing plants. IV. The toxicity to *Aphis rumicis* of certain products isolated from derris root. *Ann. Appl. Biol.*, **25**, 411–29.

Tattersfield, F. and Morris, H. M. (1924). An apparatus for testing the toxic values of contact insecticides under controlled conditions. *Bull. Ent. Res.*, **14**, 223–33.

Tattersfield, F. and Potter, C. (1943). Biological methods of determining the insecticidal values of pyrethrum preparations (particularly extracts in heavy oil). *Ann. Appl. Biol.*, **30**, 259–79.

Thompson, W. R. (1947). Use of moving averages and interpolation to estimate median-effective dose. I. Fundamental formulae, estimation and error, and relation to other methods. *Bacteriol. Rev.*, **11**, 115–45.

Thomson, G. H. (1914). The accuracy of the $\Phi(\gamma)$ process. *Brit. J. Psychol.*, **7**, 44–55.

Thomson, G. H. (1919*a*). The criterion of goodness of fit of psychophysical curves. *Biometrika*, **12**, 216–30.

Thomson, G. H. (1919*b*). A direct deduction of the constant process used in the method of right and wrong cases. *Psychol. Rev.*, **26**, 454–64.

Thomson, G. H. (1947). Review of 'Probit Analysis'. *Brit. J. Psychol. (Statistical Section)*, **1**, 71–2.

Tocher, K. D. (1949). A note on the analysis of grouped probit data. *Biometrika*, **36**, 9–17.

Trevan, J. W. (1927). The error of determination of toxicity. *Proc. Roy. Soc.* B, **101**, 483–514.

Tsutakawa, R. K. (1967). The random walk design in bio-assay. *J. Amer. Statist. Ass.*, **62**, 842–56.

Tukey, J. W. (1949). *American Statistician*, **3**, no. 4, p. 12.

Turner, N. and Bliss, C. I. (1953). Tests of synergism between nicotine and the pyrethrins. *Ann. of Appl. Biol.*, **40**, 79–90.

Urban, F. M. (1909). Die psychophysischen Massmethoden als Grundlagen empirischer Messungen. *Arch. Ges. Psychol.*, **15**, 261–355.

Urban, F. M. (1910). Die psychophysischen Massmethoden als Grundlagen empirischer Messungen (continued). *Arch. Ges. Psychol.*, **16**, 168–227.

van der Waerden, B. L. (1940a). Biologische Konzentrations auswertung. *Berichte über die Verhandlungen der Sächsischen Akademie der Wissenschaften zu Leipzig*, **92**, 41–4.

van der Waerden, B. L. (1940b). Wirksamkeits- und Konzentrationsbestimmung durch Tierversuche. *Archiv für Experimentelle Pathologie und Pharmakologie*, **195**, 389–412.

Van Soestbergen, A. A. (1956). *Over de Reactie van Sabin en Feldman*. Rotterdam.

Wadley, F. M. (1949). Dosage-mortality correlation with number treated estimated from a parallel sample. *Ann. Appl. Biol.*, **36**, 196–202.

Wetherill, G. B. (1963). Sequential estimation of quantal response curves. *J. R. Statist. Soc.*, B **25**, 1–48.

Whipple, G. C. (1916). The element of chance in sanitation. *J. Franklin Inst.*, **182**, 37–59 and 205–27.

White, R. F. and Graca, J. G. (1958). Multinomially grouped response times for the quantal response bio-assay. *Biometrics*, **14**, 462–88.

Whitlock, J. H. and Bliss, C. I. (1943). A bio-assay technique for antihelmintics. *J. Parasit.*, **29**, 48–58.

Wilson, D. C. and Sutherland, I. (1949). The age of the menarche. *British Medical Journal*, **2**, 130.

Wilson, E. B. and Worcester, J. (1943a). The determination of LD50 and its sampling error in bio-assay. *Proc. Nat. Acad. Sci.*, *Wash.*, **29**, 79–85.

Wilson, E. B. and Worcester, J. (1943b). The determination of LD50 and its sampling error in bio-assay. II. *Proc. Nat. Acad. Sci.*, *Wash.*, **29**, 114–20.

Wilson, E. B. and Worcester, J. (1943c). Bio-assay on a general curve. *Proc. Nat. Acad. Sci.*, *Wash.*, **29**, 150–4.

Wilson, E. B. and Worcester, J. (1943d). The determination of LD50 and its sampling error in bio-sasay. III. *Proc. Nat. Acad. Sci.*, *Wash.*, **29**, 257–62.

Winder, C. V. (1947). Misuse of 'deduced ratios' in the estimation of median effective doses. *Nature, Lond.*, **159**, 883.

Worcester, J. and Wilson, E. B. (1943). A table determining LD50 or the 50 % end-point. *Proc. Nat. Acad. Sci.*, *Wash.*, **29**, 207–12.

Wright, S. (1926). A frequency curve adapted to variation in percentage occurrence. *J. Amer. Statist. Ass.*, **21**, 162–78.

Yugoslav Typhoid Commission (1962). A controlled field trial of the effectiveness of phenol and alcohol typhoid vaccines. *Bull. Wld Hlth Org.*, **26**, 357–69.

Table 1 Transformation of percentages to probits

%	0.0	0.1	0.2	0.3	0.4	0.5	0.6	0.7	0.8	0.9	1	2	3	4	5
0	—	1.9098	2.1218	2.2522	2.3479	2.4242	2.4879	2.5427	2.5911	2.6344					
1	2.6737	2.7096	2.7429	2.7738	2.8027	2.8299	2.8556	2.8799	2.9031	2.9251	For more detail see				
2	2.9463	2.9665	2.9859	3.0046	3.0226	3.0400	3.0569	3.0732	3.0890	3.1043	values for 95—100				
3	3.1192	3.1337	3.1478	3.1616	3.1750	3.1881	3.2009	3.2134	3.2256	3.2376					
4	3.2493	3.2608	3.2721	3.2831	3.2940	3.3046	3.3151	3.3253	3.3354	3.3454					
5	3.3551	3.3648	3.3742	3.3836	3.3928	3.4018	3.4107	3.4195	3.4282	3.4368	9	18	27	36	45
6	3.4452	3.4536	3.4618	3.4699	3.4780	3.4859	3.4937	3.5015	3.5091	3.5167	8	16	24	32	40
7	3.5242	3.5316	3.5389	3.5462	3.5534	3.5605	3.5675	3.5745	3.5813	3.5882	7	14	21	28	36
8	3.5949	3.6016	3.6083	3.6148	3.6213	3.6278	3.6342	3.6405	3.6468	3.6531	6	13	19	26	32
9	3.6592	3.6654	3.6715	3.6775	3.6835	3.6894	3.6953	3.7012	3.7070	3.7127	6	12	18	24	30
10	3.7184	3.7241	3.7298	3.7354	3.7409	3.7464	3.7519	3.7574	3.7628	3.7681	6	11	17	22	28
11	3.7735	3.7788	3.7840	3.7893	3.7945	3.7996	3.8048	3.8099	3.8150	3.8200	5	10	16	21	26
12	3.8250	3.8300	3.8350	3.8399	3.8448	3.8497	3.8545	3.8593	3.8641	3.8689	5	10	15	20	24
13	3.8736	3.8783	3.8830	3.8877	3.8923	3.8969	3.9015	3.9061	3.9107	3.9152	5	9	14	18	23
14	3.9197	3.9242	3.9286	3.9331	3.9375	3.9419	3.9463	3.9506	3.9550	3.9593	4	9	13	18	22
15	3.9636	3.9678	3.9721	3.9763	3.9806	3.9848	3.9890	3.9931	3.9973	4.0014	4	8	13	17	21
16	4.0055	4.0096	4.0137	4.0178	4.0218	4.0259	4.0299	4.0339	4.0379	4.0419	4	8	12	16	20
17	4.0458	4.0498	4.0537	4.0576	4.0615	4.0654	4.0693	4.0731	4.0770	4.0808	4	8	12	16	19
18	4.0846	4.0884	4.0922	4.0960	4.0998	4.1035	4.1073	4.1110	4.1147	4.1184	4	8	11	15	19
19	4.1221	4.1258	4.1295	4.1331	4.1367	4.1404	4.1440	4.1476	4.1512	4.1548	4	7	11	15	18
20	4.1584	4.1619	4.1655	4.1690	4.1726	4.1761	4.1796	4.1831	4.1866	4.1901	4	7	11	14	18
21	4.1936	4.1970	4.2005	4.2039	4.2074	4.2108	4.2142	4.2176	4.2210	4.2244	3	7	10	14	17
22	4.2278	4.2312	4.2345	4.2379	4.2412	4.2446	4.2479	4.2512	4.2546	4.2579	3	7	10	13	17
23	4.2612	4.2644	4.2677	4.2710	4.2743	4.2775	4.2808	4.2840	4.2872	4.2905	3	7	10	13	16
24	4.2937	4.2969	4.3001	4.3033	4.3065	4.3097	4.3129	4.3160	4.3192	4.3224	3	6	10	13	16

Table I (*cont.*)

%	0.0	0.1	0.2	0.3	0.4	0.5	0.6	0.7	0.8	0.9	1	2	3	4	5
25	4.3255	4.3287	4.3318	4.3349	4.3380	4.3412	4.3443	4.3474	4.3505	4.3536	3	6	9	12	16
26	4.3567	4.3597	4.3628	4.3659	4.3689	4.3720	4.3750	4.3781	4.3811	4.3842	3	6	9	12	15
27	4.3872	4.3902	4.3932	4.3962	4.3992	4.4022	4.4052	4.4082	4.4112	4.4142	3	6	9	12	15
28	4.4172	4.4201	4.4231	4.4260	4.4290	4.4319	4.4349	4.4378	4.4408	4.4437	3	6	9	12	15
29	4.4466	4.4495	4.4524	4.4554	4.4583	4.4612	4.4641	4.4670	4.4698	4.4727	3	6	9	12	14
30	4.4756	4.4785	4.4813	4.4842	4.4871	4.4899	4.4928	4.4956	4.4985	4.5013	3	6	9	11	14
31	4.5041	4.5070	4.5098	4.5126	4.5155	4.5183	4.5211	4.5239	4.5267	4.5295	3	6	8	11	14
32	4.5323	4.5351	4.5379	4.5407	4.5435	4.5462	4.5490	4.5518	4.5546	4.5573	3	6	8	11	14
33	4.5601	4.5628	4.5656	4.5684	4.5711	4.5739	4.5766	4.5793	4.5821	4.5848	3	5	8	11	14
34	4.5875	4.5903	4.5930	4.5957	4.5984	4.6011	4.6039	4.6066	4.6093	4.6120	3	5	8	11	14
35	4.6147	4.6174	4.6201	4.6228	4.6255	4.6281	4.6308	4.6335	4.6362	4.6389	3	5	8	11	13
36	4.6415	4.6442	4.6469	4.6495	4.6522	4.6549	4.6575	4.6602	4.6628	4.6655	3	5	8	11	13
37	4.6681	4.6708	4.6734	4.6761	4.6787	4.6814	4.6840	4.6866	4.6893	4.6919	3	5	8	11	13
38	4.6945	4.6971	4.6998	4.7024	4.7050	4.7076	4.7102	4.7129	4.7155	4.7181	3	5	8	10	13
39	4.7207	4.7233	4.7259	4.7285	4.7311	4.7337	4.7363	4.7389	4.7415	4.7441	3	5	8	10	13
40	4.7467	4.7492	4.7518	4.7544	4.7570	4.7596	4.7622	4.7647	4.7673	4.7699	3	5	8	10	13
41	4.7725	4.7750	4.7776	4.7802	4.7827	4.7853	4.7879	4.7904	4.7930	4.7955	3	5	8	10	13
42	4.7981	4.8007	4.8032	4.8058	4.8083	4.8109	4.8134	4.8160	4.8185	4.8211	3	5	8	10	13
43	4.8236	4.8262	4.8287	4.8313	4.8338	4.8363	4.8389	4.8414	4.8440	4.8465	3	5	8	10	13
44	4.8490	4.8516	4.8541	4.8566	4.8592	4.8617	4.8642	4.8668	4.8693	4.8718	3	5	8	10	13
45	4.8743	4.8769	4.8794	4.8819	4.8844	4.8870	4.8895	4.8920	4.8945	4.8970	3	5	8	10	13
46	4.8996	4.9021	4.9046	4.9071	4.9096	4.9122	4.9147	4.9172	4.9197	4.9222	3	5	8	10	13
47	4.9247	4.9272	4.9298	4.9323	4.9348	4.9373	4.9398	4.9423	4.9448	4.9473	3	5	8	10	13
48	4.9498	4.9524	4.9549	4.9574	4.9599	4.9624	4.9649	4.9674	4.9699	4.9724	3	5	8	10	13
49	4.9749	4.9774	4.9799	4.9825	4.9850	4.9875	4.9900	4.9925	4.9950	4.9975	3	5	8	10	13

Proportional parts (top of page, by row-group):

Group 50–54	Group 55–59	Group 60–64	Group 65–69	Group 70–74
13 13 13 13 13	13 13 13 13 13	13 13 13 13 13	14 14 14 14 14	14 15 15 15 16
10 10 10 10 10	10 10 10 10 10	10 10 11 11 11	11 11 11 11 11	12 12 12 12 12
8 8 8 8 8	8 8 8 8 8	8 8 8 8 8	8 8 8 8 9	9 9 9 9 9
5 5 5 5 5	5 5 5 5 5	5 5 5 5 5	5 5 6 6 6	6 6 6 6 6
3 3 3 3 3	3 3 3 3 3	3 3 3 3 3	3 3 3 3 3	3 3 3 3 3

	0	1	2	3	4	5	6	7	8	9
50	5.0000	5.0025	5.0050	5.0075	5.0100	5.0125	5.0150	5.0175	5.0201	5.0226
51	5.0251	5.0276	5.0301	5.0326	5.0351	5.0376	5.0401	5.0426	5.0451	5.0476
52	5.0502	5.0527	5.0552	5.0577	5.0602	5.0627	5.0652	5.0677	5.0702	5.0728
53	5.0753	5.0778	5.0803	5.0828	5.0853	5.0878	5.0904	5.0929	5.0954	5.0979
54	5.1004	5.1030	5.1055	5.1080	5.1105	5.1130	5.1156	5.1181	5.1206	5.1231
55	5.1257	5.1282	5.1307	5.1332	5.1358	5.1383	5.1408	5.1434	5.1459	5.1484
56	5.1510	5.1535	5.1560	5.1586	5.1611	5.1637	5.1662	5.1687	5.1713	5.1738
57	5.1764	5.1789	5.1815	5.1840	5.1866	5.1891	5.1917	5.1942	5.1968	5.1993
58	5.2019	5.2045	5.2070	5.2096	5.2121	5.2147	5.2173	5.2198	5.2224	5.2250
59	5.2275	5.2301	5.2327	5.2353	5.2378	5.2404	5.2430	5.2456	5.2482	5.2508
60	5.2533	5.2559	5.2585	5.2611	5.2637	5.2663	5.2689	5.2715	5.2741	5.2767
61	5.2793	5.2819	5.2845	5.2871	5.2898	5.2924	5.2950	5.2976	5.3002	5.3029
62	5.3055	5.3081	5.3107	5.3134	5.3160	5.3186	5.3213	5.3239	5.3266	5.3292
63	5.3319	5.3345	5.3372	5.3398	5.3425	5.3451	5.3478	5.3505	5.3531	5.3558
64	5.3585	5.3611	5.3638	5.3665	5.3692	5.3719	5.3745	5.3772	5.3799	5.3826
65	5.3853	5.3880	5.3907	5.3934	5.3961	5.3989	5.4016	5.4043	5.4070	5.4097
66	5.4125	5.4152	5.4179	5.4207	5.4234	5.4261	5.4289	5.4316	5.4344	5.4372
67	5.4399	5.4427	5.4454	5.4482	5.4510	5.4538	5.4565	5.4593	5.4621	5.4649
68	5.4677	5.4705	5.4733	5.4761	5.4789	5.4817	5.4845	5.4874	5.4902	5.4930
69	5.4959	5.4987	5.5015	5.5044	5.5072	5.5101	5.5129	5.5158	5.5187	5.5215
70	5.5244	5.5273	5.5302	5.5330	5.5359	5.5388	5.5417	5.5446	5.5476	5.5505
71	5.5534	5.5563	5.5592	5.5622	5.5651	5.5681	5.5710	5.5740	5.5769	5.5799
72	5.5828	5.5858	5.5888	5.5918	5.5948	5.5978	5.6008	5.6038	5.6068	5.6098
73	5.6128	5.6158	5.6189	5.6219	5.6250	5.6280	5.6311	5.6341	5.6372	5.6403
74	5.6433	5.6464	5.6495	5.6526	5.6557	5.6588	5.6620	5.6651	5.6682	5.6713

Table I (*cont.*)

%	0.0	0.1	0.2	0.3	0.4	0.5	0.6	0.7	0.8	0.9	1	2	3	4	5
75	5.6745	5.6776	5.6808	5.6840	5.6871	5.6903	5.6935	5.6967	5.6999	5.7031	3	6	10	13	16
76	5.7063	5.7095	5.7128	5.7160	5.7192	5.7225	5.7257	5.7290	5.7323	5.7356	3	7	10	13	16
77	5.7388	5.7421	5.7454	5.7488	5.7521	5.7554	5.7588	5.7621	5.7655	5.7688	3	7	10	13	17
78	5.7722	5.7756	5.7790	5.7824	5.7858	5.7892	5.7926	5.7961	5.7995	5.8030	3	7	10	14	17
79	5.8064	5.8099	5.8134	5.8169	5.8204	5.8239	5.8274	5.8310	5.8345	5.8381	4	7	11	14	18
80	5.8416	5.8452	5.8488	5.8524	5.8560	5.8596	5.8633	5.8669	5.8705	5.8742	4	7	11	14	18
81	5.8779	5.8816	5.8853	5.8890	5.8927	5.8965	5.9002	5.9040	5.9078	5.9116	4	7	11	15	19
82	5.9154	5.9192	5.9230	5.9269	5.9307	5.9346	5.9385	5.9424	5.9463	5.9502	4	8	12	15	19
83	5.9542	5.9581	5.9621	5.9661	5.9701	5.9741	5.9782	5.9822	5.9863	5.9904	4	8	12	16	20
84	5.9945	5.9986	6.0027	6.0069	6.0110	6.0152	6.0194	6.0237	6.0279	6.0322	4	8	13	17	21
85	6.0364	6.0407	6.0450	6.0494	6.0537	6.0581	6.0625	6.0669	6.0714	6.0758	4	9	13	18	22
86	6.0803	6.0848	6.0893	6.0939	6.0985	6.1031	6.1077	6.1123	6.1170	6.1217	5	9	14	18	23
87	6.1264	6.1311	6.1359	6.1407	6.1455	6.1503	6.1552	6.1601	6.1650	6.1700	5	10	15	19	24
88	6.1750	6.1800	6.1850	6.1901	6.1952	6.2004	6.2055	6.2107	6.2160	6.2212	5	10	15	21	26
89	6.2265	6.2319	6.2372	6.2426	6.2481	6.2536	6.2591	6.2646	6.2702	6.2759	5	11	16	22	27
90	6.2816	6.2873	6.2930	6.2988	6.3047	6.3106	6.3165	6.3225	6.3285	6.3346	6	12	18	24	29
91	6.3408	6.3469	6.3532	6.3595	6.3658	6.3722	6.3787	6.3852	6.3917	6.3984	6	13	19	26	32
92	6.4051	6.4118	6.4187	6.4255	6.4325	6.4395	6.4466	6.4538	6.4611	6.4684	7	14	21	28	35
93	6.4758	6.4833	6.4909	6.4985	6.5063	6.5141	6.5220	6.5301	6.5382	6.5464	8	16	24	31	39
94	6.5548	6.5632	6.5718	6.5805	6.5893	6.5982	6.6072	6.6164	6.6258	6.6352	9	18	27	36	45
95	6.6449	6.6546	6.6646	6.6747	6.6849	6.6954	6.7060	6.7169	6.7279	6.7392					
	97	100	101	102	105	106	109	110	113	115					
96	6.7507	6.7624	6.7744	6.7866	6.7991	6.8119	6.8250	6.8384	6.8522	6.8663					
	117	120	122	125	128	131	134	138	141	145					
97	6.8808	6.8957	6.9110	6.9268	6.9431	6.9600	6.9774	6.9954	7.0141	7.0335					

%	0.00	0.01	0.02	0.03	0.04	0.05	0.06	0.07	0.08	0.09	1	2	3	4	5
98.0	7.0537	7.0558	7.0579	7.0600	7.0621	7.0642	7.0663	7.0684	7.0706	7.0727	2	4	6	8	11
98.1	7.0749	7.0770	7.0792	7.0814	7.0836	7.0858	7.0880	7.0902	7.0924	7.0947	2	4	7	9	11
98.2	7.0969	7.0992	7.1015	7.1038	7.1061	7.1084	7.1107	7.1130	7.1154	7.1177	2	5	7	9	12
98.3	7.1201	7.1224	7.1248	7.1272	7.1297	7.1321	7.1345	7.1370	7.1394	7.1419	2	5	7	10	12
98.4	7.1444	7.1469	7.1494	7.1520	7.1545	7.1571	7.1596	7.1622	7.1648	7.1675	3	5	8	10	13
98.5	7.1701	7.1727	7.1754	7.1781	7.1808	7.1835	7.1862	7.1890	7.1917	7.1945	3	5	8	11	14
98.6	7.1973	7.2001	7.2029	7.2058	7.2086	7.2115	7.2144	7.2173	7.2203	7.2232	3	6	9	12	14
98.7	7.2262	7.2292	7.2322	7.2353	7.2383	7.2414	7.2445	7.2476	7.2508	7.2539	3	6	9	12	15
98.8	7.2571	7.2603	7.2636	7.2668	7.2701	7.2734	7.2768	7.2801	7.2835	7.2869	3	7	10	13	17
98.9	7.2904	7.2938	7.2973	7.3009	7.3044	7.3080	7.3116	7.3152	7.3189	7.3226	4	7	11	14	18
99.0	7.3263	7.3301	7.3339	7.3378	7.3416	7.3455	7.3495	7.3535	7.3575	7.3615	4	8	12	16	20
99.1	7.3656	7.3698	7.3739	7.3781	7.3824	7.3867	7.3911	7.3954	7.3999	7.4044	4	9	13	17	22
99.2	7.4089	7.4135	7.4181	7.4228	7.4276	7.4324	7.4372	7.4422	7.4471	7.4522	5	10	14	19	24
99.3	7.4573	7.4624	7.4677	7.4730	7.4783	7.4838	7.4893	7.4949	7.5006	7.5063	5	11	16	22	27
99.4	7.5121	7.5181	7.5241	7.5302	7.5364	7.5427	7.5491	7.5556	7.5622	7.5690	6	13	19	25	32
99.5	7.5758	7.5828	7.5899	7.5972	7.6045	7.6121	7.6197	7.6276	7.6356	7.6437					
99.6	7.6521	7.6606	7.6693	7.6783	7.6874	7.6968	7.7065	7.7164	7.7266	7.7370					
99.7	7.7478	7.7589	7.7703	7.7822	7.7944	7.8070	7.8202	7.8338	7.8480	7.8627					
99.8	7.8782	7.8943	7.9112	7.9290	7.9478	7.9677	7.9889	8.0115	8.0357	8.0618					
99.9	8.0902	8.1214	8.1559	8.1947	8.2389	8.2905	8.3528	8.4316	8.5401	8.7190					

I am indebted to Professor R. A. Fisher and Dr F. Yates, and also to Messrs Oliver and Boyd, Ltd. of Edinburgh, for permission to reprint Table I from Table IX of their book *Statistical Tables for Biological, Agricultural and Medical Research*.

Table II The weighting coefficient and Q/Z

Y	Q/Z	Percentage natural response rate, C										
		0	1	2	3	4	5	6	7	8	9	10
1.1	5033.5986	0.00082	0.00000	0.00000	0.00000	0.00000	0.00000	0.00000	0.00000	0.00000	0.00000	0.00000
1.2	3425.0323	0.00118	0.00001	0.00000	0.00000	0.00000	0.00000	0.00000	0.00000	0.00000	0.00000	0.00000
1.3	2353.9045	0.00167	0.00002	0.00001	0.00001	0.00000	0.00000	0.00000	0.00000	0.00000	0.00000	0.00000
1.4	1633.9888	0.00235	0.00004	0.00002	0.00001	0.00001	0.00001	0.00001	0.00000	0.00000	0.00000	0.00000
1.5	1145.6253	0.00327	0.00007	0.00004	0.00002	0.00002	0.00001	0.00001	0.00001	0.00001	0.00001	0.00001
1.6	811.2705	0.00451	0.00015	0.00007	0.00005	0.00004	0.00003	0.00002	0.00002	0.00002	0.00002	0.00001
1.7	580.2476	0.00614	0.00028	0.00014	0.00009	0.00007	0.00006	0.00005	0.00004	0.00003	0.00003	0.00003
1.8	419.1594	0.00828	0.00053	0.00027	0.00018	0.00013	0.00011	0.00009	0.00007	0.00006	0.00006	0.00005
1.9	305.8120	0.01104	0.00097	0.00050	0.00033	0.00025	0.00020	0.00016	0.00014	0.00012	0.00011	0.00010
2.0	225.3349	0.01457	0.00172	0.00090	0.00061	0.00046	0.00036	0.00030	0.00026	0.00022	0.00020	0.00017
2.1	167.6823	0.01903	0.00297	0.00159	0.00108	0.00082	0.00065	0.00054	0.00046	0.00040	0.00035	0.00031
2.2	126.0124	0.02458	0.00496	0.00274	0.00188	0.00142	0.00114	0.00095	0.00081	0.00070	0.00062	0.00055
2.3	95.6280	0.03143	0.00803	0.00456	0.00317	0.00241	0.00194	0.00162	0.00138	0.00121	0.00106	0.00095
2.4	73.2784	0.03977	0.01256	0.00739	0.00521	0.00400	0.00324	0.00271	0.00232	0.00202	0.00179	0.00160
2.5	56.6963	0.04979	0.01895	0.01161	0.00832	0.00646	0.00525	0.00441	0.00379	0.00332	0.00294	0.00264
2.6	44.2877	0.06168	0.02763	0.01768	0.01292	0.01014	0.00831	0.00702	0.00606	0.00531	0.00472	0.00424
2.7	34.9234	0.07564	0.03895	0.02605	0.01947	0.01548	0.01280	0.01088	0.00943	0.00830	0.00740	0.00666
2.8	27.7973	0.09179	0.05316	0.03719	0.02847	0.02297	0.01918	0.01642	0.01431	0.01265	0.01131	0.01021
2.9	22.3296	0.11026	0.07043	0.05147	0.04037	0.03309	0.02794	0.02411	0.02115	0.01879	0.01687	0.01527
3.0	18.1002	0.13112	0.09080	0.06911	0.05557	0.04631	0.03957	0.03445	0.03043	0.02719	0.02452	0.02228
3.1	14.8026	0.15436	0.11419	0.09023	0.07432	0.06298	0.05449	0.04790	0.04263	0.03832	0.03473	0.03170
3.2	12.2111	0.17994	0.14046	0.11476	0.09670	0.08332	0.07300	0.06481	0.05814	0.05261	0.04795	0.04397
3.3	10.1589	0.20774	0.16935	0.14249	0.12263	0.10736	0.09525	0.08541	0.07726	0.07039	0.06453	0.05947
3.4	8.5214	0.23753	0.20057	0.17308	0.15184	0.13494	0.12116	0.10973	0.10008	0.09183	0.08469	0.07846
3.5	7.?051	0.26907	0.23373	0.20611	0.18392	0.16571	0.15050	0.13760	0.12652	0.11690	0.10848	0.10103

x												
3.6	5.5199		0.26842	0.24107	0.21836	0.19921	0.18283	0.16867	0.15631	0.14541	0.13575	0.12711
3.7	5.2705	0.33589	0.30416	0.27741	0.25456	0.23482	0.21759	0.20242	0.18896	0.17694	0.16614	0.15639
3.8	4.5571	0.37031	0.34043	0.31453	0.29186	0.27187	0.25409	0.23819	0.22387	0.21092	0.19915	0.18840
3.9	3.9675	0.40474	0.37669	0.35181	0.32960	0.30964	0.29161	0.27524	0.26031	0.24665	0.23409	0.22250
4.0	3.4771	0.43863	0.41237	0.38864	0.36707	0.34739	0.32937	0.31279	0.29749	0.28334	0.27020	0.25797
4.1	3.0665	0.47144	0.44691	0.42438	0.40362	0.38442	0.36661	0.35005	0.33461	0.32018	0.30666	0.29397
4.2	2.7206	0.50260	0.47973	0.45844	0.43858	0.42000	0.40259	0.38624	0.37085	0.35634	0.34265	0.32969
4.3	2.4276	0.53159	0.51029	0.49024	0.47134	0.45350	0.43662	0.42063	0.40546	0.39105	0.37735	0.36430
4.4	2.1780	0.55788	0.53806	0.51924	0.50134	0.48430	0.46805	0.45255	0.43774	0.42357	0.41002	0.39702
4.5	1.9640	0.58099	0.56257	0.54495	0.52806	0.51187	0.49633	0.48140	0.46705	0.45325	0.43996	0.42716
4.6	1.7797	0.60052	0.58341	0.56694	0.55106	0.53573	0.52095	0.50666	0.49286	0.47951	0.46659	0.45409
4.7	1.6202	0.61609	0.60022	0.58485	0.56995	0.55551	0.54150	0.52790	0.51470	0.50187	0.48941	0.47729
4.8	1.4813	0.62742	0.61271	0.59840	0.58446	0.57089	0.55766	0.54478	0.53221	0.51996	0.50801	0.49635
4.9	1.3599	0.63431	0.62069	0.60737	0.59436	0.58164	0.56921	0.55704	0.54514	0.53350	0.52210	0.51094
5.0	1.2533	0.63662	0.62401	0.61165	0.59954	0.58765	0.57599	0.56455	0.55332	0.54231	0.53149	0.52087
5.1	1.1593	0.63431	0.62266	0.61120	0.59994	0.58886	0.57796	0.56724	0.55669	0.54631	0.53609	0.52604
5.2	1.0759	0.62742	0.61667	0.60607	0.59562	0.58532	0.57516	0.56515	0.55527	0.54553	0.53592	0.52644
5.3	1.0018	0.61609	0.60618	0.59639	0.58672	0.57717	0.56773	0.55841	0.54919	0.54008	0.53108	0.52219
5.4	0.9357	0.60052	0.59140	0.58238	0.57346	0.56462	0.55588	0.54722	0.53866	0.53018	0.52178	0.51347
5.5	0.8764	0.58099	0.57263	0.56434	0.55612	0.54797	0.53990	0.53189	0.52396	0.51609	0.50829	0.50056
5.6	0.8230	0.55788	0.55022	0.54262	0.53507	0.52759	0.52015	0.51278	0.50545	0.49818	0.49097	0.48381
5.7	0.7749	0.53159	0.52460	0.51765	0.51075	0.50389	0.49708	0.49030	0.48357	0.47688	0.47024	0.46363
5.8	0.7313	0.50260	0.49624	0.48992	0.48363	0.47737	0.47114	0.46495	0.45879	0.45266	0.44657	0.44050
5.9	0.6917	0.47144	0.46567	0.45993	0.45422	0.44853	0.44287	0.43723	0.43162	0.42603	0.42047	0.41493
6.0	0.6557	0.43863	0.43343	0.42824	0.42308	0.41793	0.41281	0.40770	0.40261	0.39754	0.39249	0.38746
6.1	0.6227	0.40474	0.40006	0.39540	0.39075	0.38612	0.38150	0.37690	0.37231	0.36774	0.36318	0.35863
6.2	0.5926	0.37031	0.36613	0.36196	0.35781	0.35366	0.34952	0.34540	0.34128	0.33718	0.33308	0.32900
6.3	0.5649	0.33589	0.33218	0.32847	0.32477	0.32108	0.31740	0.31372	0.31005	0.30639	0.30274	0.29910
6.4	0.5394	0.30199	0.29871	0.29543	0.29216	0.28890	0.28564	0.28238	0.27914	0.27589	0.27266	0.26943
6.5	0.5158	0.26907	0.26619	0.26331	0.26044	0.25757	0.25470	0.25184	0.24899	0.24613	0.24328	0.24044

Table II (*cont.*)

Y	Q/Z	Percentage natural response rate, C										
		0	1	2	3	4	5	6	7	8	9	10
6.6	0.4940	0.23753	0.23502	0.23251	0.23001	0.22751	0.22501	0.22251	0.22001	0.21752	0.21503	0.21255
6.7	0.4739	0.20774	0.20556	0.20339	0.20122	0.19905	0.19689	0.19473	0.19256	0.19041	0.18825	0.18609
6.8	0.4551	0.17994	0.17808	0.17621	0.17435	0.17249	0.17063	0.16877	0.16691	0.16506	0.16320	0.16135
6.9	0.4376	0.15436	0.15277	0.15118	0.14960	0.14801	0.14643	0.14484	0.14326	0.14168	0.14010	0.13852
7.0	0.4214	0.13112	0.12977	0.12843	0.12709	0.12575	0.12441	0.12308	0.12174	0.12040	0.11907	0.11773
7.1	0.4062	0.11026	0.10914	0.10802	0.10689	0.10577	0.10465	0.10353	0.10241	0.10129	0.10017	0.09905
7.2	0.3919	0.09179	0.09086	0.08993	0.08900	0.08807	0.08714	0.08621	0.08528	0.08435	0.08342	0.08249
7.3	0.3786	0.07564	0.07487	0.07411	0.07334	0.07258	0.07181	0.07105	0.07029	0.06952	0.06876	0.06800
7.4	0.3661	0.06168	0.06106	0.06044	0.05982	0.05920	0.05858	0.05795	0.05733	0.05671	0.05609	0.05547
7.5	0.3543	0.04979	0.04929	0.04879	0.04828	0.04778	0.04728	0.04678	0.04628	0.04578	0.04528	0.04478
7.6	0.3432	0.03977	0.03937	0.03897	0.03857	0.03817	0.03777	0.03737	0.03697	0.03657	0.03617	0.03577
7.7	0.3327	0.03143	0.03112	0.03080	0.03049	0.03017	0.02986	0.02954	0.02922	0.02891	0.02859	0.02828
7.8	0.3228	0.02458	0.02434	0.02409	0.02384	0.02360	0.02335	0.02311	0.02286	0.02261	0.02237	0.02212
7.9	0.3134	0.01903	0.01884	0.01864	0.01845	0.01826	0.01807	0.01788	0.01769	0.01750	0.01731	0.01712
8.0	0.3046	0.01457	0.01442	0.01428	0.01413	0.01399	0.01384	0.01369	0.01355	0.01340	0.01326	0.01311
8.1	0.2962	0.01104	0.01093	0.01082	0.01071	0.01060	0.01049	0.01038	0.01027	0.01016	0.01005	0.00994
8.2	0.2882	0.00828	0.00819	0.00811	0.00803	0.00795	0.00786	0.00778	0.00770	0.00761	0.00753	0.00745
8.3	0.2806	0.00614	0.00608	0.00602	0.00596	0.00590	0.00583	0.00577	0.00571	0.00565	0.00559	0.00553
8.4	0.2734	0.00451	0.00446	0.00442	0.00437	0.00433	0.00428	0.00424	0.00419	0.00415	0.00410	0.00406
8.5	0.2666	0.00327	0.00324	0.00321	0.00318	0.00314	0.00311	0.00308	0.00305	0.00301	0.00298	0.00295
8.6	0.2600	0.00235	0.00233	0.00231	0.00228	0.00226	0.00224	0.00221	0.00219	0.00217	0.00214	0.00212
8.7	0.2538	0.00167	0.00166	0.00164	0.00162	0.00161	0.00159	0.00157	0.00156	0.00154	0.00152	0.00151
8.8	0.2478	0.00118	0.00117	0.00115	0.00114	0.00113	0.00112	0.00111	0.00110	0.00108	0.00107	0.00106
8.9	0.2421	0.00082	0.00081	0.00080	0.00080	0.00079	0.00078	0.00077	0.00076	0.00075	0.00075	0.00074
9.0	0.2367	0.00057	0.00056	0.00055	0.00055	0.00054	0.00054	0.00053	0.00053	0.00052	0.00051	0.00051

Percentage natural response rate, C

Y	Q/Z	11	12	13	14	15	16	17	18	19	20
1.1	5033.5986	0.00000	0.00000	0.00000	0.00000	0.00000	0.00000	0.00000	0.00000	0.00000	0.00000
1.2	3425.0323	0.00000	0.00000	0.00000	0.00000	0.00000	0.00000	0.00000	0.00000	0.00000	0.00000
1.3	2353.9045	0.00000	0.00000	0.00000	0.00000	0.00000	0.00000	0.00000	0.00000	0.00000	0.00000
1.4	1633.9888	0.00000	0.00000	0.00000	0.00000	0.00000	0.00000	0.00000	0.00000	0.00000	0.00000
1.5	1145.6253	0.00001	0.00001	0.00001	0.00000	0.00000	0.00000	0.00000	0.00000	0.00000	0.00000
1.6	811.2705	0.00001	0.00001	0.00001	0.00001	0.00001	0.00001	0.00001	0.00001	0.00001	0.00001
1.7	580.2476	0.00002	0.00002	0.00002	0.00002	0.00002	0.00002	0.00001	0.00001	0.00001	0.00001
1.8	419.1594	0.00005	0.00004	0.00004	0.00003	0.00003	0.00003	0.00003	0.00003	0.00002	0.00002
1.9	305.8120	0.00009	0.00008	0.00007	0.00007	0.00006	0.00006	0.00005	0.00005	0.00005	0.00004
2.0	225.3349	0.00016	0.00014	0.00013	0.00012	0.00011	0.00010	0.00010	0.00009	0.00008	0.00008
2.1	167.6823	0.00028	0.00026	0.00023	0.00022	0.00020	0.00018	0.00017	0.00016	0.00015	0.00014
2.2	126.0124	0.00050	0.00045	0.00041	0.00038	0.00035	0.00033	0.00030	0.00028	0.00026	0.00025
2.3	95.6280	0.00086	0.00078	0.00071	0.00066	0.00061	0.00056	0.00052	0.00049	0.00046	0.00043
2.4	73.2784	0.00145	0.00131	0.00120	0.00111	0.00102	0.00095	0.00088	0.00083	0.00077	0.00073
2.5	56.6963	0.00238	0.00217	0.00199	0.00183	0.00169	0.00157	0.00147	0.00137	0.00128	0.00121
2.6	44.2877	0.00384	0.00350	0.00321	0.00296	0.00274	0.00255	0.00237	0.00222	0.00208	0.00196
2.7	34.9234	0.00604	0.00551	0.00506	0.00467	0.00433	0.00403	0.00376	0.00352	0.00331	0.00311
2.8	27.7973	0.00928	0.00849	0.00781	0.00722	0.00670	0.00624	0.00583	0.00547	0.00514	0.00484
2.9	22.3296	0.01392	0.01277	0.01177	0.01090	0.01014	0.00945	0.00885	0.00830	0.00780	0.00735
3.0	18.1002	0.02038	0.01875	0.01732	0.01608	0.01497	0.01399	0.01311	0.01231	0.01159	0.01094
3.1	14.8026	0.02910	0.02685	0.02488	0.02315	0.02160	0.02022	0.01898	0.01786	0.01684	0.01590
3.2	12.2111	0.04053	0.03753	0.03488	0.03254	0.03044	0.02856	0.02686	0.02531	0.02390	0.02261
3.3	10.1589	0.05505	0.05117	0.04772	0.04465	0.04188	0.03939	0.03712	0.03506	0.03317	0.03143
3.4	8.5214	0.07297	0.06809	0.06374	0.05982	0.05628	0.05307	0.05014	0.04745	0.04498	0.04271
3.5	7.2051	0.09441	0.08848	0.08313	0.07829	0.07389	0.06987	0.06618	0.06278	0.05965	0.05674

Table II (cont.)

Y	Q/Z	Percentage natural response rate, C									
		11	12	13	14	15	16	17	18	19	20
3.6	6.1394	0.11934	0.11232	0.10595	0.10014	0.09481	0.08991	0.08540	0.08122	0.07734	0.07373
3.7	5.2705	0.14753	0.13945	0.13205	0.12525	0.11898	0.11318	0.10780	0.10279	0.09812	0.09376
3.8	4.5571	0.17854	0.16947	0.16111	0.15336	0.14616	0.13946	0.13321	0.12736	0.12187	0.11672
3.9	3.9675	0.21179	0.20185	0.19260	0.18398	0.17591	0.16836	0.16127	0.15460	0.14831	0.14237
4.0	3.4771	0.24656	0.23589	0.22589	0.21649	0.20766	0.19933	0.19146	0.18402	0.17698	0.17029
4.1	3.0665	0.28205	0.27081	0.26020	0.25017	0.24068	0.23168	0.22314	0.21501	0.20728	0.19991
4.2	2.7206	0.31742	0.30578	0.29473	0.28421	0.27420	0.26466	0.25555	0.24684	0.23852	0.23055
4.3	2.4276	0.35186	0.33998	0.32864	0.31779	0.30740	0.29744	0.28789	0.27873	0.26992	0.26145
4.4	2.1780	0.38457	0.37261	0.36112	0.35008	0.33945	0.32922	0.31937	0.30986	0.30069	0.29184
4.5	1.9640	0.41482	0.40292	0.39142	0.38032	0.36960	0.35922	0.34919	0.33947	0.33006	0.32094
4.6	1.7797	0.44198	0.43025	0.41887	0.40784	0.39713	0.38674	0.37664	0.36683	0.35729	0.34802
4.7	1.6202	0.46551	0.45404	0.44289	0.43202	0.42144	0.41113	0.40109	0.39129	0.38174	0.37242
4.8	1.4813	0.48496	0.47385	0.46299	0.45239	0.44203	0.43190	0.42199	0.41231	0.40284	0.39357
4.9	1.3599	0.50001	0.48931	0.47883	0.46855	0.45849	0.44862	0.43894	0.42945	0.42015	0.41102
5.0	1.2533	0.51044	0.50020	0.49014	0.48026	0.47055	0.46100	0.45162	0.44240	0.43333	0.42441
5.1	1.1593	0.51614	0.50639	0.49680	0.48735	0.47804	0.46887	0.45984	0.45094	0.44217	0.43354
5.2	1.0759	0.51709	0.50787	0.49876	0.48978	0.48091	0.47216	0.46353	0.45500	0.44658	0.43827
5.3	1.0018	0.51340	0.50471	0.49612	0.48762	0.47923	0.47092	0.46271	0.45459	0.44657	0.43863
5.4	0.9357	0.50524	0.49709	0.48903	0.48104	0.47313	0.46529	0.45754	0.44985	0.44224	0.43470
5.5	0.8764	0.49289	0.48529	0.47775	0.47028	0.46286	0.45551	0.44822	0.44099	0.43382	0.42671
5.6	0.8230	0.47669	0.46963	0.46262	0.45567	0.44876	0.44190	0.43509	0.42832	0.42161	0.41494
5.7	0.7749	0.45707	0.45054	0.44406	0.43761	0.43120	0.42484	0.41851	0.41222	0.40597	0.39975
5.8	0.7313	0.43447	0.42847	0.42250	0.41656	0.41066	0.40478	0.39893	0.39311	0.38733	0.38157
5.9	0.6917	0.40942	0.40393	0.39847	0.39302	0.38761	0.38221	0.37684	0.37149	0.36617	0.36087
6.0	0.6557	0.38245	0.37745	0.37248	0.36752	0.36258	0.35766	0.35275	0.34787	0.34300	0.33815

x											
6.2	0.5926	0.32493	0.32087	0.31681	0.31277	0.30874	0.30472	0.30071	0.29671	0.29272	0.28874
6.3	0.5649	0.29546	0.29183	0.28821	0.28460	0.28099	0.27739	0.27380	0.27022	0.26664	0.26308
6.4	0.5394	0.26620	0.26298	0.25977	0.25656	0.25335	0.25016	0.24696	0.24378	0.24060	0.23742
6.5	0.5158	0.23760	0.23476	0.23193	0.22910	0.22628	0.22346	0.22064	0.21783	0.21502	0.21222
6.6	0.4940	0.21007	0.20759	0.20511	0.20264	0.20016	0.19770	0.19523	0.19277	0.19031	0.18785
6.7	0.4739	0.18394	0.18179	0.17964	0.17749	0.17535	0.17320	0.17106	0.16892	0.16679	0.16465
6.8	0.4551	0.15950	0.15765	0.15580	0.15395	0.15210	0.15026	0.14841	0.14657	0.14473	0.14289
6.9	0.4376	0.13694	0.13536	0.13378	0.13220	0.13063	0.12905	0.12748	0.12591	0.12433	0.12276
7.0	0.4214	0.11639	0.11506	0.11373	0.11239	0.11106	0.10973	0.10840	0.10707	0.10574	0.10441
7.1	0.4062	0.09794	0.09682	0.09570	0.09458	0.09347	0.09235	0.09123	0.09012	0.08900	0.08789
7.2	0.3919	0.08157	0.08064	0.07971	0.07878	0.07786	0.07693	0.07600	0.07508	0.07415	0.07323
7.3	0.3786	0.06724	0.06647	0.06571	0.06495	0.06419	0.06342	0.06266	0.06190	0.06114	0.06038
7.4	0.3661	0.05485	0.05423	0.05361	0.05299	0.05237	0.05175	0.05113	0.05051	0.04989	0.04927
7.5	0.3543	0.04428	0.04378	0.04328	0.04278	0.04228	0.04178	0.04128	0.04078	0.04028	0.03978
7.6	0.3432	0.03537	0.03498	0.03458	0.03418	0.03378	0.03338	0.03298	0.03258	0.03218	0.03178
7.7	0.3327	0.02796	0.02765	0.02733	0.02702	0.02670	0.02639	0.02607	0.02576	0.02544	0.02513
7.8	0.3228	0.02187	0.02163	0.02138	0.02113	0.02089	0.02064	0.02040	0.02015	0.01990	0.01966
7.9	0.3134	0.01693	0.01674	0.01655	0.01636	0.01617	0.01598	0.01579	0.01560	0.01541	0.01522
8.0	0.3046	0.01297	0.01282	0.01267	0.01253	0.01238	0.01224	0.01209	0.01194	0.01180	0.01165
8.1	0.2962	0.00982	0.00971	0.00960	0.00949	0.00938	0.00927	0.00916	0.00905	0.00894	0.00883
8.2	0.2882	0.00737	0.00728	0.00720	0.00712	0.00704	0.00695	0.00687	0.00679	0.00670	0.00662
8.3	0.2806	0.00547	0.00540	0.00534	0.00528	0.00522	0.00516	0.00510	0.00504	0.00497	0.00491
8.4	0.2734	0.00401	0.00397	0.00392	0.00388	0.00383	0.00379	0.00374	0.00370	0.00365	0.00361
8.5	0.2666	0.00291	0.00288	0.00285	0.00282	0.00278	0.00275	0.00272	0.00269	0.00265	0.00262
8.6	0.2600	0.00209	0.00207	0.00205	0.00202	0.00200	0.00198	0.00195	0.00193	0.00191	0.00188
8.7	0.2538	0.00149	0.00147	0.00146	0.00144	0.00142	0.00141	0.00139	0.00137	0.00136	0.00134
8.8	0.2478	0.00105	0.00104	0.00103	0.00101	0.00100	0.00099	0.00098	0.00097	0.00095	0.00094
8.9	0.2421	0.00073	0.00072	0.00071	0.00071	0.00070	0.00069	0.00068	0.00067	0.00066	0.00066
9.0	0.2367	0.00050	0.00050	0.00049	0.00049	0.00048	0.00048	0.00047	0.00046	0.00046	0.00045

Table II (*cont.*)

Y	Q/Z	Percentage natural response rate, C									
		21	22	23	24	25	26	27	28	29	30
1.1	5033.5986	0.00000	0.00000	0.00000	0.00000	0.00000	0.00000	0.00000	0.00000	0.00000	0.00000
1.2	3425.0323	0.00000	0.00000	0.00000	0.00000	0.00000	0.00000	0.00000	0.00000	0.00000	0.00000
1.3	2353.9045	0.00000	0.00000	0.00000	0.00000	0.00000	0.00000	0.00000	0.00000	0.00000	0.00000
1.4	1633.9888	0.00000	0.00000	0.00000	0.00000	0.00000	0.00000	0.00000	0.00000	0.00000	0.00000
1.5	1145.6253	0.00000	0.00000	0.00000	0.00000	0.00000	0.00000	0.00000	0.00000	0.00000	0.00000
1.6	811.2705	0.00001	0.00001	0.00001	0.00000	0.00000	0.00000	0.00000	0.00000	0.00000	0.00000
1.7	580.2476	0.00001	0.00001	0.00001	0.00001	0.00001	0.00001	0.00001	0.00001	0.00001	0.00001
1.8	419.1594	0.00002	0.00002	0.00002	0.00002	0.00002	0.00002	0.00002	0.00001	0.00001	0.00001
1.9	305.8120	0.00004	0.00004	0.00004	0.00003	0.00003	0.00003	0.00003	0.00003	0.00003	0.00002
2.0	225.3349	0.00007	0.00007	0.00007	0.00006	0.00006	0.00006	0.00005	0.00005	0.00005	0.00005
2.1	167.6823	0.00013	0.00013	0.00012	0.00011	0.00011	0.00010	0.00010	0.00009	0.00009	0.00008
2.2	126.0124	0.00023	0.00022	0.00021	0.00020	0.00019	0.00018	0.00017	0.00016	0.00015	0.00015
2.3	95.6280	0.00040	0.00038	0.00036	0.00034	0.00032	0.00031	0.00029	0.00028	0.00026	0.00025
2.4	73.2784	0.00069	0.00065	0.00061	0.00058	0.00055	0.00052	0.00049	0.00047	0.00045	0.00043
2.5	56.6963	0.00114	0.00107	0.00101	0.00096	0.00091	0.00086	0.00082	0.00078	0.00075	0.00071
2.6	44.2877	0.00185	0.00174	0.00165	0.00156	0.00148	0.00141	0.00134	0.00127	0.00121	0.00116
2.7	34.9234	0.00293	0.00277	0.00262	0.00248	0.00236	0.00224	0.00213	0.00203	0.00194	0.00185
2.8	27.7973	0.00456	0.00431	0.00408	0.00387	0.00368	0.00349	0.00333	0.00317	0.00302	0.00288
2.9	22.3296	0.00694	0.00657	0.00622	0.00590	0.00561	0.00533	0.00508	0.00484	0.00462	0.00441
3.0	18.1002	0.01034	0.00979	0.00928	0.00881	0.00838	0.00797	0.00760	0.00725	0.00692	0.00661
3.1	14.8026	0.01505	0.01426	0.01354	0.01287	0.01224	0.01166	0.01112	0.01061	0.01014	0.00969
3.2	12.2111	0.02143	0.02033	0.01932	0.01838	0.01751	0.01669	0.01593	0.01522	0.01455	0.01392
3.3	10.1589	0.02983	0.02834	0.02697	0.02569	0.02450	0.02338	0.02234	0.02136	0.02044	0.01957
3.4	8.5214	0.04060	0.03864	0.03682	0.03512	0.03354	0.03205	0.03065	0.02934	0.02810	0.02693
3.5	7.2051	0.05404	0.05153	0.04918	0.04698	0.04492	0.04299	0.04117	0.03945	0.03782	0.03629

3.7	5.2705	0.08967	0.08582	0.08221	0.07881	0.07559	0.07255	0.06967	0.06695	0.06435	0.06189
3.8	4.5571	0.11187	0.10730	0.10298	0.09890	0.09503	0.09136	0.08787	0.08455	0.08139	0.07838
3.9	3.9675	0.13676	0.13145	0.12641	0.12163	0.11708	0.11275	0.10862	0.10468	0.10091	0.09732
4.0	3.4771	0.16394	0.15791	0.15216	0.14668	0.14145	0.13645	0.13167	0.12710	0.12271	0.11851
4.1	3.0665	0.19288	0.18616	0.17974	0.17360	0.16771	0.16207	0.15665	0.15145	0.14645	0.14164
4.2	2.7206	0.22291	0.21559	0.20856	0.20180	0.19531	0.18906	0.18304	0.17725	0.17166	0.16626
4.3	2.4276	0.25331	0.24546	0.23790	0.23061	0.22358	0.21679	0.21023	0.20389	0.19776	0.19182
4.4	2.1780	0.28329	0.27503	0.26704	0.25930	0.25131	0.24456	0.23753	0.23072	0.22411	0.21769
4.5	1.9640	0.31210	0.30352	0.29520	0.28712	0.27927	0.27165	0.26424	0.25703	0.25001	0.24319
4.6	1.7797	0.33900	0.33022	0.32167	0.31335	0.30524	0.29734	0.28963	0.28212	0.27479	0.26764
4.7	1.6202	0.36332	0.35444	0.34578	0.33731	0.32904	0.32095	0.31305	0.30533	0.29777	0.29038
4.8	1.4813	0.38450	0.37562	0.36693	0.35841	0.35008	0.34191	0.33390	0.32605	0.31836	0.31082
4.9	1.3599	0.40206	0.39327	0.38464	0.37617	0.36735	0.35968	0.35166	0.34378	0.33604	0.32843
5.0	1.2533	0.41564	0.40702	0.39853	0.39019	0.38197	0.37389	0.36593	0.35810	0.35039	0.34280
5.1	1.1593	0.42502	0.41663	0.40836	0.40020	0.39216	0.38423	0.37641	0.36870	0.36109	0.35359
5.2	1.0759	0.43007	0.42196	0.41396	0.40606	0.39825	0.39054	0.38292	0.37540	0.36796	0.36062
5.3	1.0018	0.43077	0.42300	0.41532	0.40772	0.40020	0.39276	0.38540	0.37812	0.37091	0.36378
5.4	0.9357	0.42724	0.41984	0.41252	0.40526	0.39807	0.39094	0.38388	0.37689	0.36996	0.36309
5.5	0.8764	0.41966	0.41266	0.40572	0.39884	0.39201	0.38524	0.37852	0.37185	0.36524	0.35868
5.6	0.8230	0.40832	0.40174	0.39521	0.38873	0.38229	0.37590	0.36954	0.36324	0.35697	0.35075
5.7	0.7749	0.39357	0.38743	0.38133	0.37526	0.36923	0.36323	0.35727	0.35134	0.34545	0.33959
5.8	0.7313	0.37584	0.37014	0.36447	0.35883	0.35322	0.34763	0.34207	0.33655	0.33104	0.32557
5.9	0.6917	0.35559	0.35033	0.34510	0.33989	0.33470	0.32954	0.32439	0.31927	0.31417	0.30909
6.0	0.6557	0.33332	0.32850	0.32370	0.31892	0.31416	0.30942	0.30469	0.29997	0.29528	0.29060
6.1	0.6227	0.30954	0.30516	0.30079	0.29643	0.29209	0.28776	0.28344	0.27914	0.27485	0.27057
6.2	0.5926	0.28477	0.28081	0.27686	0.27292	0.26899	0.26507	0.26116	0.25726	0.25337	0.24949
6.3	0.5649	0.25951	0.25596	0.25242	0.24888	0.24535	0.24182	0.23831	0.23480	0.23130	0.22780
6.4	0.5394	0.23425	0.23109	0.22793	0.22477	0.22163	0.21848	0.21535	0.21221	0.20909	0.20597
6.5	0.5158	0.20942	0.20662	0.20383	0.20104	0.19825	0.19547	0.19270	0.18992	0.18715	0.18439

Table II (*cont.*)

		Percentage natural response rate, C									
Y	Q/Z	21	22	23	24	25	26	27	28	29	30
6.6	0.4940	0.18540	0.18294	0.18049	0.17805	0.17561	0.17317	0.17073	0.16829	0.16586	0.16343
6.7	0.4739	0.16252	0.16039	0.15826	0.15613	0.15401	0.15188	0.14976	0.14764	0.14552	0.14341
6.8	0.4551	0.14105	0.13921	0.13738	0.13554	0.13371	0.13188	0.13005	0.12822	0.12639	0.12457
6.9	0.4376	0.12119	0.11962	0.11806	0.11649	0.11492	0.11336	0.11179	0.11023	0.10866	0.10710
7.0	0.4214	0.10308	0.10175	0.10042	0.09909	0.09777	0.09644	0.09512	0.09379	0.09247	0.09114
7.1	0.4062	0.08677	0.08566	0.08455	0.08343	0.08232	0.08121	0.08010	0.07899	0.07787	0.07676
7.2	0.3919	0.07230	0.07137	0.07045	0.06952	0.06860	0.06768	0.06675	0.06583	0.06491	0.06398
7.3	0.3786	0.05962	0.05886	0.05809	0.05733	0.05657	0.05581	0.05505	0.05429	0.05353	0.05277
7.4	0.3661	0.04865	0.04803	0.04741	0.04679	0.04617	0.04555	0.04493	0.04431	0.04369	0.04307
7.5	0.3543	0.03928	0.03878	0.03828	0.03778	0.03728	0.03678	0.03628	0.03578	0.03528	0.03479
7.6	0.3432	0.03138	0.03099	0.03059	0.03019	0.02979	0.02939	0.02899	0.02859	0.02820	0.02780
7.7	0.3327	0.02481	0.02450	0.02418	0.02387	0.02355	0.02324	0.02292	0.02261	0.02229	0.02198
7.8	0.3228	0.01941	0.01916	0.01892	0.01867	0.01843	0.01818	0.01793	0.01769	0.01744	0.01720
7.9	0.3134	0.01502	0.01483	0.01464	0.01445	0.01426	0.01407	0.01388	0.01369	0.01350	0.01331
8.0	0.3046	0.01151	0.01136	0.01122	0.01107	0.01092	0.01078	0.01063	0.01049	0.01034	0.01019
8.1	0.2962	0.00872	0.00861	0.00850	0.00839	0.00828	0.00817	0.00806	0.00795	0.00784	0.00773
8.2	0.2882	0.00654	0.00646	0.00637	0.00629	0.00621	0.00612	0.00604	0.00596	0.00588	0.00579
8.3	0.2806	0.00485	0.00479	0.00473	0.00467	0.00461	0.00454	0.00448	0.00442	0.00436	0.00430
8.4	0.2734	0.00356	0.00352	0.00347	0.00343	0.00338	0.00334	0.00329	0.00325	0.00320	0.00316
8.5	0.2666	0.00259	0.00255	0.00252	0.00249	0.00246	0.00242	0.00239	0.00236	0.00232	0.00229
8.6	0.2600	0.00186	0.00184	0.00181	0.00179	0.00177	0.00174	0.00172	0.00169	0.00167	0.00165
8.7	0.2538	0.00132	0.00131	0.00129	0.00127	0.00126	0.00124	0.00122	0.00121	0.00119	0.00117
8.8	0.2478	0.00093	0.00092	0.00091	0.00090	0.00088	0.00087	0.00086	0.00085	0.00084	0.00082
8.9	0.2421	0.00065	0.00064	0.00063	0.00062	0.00062	0.00061	0.00060	0.00059	0.00058	0.00057
9.0	0.2367	0.00045	0.00044	0.00044	0.00043	0.00042	0.00042	0.00041	0.00041	0.00040	0.00040

Percentage natural response rate, C

Y	Q/Z	31	32	33	34	35	36	37	38	39	40
1.1	5033.5986	0.00000	0.00000	0.00000	0.00000	0.00000	0.00000	0.00000	0.00000	0.00000	0.00000
1.2	3425.0323	0.00000	0.00000	0.00000	0.00000	0.00000	0.00000	0.00000	0.00000	0.00000	0.00000
1.3	2353.9045	0.00000	0.00000	0.00000	0.00000	0.00000	0.00000	0.00000	0.00000	0.00000	0.00000
1.4	1633.9888	0.00000	0.00000	0.00000	0.00000	0.00000	0.00000	0.00000	0.00000	0.00000	0.00000
1.5	1145.6253	0.00000	0.00000	0.00000	0.00000	0.00000	0.00000	0.00000	0.00000	0.00000	0.00000
1.6	811.2705	0.00000	0.00000	0.00000	0.00000	0.00000	0.00000	0.00000	0.00000	0.00000	0.00000
1.7	580.2476	0.00001	0.00001	0.00001	0.00001	0.00001	0.00001	0.00001	0.00000	0.00000	0.00000
1.8	419.1594	0.00001	0.00001	0.00001	0.00001	0.00001	0.00001	0.00001	0.00001	0.00001	0.00001
1.9	305.8120	0.00002	0.00002	0.00002	0.00002	0.00002	0.00002	0.00002	0.00002	0.00002	0.00002
2.0	225.3349	0.00004	0.00004	0.00004	0.00004	0.00004	0.00003	0.00003	0.00003	0.00003	0.00003
2.1	167.6823	0.00008	0.00008	0.00007	0.00007	0.00007	0.00006	0.00006	0.00006	0.00006	0.00005
2.2	126.0124	0.00014	0.00013	0.00013	0.00012	0.00012	0.00011	0.00011	0.00010	0.00010	0.00009
2.3	95.6280	0.00024	0.00023	0.00022	0.00021	0.00020	0.00019	0.00018	0.00018	0.00017	0.00016
2.4	73.2784	0.00041	0.00039	0.00037	0.00036	0.00034	0.00033	0.00031	0.00030	0.00029	0.00028
2.5	56.6963	0.00068	0.00065	0.00062	0.00059	0.00057	0.00054	0.00052	0.00050	0.00048	0.00046
2.6	44.2877	0.00111	0.00106	0.00101	0.00097	0.00093	0.00089	0.00085	0.00081	0.00078	0.00075
2.7	34.9234	0.00176	0.00169	0.00161	0.00154	0.00148	0.00142	0.00136	0.00130	0.00125	0.00120
2.8	27.7973	0.00276	0.00263	0.00252	0.00241	0.00231	0.00221	0.00212	0.00204	0.00195	0.00188
2.9	22.3296	0.00422	0.00403	0.00386	0.00370	0.00354	0.00339	0.00325	0.00312	0.00300	0.00288
3.0	18.1002	0.00632	0.00605	0.00579	0.00555	0.00532	0.00510	0.00489	0.00469	0.00451	0.00433
3.1	14.8026	0.00927	0.00888	0.00850	0.00815	0.00782	0.00750	0.00720	0.00691	0.00664	0.00637
3.2	12.2111	0.01333	0.01276	0.01223	0.01173	0.01126	0.01080	0.01037	0.00996	0.00957	0.00920
3.3	10.1589	0.01875	0.01797	0.01724	0.01654	0.01588	0.01525	0.01465	0.01408	0.01354	0.01302
3.4	8.5214	0.02582	0.02478	0.02378	0.02284	0.02194	0.02109	0.02027	0.01949	0.01875	0.01804
3.5	7.2051	0.03483	0.03345	0.03214	0.03089	0.02970	0.02856	0.02748	0.02645	0.02546	0.02451

Table II (cont.)

		Percentage natural response rate, C									
Y	Q/Z	31	32	33	34	35	36	37	38	39	40
3.6	6.1394	0.04601	0.04423	0.04254	0.04093	0.03938	0.03791	0.03651	0.03516	0.03387	0.03263
3.7	5.2705	0.05954	0.05731	0.05517	0.05313	0.05118	0.04932	0.04753	0.04581	0.04417	0.04259
3.8	4.5571	0.07551	0.07276	0.07013	0.06761	0.06520	0.06289	0.06067	0.05853	0.05648	0.05451
3.9	3.9675	0.09387	0.09057	0.08741	0.08437	0.08145	0.07865	0.07595	0.07335	0.07085	0.06844
4.0	3.4771	0.11447	0.11059	0.10687	0.10328	0.09983	0.09650	0.09329	0.09020	0.08721	0.08432
4.1	3.0665	0.13701	0.13255	0.12825	0.12410	0.12010	0.11623	0.11249	0.10888	0.10538	0.10200
4.2	2.7206	0.16106	0.15603	0.15116	0.14646	0.14191	0.13751	0.13324	0.12910	0.12509	0.12120
4.3	2.4276	0.18608	0.18051	0.17512	0.16989	0.16481	0.15989	0.15511	0.15046	0.14595	0.14156
4.4	2.1780	0.21146	0.20541	0.19953	0.19382	0.18826	0.18285	0.17758	0.17246	0.16747	0.16261
4.5	1.9640	0.23655	0.23008	0.22377	0.21763	0.21164	0.20580	0.20010	0.19454	0.18911	0.18382
4.6	1.7797	0.26066	0.25384	0.24719	0.24069	0.23433	0.22812	0.22205	0.21611	0.21031	0.20462
4.7	1.6202	0.28315	0.27607	0.26914	0.26236	0.25572	0.24921	0.24283	0.23659	0.23046	0.22443
4.8	1.4813	0.30342	0.29617	0.28905	0.28207	0.27521	0.26848	0.26188	0.25539	0.24902	0.24276
4.9	1.3599	0.32096	0.31361	0.30638	0.29928	0.29229	0.28542	0.27866	0.27201	0.26547	0.25904
5.0	1.2533	0.33532	0.32796	0.32070	0.31356	0.30652	0.29959	0.29275	0.28602	0.27938	0.27284
5.1	1.1593	0.34619	0.33889	0.33168	0.32457	0.31756	0.31063	0.30380	0.29705	0.29039	0.28381
5.2	1.0759	0.35336	0.34618	0.33910	0.33209	0.32516	0.31832	0.31155	0.30486	0.29824	0.29170
5.3	1.0018	0.35672	0.34974	0.34282	0.33598	0.32921	0.32250	0.31587	0.30930	0.30279	0.29635
5.4	0.9357	0.35629	0.34954	0.34286	0.33624	0.32967	0.32317	0.31672	0.31032	0.30399	0.29770
5.5	0.8764	0.35217	0.34571	0.33930	0.33294	0.32663	0.32037	0.31416	0.30799	0.30187	0.29580
5.6	0.8230	0.34457	0.33843	0.33233	0.32628	0.32026	0.31428	0.30835	0.30245	0.29659	0.29077
5.7	0.7749	0.33377	0.32798	0.32222	0.31650	0.31081	0.30515	0.29953	0.29393	0.28837	0.28284
5.8	0.7313	0.32012	0.31470	0.30931	0.30394	0.29860	0.29329	0.28800	0.28273	0.27750	0.27229
5.9	0.6917	0.30403	0.29899	0.29398	0.28898	0.28401	0.27906	0.27413	0.26921	0.26432	0.25945
6.0	0.6557	0.28594	0.28129	0.27666	0.27205	0.26746	0.26288	0.25831	0.25377	0.24923	0.24472

6.1	0.6227	0.26631	0.26206	0.25782	0.25359	0.24938	0.24518	0.24099	0.23681	0.23265	0.22850
6.2	0.5926	0.24561	0.24175	0.23790	0.23406	0.23022	0.22640	0.22259	0.21878	0.21499	0.21120
6.3	0.5649	0.22431	0.22083	0.21736	0.21390	0.21044	0.20699	0.20354	0.20010	0.19667	0.19325
6.4	0.5394	0.20285	0.19974	0.19663	0.19353	0.19044	0.18735	0.18426	0.18119	0.17811	0.17504
6.5	0.5158	0.18163	0.17887	0.17612	0.17337	0.17062	0.16788	0.16514	0.16240	0.15967	0.15695
6.6	0.4940	0.16101	0.15858	0.15616	0.15374	0.15133	0.14891	0.14650	0.14410	0.14169	0.13929
6.7	0.4739	0.14129	0.13918	0.13707	0.13496	0.13286	0.13075	0.12865	0.12655	0.12445	0.12236
6.8	0.4551	0.12274	0.12092	0.11910	0.11728	0.11546	0.11364	0.11182	0.11001	0.10819	0.10638
6.9	0.4376	0.10554	0.10398	0.10242	0.10086	0.09931	0.09775	0.09620	0.09464	0.09309	0.09153
7.0	0.4214	0.08982	0.08850	0.08718	0.08586	0.08454	0.08322	0.08190	0.08058	0.07926	0.07794
7.1	0.4062	0.07565	0.07454	0.07343	0.07232	0.07122	0.07011	0.06900	0.06789	0.06679	0.06568
7.2	0.3919	0.06306	0.06214	0.06121	0.06029	0.05937	0.05845	0.05753	0.05661	0.05569	0.05476
7.3	0.3786	0.05201	0.05125	0.05049	0.04974	0.04898	0.04822	0.04746	0.04670	0.04594	0.04519
7.4	0.3661	0.04245	0.04183	0.04122	0.04060	0.03998	0.03936	0.03874	0.03812	0.03751	0.03689
7.5	0.3543	0.03429	0.03379	0.03329	0.03279	0.03229	0.03179	0.03129	0.03079	0.03030	0.02980
7.6	0.3432	0.02740	0.02700	0.02660	0.02620	0.02581	0.02541	0.02501	0.02461	0.02421	0.02382
7.7	0.3327	0.02166	0.02135	0.02104	0.02072	0.02041	0.02009	0.01978	0.01946	0.01915	0.01883
7.8	0.3228	0.01695	0.01670	0.01646	0.01621	0.01597	0.01572	0.01547	0.01523	0.01498	0.01474
7.9	0.3134	0.01312	0.01293	0.01274	0.01255	0.01236	0.01217	0.01198	0.01179	0.01160	0.01141
8.0	0.3046	0.01005	0.00990	0.00976	0.00961	0.00947	0.00932	0.00917	0.00903	0.00888	0.00874
8.1	0.2962	0.00762	0.00750	0.00739	0.00728	0.00717	0.00706	0.00695	0.00684	0.00673	0.00662
8.2	0.2882	0.00571	0.00563	0.00554	0.00546	0.00538	0.00530	0.00521	0.00513	0.00505	0.00497
8.3	0.2806	0.00424	0.00418	0.00411	0.00405	0.00399	0.00393	0.00387	0.00381	0.00375	0.00368
8.4	0.2734	0.00311	0.00307	0.00302	0.00297	0.00293	0.00288	0.00284	0.00279	0.00275	0.00270
8.5	0.2666	0.00226	0.00223	0.00219	0.00216	0.00213	0.00210	0.00206	0.00203	0.00200	0.00196
8.6	0.2600	0.00162	0.00160	0.00158	0.00155	0.00153	0.00151	0.00148	0.00146	0.00144	0.00141
8.7	0.2538	0.00116	0.00114	0.00112	0.00110	0.00109	0.00107	0.00105	0.00104	0.00102	0.00100
8.8	0.2478	0.00081	0.00080	0.00079	0.00078	0.00077	0.00075	0.00074	0.00073	0.00072	0.00071
8.9	0.2421	0.00057	0.00056	0.00055	0.00054	0.00053	0.00053	0.00052	0.00051	0.00050	0.00049
9.0	0.2367	0.00039	0.00038	0.00038	0.00037	0.00037	0.00036	0.00036	0.00035	0.00034	0.00034

Table II (*cont.*)

Y	Q/Z	Percentage natural response rate, C									
		41	42	43	44	45	46	47	48	49	50
1.1	5033.5986	0.00000	0.00000	0.00000	0.00000	0.00000	0.00000	0.00000	0.00000	0.00000	0.00000
1.2	3425.0323	0.00000	0.00000	0.00000	0.00000	0.00000	0.00000	0.00000	0.00000	0.00000	0.00000
1.3	2353.9045	0.00000	0.00000	0.00000	0.00000	0.00000	0.00000	0.00000	0.00000	0.00000	0.00000
1.4	1633.9888	0.00000	0.00000	0.00000	0.00000	0.00000	0.00000	0.00000	0.00000	0.00000	0.00000
1.5	1145.6253	0.00000	0.00000	0.00000	0.00000	0.00000	0.00000	0.00000	0.00000	0.00000	0.00000
1.6	811.2705	0.00000	0.00000	0.00000	0.00000	0.00000	0.00000	0.00000	0.00000	0.00000	0.00000
1.7	580.2476	0.00000	0.00000	0.00000	0.00000	0.00000	0.00000	0.00000	0.00000	0.00000	0.00000
1.8	419.1594	0.00001	0.00001	0.00001	0.00001	0.00001	0.00001	0.00001	0.00001	0.00001	0.00001
1.9	305.8120	0.00002	0.00001	0.00001	0.00001	0.00001	0.00001	0.00001	0.00001	0.00001	0.00001
2.0	225.3349	0.00003	0.00003	0.00003	0.00002	0.00002	0.00002	0.00002	0.00002	0.00002	0.00002
2.1	167.6823	0.00005	0.00005	0.00005	0.00005	0.00004	0.00004	0.00004	0.00004	0.00004	0.00004
2.2	126.0124	0.00009	0.00009	0.00008	0.00008	0.00008	0.00007	0.00007	0.00007	0.00007	0.00006
2.3	95.6280	0.00016	0.00015	0.00014	0.00014	0.00013	0.00013	0.00012	0.00012	0.00011	0.00011
2.4	73.2784	0.00026	0.00025	0.00024	0.00023	0.00023	0.00022	0.00021	0.00020	0.00019	0.00018
2.5	56.6963	0.00044	0.00042	0.00041	0.00039	0.00038	0.00036	0.00035	0.00033	0.00032	0.00031
2.6	44.2877	0.00072	0.00069	0.00066	0.00064	0.00061	0.00059	0.00056	0.00054	0.00052	0.00050
2.7	34.9234	0.00115	0.00110	0.00106	0.00102	0.00098	0.00094	0.00090	0.00087	0.00083	0.00080
2.8	27.7973	0.00180	0.00173	0.00166	0.00160	0.00153	0.00147	0.00142	0.00136	0.00131	0.00126
2.9	22.3296	0.00276	0.00265	0.00255	0.00245	0.00236	0.00226	0.00218	0.00209	0.00201	0.00194
3.0	18.1002	0.00416	0.00399	0.00384	0.00369	0.00355	0.00341	0.00328	0.00315	0.00303	0.00292
3.1	14.8026	0.00613	0.00589	0.00566	0.00544	0.00523	0.00503	0.00484	0.00466	0.00448	0.00431
3.2	12.2111	0.00885	0.00851	0.00818	0.00787	0.00757	0.00728	0.00701	0.00674	0.00649	0.00624
3.3	10.1589	0.01252	0.01204	0.01159	0.01115	0.01073	0.01033	0.00994	0.00957	0.00921	0.00886
3.4	8.5214	0.01736	0.01671	0.01609	0.01549	0.01491	0.01436	0.01382	0.01331	0.01282	0.01234
3.5	7.2051	0.02360	0.02273	0.02189	0.02109	0.02031	0.01957	0.01885	0.01816	0.01749	0.01685

x											
										0.02342	0.02257
3.7	5.2705	0.04107	0.03961	0.03820	0.03684	0.03554	0.03427	0.03306	0.03188	0.03074	0.02964
3.8	4.5571	0.05261	0.05078	0.04901	0.04730	0.04566	0.04407	0.04253	0.04105	0.03961	0.03821
3.9	3.9675	0.06611	0.06386	0.06169	0.05959	0.05757	0.05560	0.05370	0.05186	0.05008	0.04835
4.0	3.4771	0.08153	0.07883	0.07622	0.07369	0.07124	0.06887	0.06657	0.06433	0.06217	0.06006
4.1	3.0665	0.09872	0.09554	0.09246	0.08948	0.08658	0.08376	0.08103	0.07838	0.07579	0.07328
4.2	2.7206	0.11743	0.11376	0.11020	0.10674	0.10337	0.10010	0.09692	0.09382	0.09080	0.08786
4.3	2.4276	0.13729	0.13314	0.12910	0.12516	0.12133	0.11759	0.11395	0.11040	0.10694	0.10357
4.4	2.1780	0.15787	0.15325	0.14874	0.14434	0.14005	0.13587	0.13178	0.12778	0.12388	0.12007
4.5	1.9640	0.17864	0.17359	0.16865	0.16382	0.15910	0.15448	0.14997	0.14555	0.14122	0.13699
4.6	1.7797	0.19906	0.19362	0.18829	0.18307	0.17796	0.17295	0.16804	0.16323	0.15852	0.15390
4.7	1.6202	0.21857	0.21280	0.20713	0.20158	0.19612	0.19077	0.18552	0.18036	0.17530	0.17032
4.8	1.4813	0.23662	0.23058	0.22464	0.21881	0.21307	0.20744	0.20189	0.19644	0.19108	0.18581
4.9	1.3599	0.25270	0.24647	0.24033	0.23428	0.22833	0.22247	0.21670	0.21102	0.20542	0.19990
5.0	1.2533	0.26639	0.26003	0.25376	0.24757	0.24148	0.23546	0.22953	0.22368	0.21790	0.21221
5.1	1.1593	0.27732	0.27091	0.26458	0.25832	0.25215	0.24605	0.24002	0.23407	0.22819	0.22237
5.2	1.0759	0.28524	0.27884	0.27252	0.26626	0.26008	0.25396	0.24790	0.24192	0.23599	0.23013
5.3	1.0018	0.28998	0.28366	0.27741	0.27122	0.26509	0.25901	0.25300	0.24704	0.24114	0.23530
5.4	0.9357	0.29148	0.28530	0.27918	0.27311	0.26709	0.26113	0.25521	0.24935	0.24353	0.23776
5.5	0.8764	0.28977	0.28379	0.27785	0.27196	0.26611	0.26031	0.25454	0.24882	0.24314	0.23751
5.6	0.8230	0.28499	0.27925	0.27354	0.26787	0.26224	0.25664	0.25108	0.24555	0.24006	0.23461
5.7	0.7749	0.27734	0.27187	0.26644	0.26103	0.25565	0.25031	0.24499	0.23970	0.23444	0.22921
5.8	0.7313	0.26710	0.26194	0.25680	0.25169	0.24660	0.24154	0.23650	0.23149	0.22650	0.22153
5.9	0.6917	0.25460	0.24977	0.24496	0.24017	0.23540	0.23064	0.22591	0.22120	0.21650	0.21183
6.0	0.6557	0.24022	0.23573	0.23127	0.22681	0.22238	0.21795	0.21355	0.20916	0.20478	0.20042
6.1	0.6227	0.22436	0.22023	0.21611	0.21201	0.20792	0.20384	0.19977	0.19572	0.19167	0.18764
6.2	0.5926	0.20742	0.20366	0.19990	0.19615	0.19241	0.18868	0.18496	0.18125	0.17755	0.17385
6.3	0.5649	0.18984	0.18643	0.18302	0.17963	0.17624	0.17286	0.16949	0.16612	0.16276	0.15940
6.4	0.5394	0.17198	0.16892	0.16587	0.16282	0.15978	0.15674	0.15371	0.15068	0.14766	0.14464
6.5	0.5158	0.15422	0.15150	0.14879	0.14608	0.14337	0.14066	0.13796	0.13527	0.13257	0.12989

Table II (*cont.*)

Y	Q/Z	Percentage natural response rate, C									
		41	42	43	44	45	46	47	48	49	50
6.6	0.4940	0.13689	0.13450	0.13210	0.12971	0.12732	0.12494	0.12255	0.12017	0.11780	0.11542
6.7	0.4739	0.12026	0.11817	0.11608	0.11399	0.11191	0.10982	0.10774	0.10566	0.10358	0.10150
6.8	0.4551	0.10457	0.10276	0.10095	0.09914	0.09734	0.09553	0.09373	0.09193	0.09013	0.08833
6.9	0.4376	0.08998	0.08843	0.08688	0.08533	0.08378	0.08224	0.08069	0.07914	0.07760	0.07606
7.0	0.4214	0.07663	0.07531	0.07399	0.07268	0.07137	0.07005	0.06874	0.06743	0.06611	0.06480
7.1	0.4062	0.06457	0.06347	0.06236	0.06126	0.06015	0.05905	0.05794	0.05684	0.05574	0.05463
7.2	0.3919	0.05384	0.05292	0.05200	0.05109	0.05017	0.04925	0.04833	0.04741	0.04649	0.04557
7.3	0.3786	0.04443	0.04367	0.04291	0.04215	0.04140	0.04064	0.03988	0.03913	0.03837	0.03761
7.4	0.3661	0.03627	0.03565	0.03504	0.03442	0.03380	0.03318	0.03257	0.03195	0.03133	0.03072
7.5	0.3543	0.02930	0.02880	0.02830	0.02780	0.02731	0.02681	0.02631	0.02581	0.02531	0.02482
7.6	0.3432	0.02342	0.02302	0.02262	0.02222	0.02183	0.02143	0.02103	0.02063	0.02023	0.01984
7.7	0.3327	0.01852	0.01820	0.01789	0.01757	0.01726	0.01695	0.01663	0.01632	0.01600	0.01569
7.8	0.3228	0.01449	0.01424	0.01400	0.01375	0.01351	0.01326	0.01301	0.01277	0.01252	0.01228
7.9	0.3134	0.01122	0.01103	0.01084	0.01065	0.01046	0.01027	0.01007	0.00988	0.00969	0.00950
8.0	0.3046	0.00859	0.00845	0.00830	0.00815	0.00801	0.00786	0.00772	0.00757	0.00743	0.00728
8.1	0.2962	0.00651	0.00640	0.00629	0.00618	0.00607	0.00596	0.00585	0.00574	0.00563	0.00552
8.2	0.2882	0.00488	0.00480	0.00472	0.00463	0.00455	0.00447	0.00439	0.00430	0.00422	0.00414
8.3	0.2806	0.00362	0.00356	0.00350	0.00344	0.00338	0.00332	0.00325	0.00319	0.00313	0.00307
8.4	0.2734	0.00266	0.00261	0.00257	0.00252	0.00248	0.00243	0.00239	0.00234	0.00230	0.00225
8.5	0.2666	0.00193	0.00190	0.00187	0.00183	0.00180	0.00177	0.00174	0.00170	0.00167	0.00164
8.6	0.2600	0.00139	0.00137	0.00134	0.00132	0.00129	0.00127	0.00125	0.00122	0.00120	0.00118
8.7	0.2538	0.00099	0.00097	0.00095	0.00094	0.00092	0.00090	0.00089	0.00087	0.00085	0.00084
8.8	0.2478	0.00070	0.00068	0.00067	0.00066	0.00065	0.00064	0.00062	0.00061	0.00060	0.00059
8.9	0.2421	0.00048	0.00048	0.00047	0.00046	0.00045	0.00044	0.00043	0.00043	0.00042	0.00041
9.0	0.2367	0.00033	0.00033	0.00032	0.00032	0.00031	0.00031	0.00030	0.00029	0.00029	0.00028

Percentage natural response rate, C

Y	Q/Z	52	54	56	58	60	62	64	66	68	70
1.1	5033.5986	0.00000	0.00000	0.00000	0.00000	0.00000	0.00000	0.00000	0.00000	0.00000	0.00000
1.2	3425.0323	0.00000	0.00000	0.00000	0.00000	0.00000	0.00000	0.00000	0.00000	0.00000	0.00000
1.3	2353.9045	0.00000	0.00000	0.00000	0.00000	0.00000	0.00000	0.00000	0.00000	0.00000	0.00000
1.4	1633.9888	0.00000	0.00000	0.00000	0.00000	0.00000	0.00000	0.00000	0.00000	0.00000	0.00000
1.5	1145.6253	0.00000	0.00000	0.00000	0.00000	0.00000	0.00000	0.00000	0.00000	0.00000	0.00000
1.6	811.2705	0.00000	0.00000	0.00000	0.00000	0.00000	0.00000	0.00000	0.00000	0.00000	0.00000
1.7	580.2476	0.00000	0.00000	0.00000	0.00000	0.00000	0.00000	0.00000	0.00000	0.00000	0.00000
1.8	419.1594	0.00001	0.00000	0.00000	0.00000	0.00000	0.00000	0.00000	0.00000	0.00000	0.00000
1.9	305.8120	0.00001	0.00001	0.00001	0.00001	0.00001	0.00001	0.00001	0.00001	0.00001	0.00000
2.0	225.3349	0.00002	0.00002	0.00002	0.00001	0.00001	0.00001	0.00001	0.00001	0.00001	0.00001
2.1	167.6823	0.00003	0.00003	0.00003	0.00003	0.00002	0.00002	0.00002	0.00002	0.00002	0.00002
2.2	126.0124	0.00006	0.00005	0.00005	0.00005	0.00004	0.00004	0.00004	0.00003	0.00003	0.00003
2.3	95.6280	0.00010	0.00009	0.00009	0.00008	0.00007	0.00007	0.00006	0.00006	0.00005	0.00005
2.4	73.2784	0.00017	0.00016	0.00015	0.00013	0.00012	0.00011	0.00010	0.00010	0.00009	0.00008
2.5	56.6963	0.00028	0.00026	0.00024	0.00022	0.00021	0.00019	0.00017	0.00016	0.00015	0.00013
2.6	44.2877	0.00046	0.00043	0.00039	0.00036	0.00034	0.00031	0.00028	0.00026	0.00024	0.00022
2.7	34.9234	0.00074	0.00068	0.00063	0.00058	0.00054	0.00049	0.00045	0.00042	0.00038	0.00035
2.8	27.7973	0.00116	0.00107	0.00099	0.00091	0.00084	0.00078	0.00071	0.00065	0.00060	0.00054
2.9	22.3296	0.00179	0.00165	0.00153	0.00141	0.00130	0.00119	0.00110	0.00101	0.00092	0.00084
3.0	18.1002	0.00270	0.00249	0.00230	0.00213	0.00196	0.00180	0.00166	0.00152	0.00139	0.00127
3.1	14.8026	0.00399	0.00369	0.00341	0.00314	0.00290	0.00267	0.00245	0.00225	0.00206	0.00188
3.2	12.2111	0.00578	0.00534	0.00494	0.00456	0.00421	0.00388	0.00356	0.00327	0.00299	0.00273
3.3	10.1589	0.00821	0.00760	0.00703	0.00649	0.00599	0.00552	0.00508	0.00466	0.00427	0.00389
3.4	8.5214	0.01144	0.01059	0.00981	0.00907	0.00837	0.00772	0.00710	0.00652	0.00597	0.00545
3.5	7.2051	0.01563	0.01449	0.01342	0.01242	0.01147	0.01058	0.00975	0.00895	0.00820	0.00749

Table II (*cont.*)

Y	Q/Z	Percentage natural response rate, C									
		52	54	56	58	60	62	64	66	68	70
3.6	6.1394	0.02095	0.01944	0.01802	0.01668	0.01543	0.01424	0.01312	0.01206	0.01106	0.01010
3.7	5.2705	0.02755	0.02559	0.02374	0.02200	0.02036	0.01881	0.01735	0.01595	0.01463	0.01338
3.8	4.5571	0.03556	0.03306	0.03070	0.02848	0.02638	0.02440	0.02251	0.02072	0.01902	0.01740
3.9	3.9675	0.04504	0.04193	0.03899	0.03620	0.03357	0.03107	0.02870	0.02644	0.02429	0.02224
4.0	3.4771	0.05603	0.05222	0.04862	0.04520	0.04196	0.03887	0.03594	0.03314	0.03047	0.02793
4.1	3.0665	0.06847	0.06390	0.05956	0.05545	0.05153	0.04779	0.04423	0.04083	0.03758	0.03447
4.2	2.7206	0.08221	0.07684	0.07172	0.06685	0.06220	0.05776	0.05352	0.04946	0.04557	0.04184
4.3	2.4276	0.09705	0.09084	0.08492	0.07926	0.07384	0.06865	0.06368	0.05892	0.05434	0.04995
4.4	2.1780	0.11270	0.10565	0.09890	0.09244	0.08623	0.08028	0.07456	0.06906	0.06377	0.05867
4.5	1.9640	0.12879	0.12092	0.11336	0.10610	0.09912	0.09240	0.08592	0.07968	0.07366	0.06785
4.6	1.7797	0.14491	0.13627	0.12794	0.11992	0.11218	0.10471	0.09750	0.09053	0.08379	0.07727
4.7	1.6202	0.16064	0.15129	0.14225	0.13352	0.12507	0.11690	0.10899	0.10132	0.09389	0.08669
4.8	1.4813	0.17551	0.16554	0.15588	0.14652	0.13744	0.12863	0.12007	0.11177	0.10370	0.09585
4.9	1.3599	0.18911	0.17863	0.16844	0.15854	0.14891	0.13954	0.13043	0.12155	0.11291	0.10449
5.0	1.2533	0.20104	0.19016	0.17956	0.16923	0.15915	0.14933	0.13975	0.13039	0.12126	0.11234
5.1	1.1593	0.21096	0.19981	0.18891	0.17827	0.16787	0.15769	0.14775	0.13802	0.12850	0.11918
5.2	1.0759	0.21860	0.20730	0.19624	0.18541	0.17479	0.16439	0.15419	0.14420	0.13440	0.12478
5.3	1.0018	0.22377	0.21246	0.20135	0.19045	0.17975	0.16923	0.15891	0.14876	0.13879	0.12899
5.4	0.9357	0.22636	0.21516	0.20413	0.19328	0.18261	0.17210	0.16176	0.15158	0.14156	0.13169
5.5	0.8764	0.22636	0.21536	0.20453	0.19385	0.18332	0.17293	0.16270	0.15260	0.14264	0.13281
5.6	0.8230	0.22380	0.21313	0.20259	0.19219	0.18191	0.17175	0.16172	0.15181	0.14202	0.13235
5.7	0.7749	0.21884	0.20858	0.19843	0.18839	0.17846	0.16863	0.15891	0.14929	0.13977	0.13035
5.8	0.7313	0.21166	0.20189	0.19221	0.18262	0.17312	0.16371	0.15438	0.14514	0.13598	0.12690
5.9	0.6917	0.20253	0.19331	0.18417	0.17509	0.16609	0.15717	0.14831	0.13952	0.13080	0.12214
6.0	0.6557	0.19174	0.18312	0.17456	0.16606	0.15762	0.14923	0.14090	0.13263	0.12441	0.11624

x										
6.1	0.6227	0.17961	0.17163	0.16370	0.15580	0.14796	0.14016	0.13240	0.12469	0.11702	0.10940
6.2	0.5926	0.16649	0.15917	0.15188	0.14462	0.13740	0.13022	0.12307	0.11595	0.10887	0.10182
6.3	0.5649	0.15272	0.14606	0.13942	0.13282	0.12624	0.11969	0.11316	0.10666	0.10018	0.09374
6.4	0.5394	0.13862	0.13262	0.12665	0.12069	0.11475	0.10883	0.10293	0.09705	0.09119	0.08535
6.5	0.5158	0.12452	0.11916	0.11383	0.10850	0.10319	0.09790	0.09262	0.08736	0.08210	0.07687
6.6	0.4940	0.11068	0.10595	0.10123	0.09652	0.09182	0.08713	0.08245	0.07779	0.07313	0.06848
6.7	0.4739	0.09735	0.09321	0.08908	0.08495	0.08033	0.07672	0.07262	0.06852	0.06443	0.06035
6.8	0.4551	0.08473	0.08114	0.07756	0.07398	0.07040	0.06683	0.06327	0.05971	0.05616	0.05261
6.9	0.4376	0.07297	0.06989	0.06681	0.06374	0.06067	0.05760	0.05454	0.05148	0.04842	0.04537
7.0	0.4214	0.06218	0.05956	0.05695	0.05433	0.05172	0.04911	0.04651	0.04390	0.04130	0.03870
7.1	0.4062	0.05243	0.05023	0.04803	0.04583	0.04363	0.04143	0.03924	0.03704	0.03485	0.03266
7.2	0.3919	0.04374	0.04190	0.04007	0.03824	0.03641	0.03458	0.03275	0.03092	0.02909	0.02727
7.3	0.3786	0.03610	0.03459	0.03308	0.03157	0.03006	0.02855	0.02704	0.02553	0.02403	0.02252
7.4	0.3661	0.02948	0.02825	0.02702	0.02578	0.02455	0.02332	0.02209	0.02086	0.01963	0.01840
7.5	0.3543	0.02382	0.02283	0.02183	0.02084	0.01984	0.01885	0.01785	0.01686	0.01586	0.01487
7.6	0.3432	0.01904	0.01825	0.01745	0.01666	0.01586	0.01507	0.01427	0.01348	0.01269	0.01189
7.7	0.3327	0.01506	0.01443	0.01380	0.01317	0.01255	0.01192	0.01129	0.01066	0.01003	0.00941
7.8	0.3228	0.01178	0.01129	0.01080	0.01031	0.00982	0.00933	0.00884	0.00834	0.00785	0.00736
7.9	0.3134	0.00912	0.00874	0.00836	0.00798	0.00769	0.00722	0.00684	0.00646	0.00608	0.00570
8.0	0.3046	0.00699	0.00670	0.00641	0.00611	0.00582	0.00553	0.00524	0.00495	0.00466	0.00437
8.1	0.2962	0.00530	0.00508	0.00486	0.00463	0.00441	0.00419	0.00397	0.00375	0.00353	0.00331
8.2	0.2882	0.00397	0.00381	0.00364	0.00348	0.00331	0.00314	0.00298	0.00281	0.00265	0.00248
8.3	0.2806	0.00295	0.00282	0.00270	0.00258	0.00246	0.00233	0.00221	0.00209	0.00196	0.00184
8.4	0.2734	0.00216	0.00207	0.00198	0.00189	0.00180	0.00171	0.00162	0.00153	0.00144	0.00135
8.5	0.2666	0.00157	0.00151	0.00144	0.00138	0.00131	0.00124	0.00118	0.00111	0.00105	0.00098
8.6	0.2600	0.00113	0.00108	0.00104	0.00099	0.00094	0.00089	0.00085	0.00080	0.00075	0.00071
8.7	0.2538	0.00080	0.00077	0.00074	0.00070	0.00067	0.00064	0.00060	0.00057	0.00054	0.00050
8.8	0.2478	0.00057	0.00054	0.00052	0.00049	0.00047	0.00045	0.00042	0.00040	0.00038	0.00035
8.9	0.2421	0.00039	0.00038	0.00036	0.00034	0.00033	0.00031	0.00030	0.00028	0.00026	0.00025
9.0	0.2367	0.00027	0.00026	0.00025	0.00024	0.00023	0.00021	0.00020	0.00019	0.00018	0.00017

Table II (cont.)

Y	Q/Z	Percentage natural response rate, C									
		72	74	76	78	80	82	84	86	88	90
1.1	5033.5986	0.00000	0.00000	0.00000	0.00000	0.00000	0.00000	0.00000	0.00000	0.00000	0.00000
1.2	3425.0323	0.00000	0.00000	0.00000	0.00000	0.00000	0.00000	0.00000	0.00000	0.00000	0.00000
1.3	2353.9045	0.00000	0.00000	0.00000	0.00000	0.00000	0.00000	0.00000	0.00000	0.00000	0.00000
1.4	1633.9888	0.00000	0.00000	0.00000	0.00000	0.00000	0.00000	0.00000	0.00000	0.00000	0.00000
1.5	1145.6253	0.00000	0.00000	0.00000	0.00000	0.00000	0.00000	0.00000	0.00000	0.00000	0.00000
1.6	811.2705	0.00000	0.00000	0.00000	0.00000	0.00000	0.00000	0.00000	0.00000	0.00000	0.00000
1.7	580.2476	0.00000	0.00000	0.00000	0.00000	0.00000	0.00000	0.00000	0.00000	0.00000	0.00000
1.8	419.1594	0.00000	0.00000	0.00000	0.00000	0.00000	0.00000	0.00000	0.00000	0.00000	0.00000
1.9	305.8120	0.00000	0.00000	0.00000	0.00000	0.00000	0.00000	0.00000	0.00000	0.00000	0.00000
2.0	225.3349	0.00001	0.00001	0.00001	0.00001	0.00000	0.00000	0.00000	0.00000	0.00000	0.00000
2.1	167.6823	0.00001	0.00001	0.00001	0.00001	0.00001	0.00001	0.00001	0.00001	0.00001	0.00000
2.2	126.0124	0.00002	0.00002	0.00002	0.00002	0.00002	0.00001	0.00001	0.00001	0.00001	0.00001
2.3	95.6280	0.00004	0.00004	0.00003	0.00003	0.00003	0.00002	0.00002	0.00002	0.00001	0.00001
2.4	73.2784	0.00007	0.00007	0.00006	0.00005	0.00005	0.00004	0.00004	0.00003	0.00003	0.00002
2.5	56.6963	0.00012	0.00011	0.00010	0.00009	0.00008	0.00007	0.00006	0.00005	0.00004	0.00003
2.6	44.2877	0.00020	0.00018	0.00016	0.00014	0.00013	0.00011	0.00010	0.00008	0.00007	0.00006
2.7	34.9234	0.00031	0.00028	0.00026	0.00023	0.00020	0.00018	0.00015	0.00013	0.00011	0.00009
2.8	27.7973	0.00049	0.00045	0.00040	0.00036	0.00032	0.00028	0.00024	0.00021	0.00017	0.00014
2.9	22.3296	0.00076	0.00069	0.00062	0.00055	0.00049	0.00043	0.00037	0.00032	0.00027	0.00022
3.0	18.1002	0.00115	0.00104	0.00094	0.00084	0.00074	0.00065	0.00057	0.00048	0.00041	0.00033
3.1	14.8026	0.00170	0.00154	0.00139	0.00124	0.00110	0.00097	0.00084	0.00072	0.00060	0.00049
3.2	12.2111	0.00248	0.00224	0.00202	0.00181	0.00160	0.00141	0.00122	0.00105	0.00088	0.00072
3.3	10.1589	0.00354	0.00320	0.00288	0.00258	0.00229	0.00201	0.00175	0.00150	0.00125	0.00102
3.4	8.5214	0.00496	0.00449	0.00404	0.00362	0.00321	0.00282	0.00245	0.00210	0.00176	0.00144
3.5	7.2051	0.00681	0.00617	0.00556	0.00498	0.00442	0.00389	0.00338	0.00289	0.00243	0.00198

x											
3.6	6.1394	0.00269	0.00329	0.00392	0.00457	0.00526	0.00598	0.00673	0.00751	0.00833	0.00920
3.7	5.2705	0.00357	0.00438	0.00521	0.00608	0.00699	0.00794	0.00893	0.00996	0.01105	0.01219
3.8	4.5571	0.00467	0.00572	0.00681	0.00794	0.00912	0.01035	0.01164	0.01298	0.01439	0.01586
3.9	3.9675	0.00601	0.00735	0.00875	0.01020	0.01170	0.01328	0.01492	0.01663	0.01841	0.02028
4.0	3.4771	0.00760	0.00929	0.01104	0.01287	0.01476	0.01673	0.01879	0.02093	0.02316	0.02549
4.1	3.0665	0.00945	0.01154	0.01371	0.01597	0.01831	0.02074	0.02327	0.02590	0.02864	0.03149
4.2	2.7206	0.01156	0.01411	0.01676	0.01950	0.02233	0.02528	0.02834	0.03152	0.03482	0.03826
4.3	2.4276	0.01392	0.01698	0.02015	0.02342	0.02681	0.03032	0.03396	0.03774	0.04165	0.04572
4.4	2.1780	0.01650	0.02011	0.02384	0.02770	0.03168	0.03580	0.04006	0.04446	0.04903	0.05377
4.5	1.9640	0.01926	0.02346	0.02779	0.03225	0.03685	0.04161	0.04651	0.05158	0.05682	0.06224
4.6	1.7797	0.02214	0.02695	0.03190	0.03699	0.04223	0.04763	0.05319	0.05893	0.06485	0.07096
4.7	1.6202	0.02509	0.03051	0.03608	0.04180	0.04767	0.05372	0.05994	0.06633	0.07292	0.07970
4.8	1.4813	0.02802	0.03404	0.04022	0.04655	0.05305	0.05971	0.06656	0.07359	0.08081	0.08822
4.9	1.3599	0.03085	0.03745	0.04421	0.05112	0.05820	0.06544	0.07287	0.08048	0.08828	0.09628
5.0	1.2533	0.03351	0.04064	0.04792	0.05536	0.06296	0.07074	0.07868	0.08681	0.09513	0.10364
5.1	1.1593	0.03589	0.04349	0.05124	0.05914	0.06720	0.07543	0.08382	0.09238	0.10113	0.11006
5.2	1.0759	0.03794	0.04593	0.05407	0.06235	0.07078	0.07937	0.08811	0.09702	0.10610	0.11535
5.3	1.0018	0.03958	0.04788	0.05631	0.06488	0.07358	0.08244	0.09144	0.10059	0.10990	0.11936
5.4	0.9357	0.04076	0.04927	0.05790	0.06665	0.07553	0.08454	0.09369	0.10298	0.11240	0.12197
5.5	0.8764	0.04145	0.05006	0.05878	0.06762	0.07656	0.08563	0.09482	0.10413	0.11356	0.12312
5.6	0.8230	0.04163	0.05024	0.05895	0.06775	0.07666	0.08567	0.09479	0.10402	0.11335	0.12279
5.7	0.7749	0.04130	0.04980	0.05839	0.06707	0.07584	0.08469	0.09364	0.10267	0.11180	0.12103
5.8	0.7313	0.04047	0.04877	0.05715	0.06560	0.07413	0.08273	0.09141	0.10016	0.10900	0.11791
5.9	0.6917	0.03919	0.04720	0.05528	0.06341	0.07161	0.07987	0.08820	0.09659	0.10504	0.11356
6.0	0.6557	0.03750	0.04514	0.05284	0.06058	0.06838	0.07623	0.08412	0.09208	0.10008	0.10813
6.1	0.6227	0.03546	0.04267	0.04992	0.05721	0.06454	0.07192	0.07933	0.08678	0.09428	0.10182
6.2	0.5926	0.03315	0.03987	0.04663	0.05342	0.06023	0.06708	0.07397	0.08088	0.08783	0.09481
6.3	0.5649	0.03063	0.03683	0.04306	0.04930	0.05558	0.06187	0.06820	0.07454	0.08091	0.08731
6.4	0.5394	0.02799	0.03364	0.03931	0.04500	0.05071	0.05643	0.06218	0.06794	0.07372	0.07953
6.5	0.5158	0.02528	0.03037	0.03548	0.04061	0.04575	0.05090	0.05606	0.06124	0.06644	0.07165

308

Table II (cont.)

Y	Q/Z	Percentage natural response rate, C									
		72	74	76	78	80	82	84	86	88	90
6.6	0.4940	0.06384	0.05922	0.05460	0.05000	0.04540	0.04082	0.03624	0.03168	0.02712	0.02258
6.7	0.4739	0.05628	0.05221	0.04815	0.04410	0.04005	0.03601	0.03198	0.02796	0.02395	0.01994
6.8	0.4551	0.04907	0.04553	0.04200	0.03847	0.03495	0.03143	0.02792	0.02441	0.02091	0.01741
6.9	0.4376	0.04232	0.03927	0.03623	0.03319	0.03016	0.02713	0.02410	0.02107	0.01805	0.01504
7.0	0.4214	0.03611	0.03351	0.03092	0.02833	0.02574	0.02316	0.02058	0.01800	0.01542	0.01284
7.1	0.4062	0.03047	0.02829	0.02610	0.02392	0.02174	0.01956	0.01738	0.01520	0.01302	0.01085
7.2	0.3919	0.02544	0.02362	0.02180	0.01997	0.01815	0.01633	0.01451	0.01270	0.01088	0.00906
7.3	0.3786	0.02101	0.01951	0.01800	0.01650	0.01500	0.01349	0.01199	0.01049	0.00899	0.00749
7.4	0.3661	0.01717	0.01594	0.01471	0.01348	0.01226	0.01103	0.00980	0.00857	0.00735	0.00612
7.5	0.3543	0.01388	0.01289	0.01189	0.01090	0.00991	0.00892	0.00792	0.00693	0.00594	0.00495
7.6	0.3432	0.01110	0.01030	0.00951	0.00872	0.00792	0.00713	0.00634	0.00555	0.00475	0.00396
7.7	0.3327	0.00878	0.00815	0.00752	0.00690	0.00627	0.00564	0.00501	0.00439	0.00376	0.00313
7.8	0.3228	0.00687	0.00638	0.00589	0.00540	0.00491	0.00442	0.00392	0.00343	0.00294	0.00245
7.9	0.3134	0.00532	0.00494	0.00456	0.00418	0.00380	0.00342	0.00304	0.00266	0.00228	0.00190
8.0	0.3046	0.00408	0.00378	0.00349	0.00320	0.00291	0.00262	0.00233	0.00204	0.00175	0.00146
8.1	0.2962	0.00309	0.00287	0.00265	0.00243	0.00221	0.00199	0.00176	0.00154	0.00132	0.00110
8.2	0.2882	0.00232	0.00215	0.00199	0.00182	0.00165	0.00149	0.00132	0.00116	0.00099	0.00083
8.3	0.2806	0.00172	0.00160	0.00147	0.00135	0.00123	0.00110	0.00098	0.00086	0.00074	0.00061
8.4	0.2734	0.00126	0.00117	0.00108	0.00099	0.00090	0.00081	0.00072	0.00063	0.00054	0.00045
8.5	0.2666	0.00092	0.00085	0.00079	0.00072	0.00065	0.00059	0.00052	0.00046	0.00039	0.00033
8.6	0.2600	0.00066	0.00061	0.00056	0.00052	0.00047	0.00042	0.00038	0.00033	0.00028	0.00024
8.7	0.2538	0.00047	0.00044	0.00040	0.00037	0.00033	0.00030	0.00027	0.00023	0.00020	0.00017
8.8	0.2478	0.00033	0.00031	0.00028	0.00026	0.00024	0.00021	0.00019	0.00016	0.00014	0.00012
8.9	0.2421	0.00023	0.00021	0.00020	0.00018	0.00016	0.00015	0.00013	0.00011	0.00010	0.00008
9.0	0.2367	0.00016	0.00015	0.00014	0.00012	0.00011	0.00010	0.00009	0.00008	0.00007	0.00006

Table III Maximum and minimum working probits and ranges

Minimum working probits			Maximum working probits	
Expected probit Y	$y_0 = Y - P/Z$	Range $1/Z$	$y_1 = Y + Q/Z$	Expected probit Y
1.1	0.8579	5033.8402	9.1421	8.9
1.2	0.9522	3425.2797	9.0478	8.8
1.3	1.0462	2354.1580	8.9538	8.7
1.4	1.1400	1634.2486	8.8600	8.6
1.5	1.2334	1145.8917	8.7666	8.5
1.6	1.3266	811.5439	8.6734	8.4
1.7	1.4194	580.5282	8.5806	8.3
1.8	1.5118	419.4475	8.4882	8.2
1.9	1.6038	306.1081	8.3962	8.1
2.0	1.6954	225.6395	8.3046	8.0
2.1	1.7866	167.9957	8.2134	7.9
2.2	1.8772	126.3352	8.1228	7.8
2.3	1.9673	95.9607	8.0327	7.7
2.4	2.0568	73.6216	7.9432	7.6
2.5	2.1457	57.0506	7.8543	7.5
2.6	2.2339	44.6538	7.7661	7.4
2.7	2.3214	35.3020	7.6786	7.3
2.8	2.4081	28.1892	7.5919	7.2
2.9	2.4938	22.7357	7.5062	7.1
3.0	2.5786	18.5216	7.4214	7.0
3.1	2.6624	15.2402	7.3376	6.9
3.2	2.7449	12.6662	7.2551	6.8
3.3	2.8261	10.6327	7.1739	6.7
3.4	2.9060	9.0154	7.0940	6.6
3.5	2.9842	7.7210	7.0158	6.5
3.6	3.0606	6.6788	6.9394	6.4
3.7	3.1351	5.8354	6.8649	6.3
3.8	3.2074	5.1497	6.7926	6.2
3.9	3.2773	4.5903	6.7227	6.1
4.0	3.3443	4.1327	6.6557	6.0

Table III (*cont.*)

Minimum working probits			Maximum working probits	
Expected probit Y	$y_0 = Y - P/Z$	Range $1/Z$	$y_1 = Y + Q/Z$	Expected probit Y
4.1	3.4083	3.7582	6.5917	5.9
4.2	3.4687	3.4519	6.5313	5.8
4.3	3.5251	3.2025	6.4749	5.7
4.4	3.5770	3.0010	6.4230	5.6
4.5	3.6236	2.8404	6.3764	5.5
4.6	3.6643	2.7154	6.3357	5.4
4.7	3.6982	2.6220	6.3018	5.3
4.8	3.7241	2.5573	6.2759	5.2
4.9	3.7407	2.5192	6.2593	5.1
5.0	3.7467	2.5066	6.2533	5.0
5.1	3.7401	2.5192	6.2599	4.9
5.2	3.7187	2.5573	6.2813	4.8
5.3	3.6798	2.6220	6.3202	4.7
5.4	3.6203	2.7154	6.3797	4.6
5.5	3.5360	2.8404	6.4640	4.5
5.6	3.4220	3.0010	6.5780	4.4
5.7	3.2724	3.2025	6.7276	4.3
5.8	3.0794	3.4519	6.9206	4.2
5.9	2.8335	3.7582	7.1665	4.1
6.0	2.5229	4.1327	7.4771	4.0
6.1	2.1325	4.5903	7.8675	3.9
6.2	1.6429	5.1497	8.3571	3.8
6.3	1.0295	5.8354	8.9705	3.7
6.4	0.2606	6.6788	9.7394	3.6
6.5	-0.7051	7.7210	10.7051	3.5

Table IV Working probits

($Y = 2.0$–2.9; 0–50% response)

% response	Expected probit, Y									
	2.0	2.1	2.2	2.3	2.4	2.5	2.6	2.7	2.8	2.9
0	1.695	1.787	1.877	1.967	2.057	2.146	2.234	2.321	2.408	2.494
1	3.951	3.467	3.141	2.927	2.793	2.716	2.681	2.674	2.690	2.721
2	6.207	5.147	4.404	3.886	3.529	3.287	3.127	3.027	.972	.949
3	8.463	6.827	5.667	4.846	4.265	.857	.574	.380	3.254	3.176
4	—	8.507	6.931	5.806	5.002	4.428	4.020	.733	.536	.403
5	—	—	8.194	6.765	.938	.998	.467	4.086	.818	.631
6	—	—	9.458	7.725	6.474	5.569	4.913	4.440	4.099	3.858
7	—	—	—	8.684	7.210	6.139	5.360	.793	.381	4.085
8	—	—	—	9.644	.946	.710	.806	5.146	.663	.313
9	—	—	—	—	8.683	7.280	6.253	.499	.945	.540
10	—	—	—	—	9.419	.851	.699	.852	5.227	.707
11	—	—	—	—	—	8.421	7.146	6.205	5.509	4.995
12	—	—	—	—	—	.992	.592	.558	.791	5.222
13	—	—	—	—	—	9.562	8.039	.911	6.073	.449
14	—	—	—	—	—	—	.486	7.264	.355	.677
15	—	—	—	—	—	—	.932	.617	.636	.904
16	—	—	—	—	—	—	9.379	7.970	6.918	6.132
17	—	—	—	—	—	—	.825	8.323	7.200	.359
18	—	—	—	—	—	—	—	.676	.482	.586
19	—	—	—	—	—	—	—	9.029	.761	.814
20	—	—	—	—	—	—	—	.382	8.046	7.041
21	—	—	—	—	—	—	—	9.735	8.328	7.268
22	—	—	—	—	—	—	—	—	.610	.496
23	—	—	—	—	—	—	—	—	.892	.723
24	—	—	—	—	—	—	—	—	9.173	.950
25	—	—	—	—	—	—	—	—	.455	8.178
26	—	—	—	—	—	—	—	—	9.737	8.405
27	—	—	—	—	—	—	—	—	—	.633
28	—	—	—	—	—	—	—	—	—	.860
29	—	—	—	—	—	—	—	—	—	9.087
30	—	—	—	—	—	—	—	—	—	.315
31	—	—	—	—	—	—	—	—	—	9.542
32	—	—	—	—	—	—	—	—	—	.769
33	—	—	—	—	—	—	—	—	—	.997
34	—	—	—	—	—	—	—	—	—	—
35	—	—	—	—	—	—	—	—	—	—

Table IV (cont.)

(Y = 3.0–3.9; 0–50% response)

% response	Expected probit, Y									
	3.0	3.1	3.2	3.3	3.4	3.5	3.6	3.7	3.8	3.9
0	2.579	2.662	2.745	2.826	2.906	2.984	3.061	3.135	3.207	3.277
1	2.764	2.815	2.872	2.932	2.996	3.061	3.127	3.193	3.259	3.323
2	.949	.967	.998	3.039	3.086	.139	.194	.252	.310	.369
3	3.134	3.120	3.125	.145	.176	.216	.261	.310	.362	.415
4	.319	.272	.252	.251	.267	.293	.328	.369	.413	.461
5	.505	.424	.378	.358	.357	.370	.395	.427	.465	.507
6	3.690	3.577	3.505	3.464	3.447	3.447	3.461	3.485	3.516	3.553
7	.875	.729	.632	.570	.537	.525	.528	.544	.568	.599
8	4.060	.882	.758	.677	.627	.602	.595	.602	.619	.645
9	.246	4.034	.885	.783	.717	.679	.662	.660	.671	.690
10	.431	.186	4.012	.889	.808	.756	.728	.719	.722	.736
11	4.616	4.339	4.138	3.996	3.898	3.834	3.795	3.777	3.774	3.782
12	.801	.491	.265	4.102	.988	.911	.862	.835	.825	.828
13	.986	.644	.391	.208	4.078	.988	.929	.894	.877	.874
14	5.172	.796	.518	.315	.168	4.065	.996	.952	.928	.920
15	.357	.948	.645	.421	.258	.142	4.062	4.010	.980	.966
16	5.542	5.101	4.771	4.527	4.348	4.220	4.129	4.069	4.031	4.012
17	.727	.253	.898	.634	.439	.297	.196	.127	.083	.058
18	.913	.406	5.025	.740	.529	.374	.263	.185	.134	.104
19	6.098	.558	.151	.846	.619	.451	.330	.244	.186	.149
20	.283	.710	.278	.953	.709	.528	.396	.302	.237	.195
21	6.468	5.863	5.405	5.059	4.799	4.606	4.463	4.361	4.289	4.241
22	.653	6.015	.531	.165	.889	.683	.530	.419	.340	.287
23	.839	.168	.658	.272	.979	.760	.597	.477	.392	.333
24	7.024	.320	.785	.378	5.070	.837	.664	.536	.443	.379
25	.209	.472	.911	.484	.160	.914	.730	.594	.495	.425
26	7.394	6.625	6.038	5.591	5.250	4.992	4.797	4.652	4.546	4.471
27	.580	.777	.165	.697	.340	5.069	.864	.711	.598	.517
28	.765	.930	.291	.803	.430	.146	.931	.769	.649	.563
29	.950	7.082	.418	.910	.520	.223	.997	.827	.701	.608
30	8.135	.234	.545	6.016	.610	.300	5.064	.886	.752	.654
31	8.320	7.387	6.671	6.122	5.701	5.378	5.131	4.944	4.804	4.700
32	.506	.539	.798	.229	.791	.455	.198	5.002	.855	.746
33	.691	.692	.925	.335	.881	.532	.265	.061	.907	.792
34	.876	.844	7.051	.441	.971	.609	.331	.119	.958	.838
35	9.061	.996	.178	.548	6.061	.687	.398	.177	5.010	.884
36	9.247	8.149	7.305	6.654	6.151	5.764	5.465	5.236	5.061	4.930
37	.432	.301	.431	.760	.242	.841	.532	.294	.113	.976
38	.617	.454	.558	.867	.332	.918	.599	.353	.164	5.022
39	.802	.606	.685	.973	.422	.995	.665	.411	.216	.068
40	.987	.758	.811	7.079	.512	6.073	.732	.469	.267	.113
41	—	8.911	7.938	7.186	6.602	6.150	5.799	5.528	5.319	5.159
42	—	9.063	8.065	.292	.692	.227	.866	.586	.370	.205
43	—	.216	.191	.398	.782	.304	.932	.644	.422	.251
44	—	.368	.318	.505	.873	.381	.999	.703	.473	.297
45	—	.520	.445	.611	.963	.459	6.066	.761	.525	.343
46	—	9.673	8.571	7.717	7.053	6.536	6.133	5.819	5.576	5.389
47	—	.825	.698	.824	.143	.613	.200	.878	.628	.435
48	—	.978	.825	.930	.233	.690	.266	.936	.679	.481
49	—	—	.951	8.036	.323	.767	.333	.994	.731	.527
50	—	—	9.078	.143	.414	.845	.400	6.053	.782	.572

Table IV (*cont.*)

($Y = 4.0$–4.9; 0–50% response)

% response	4.0	4.1	4.2	4.3	4.4	4.5	4.6	4.7	4.8	4.9
				Expected probit, Y						
0	3.344	3.408	3.469	3.525	3.577	3.624	3.664	3.698	3.724	3.741
1	3.386	3.446	3.503	3.557	3.607	3.652	3.691	3.724	3.750	3.766
2	.427	.487	.538	.589	.637	.680	.710	.751	.775	.791
3	.468	.521	.572	.621	.667	.709	.746	.777	.801	.816
4	.510	.559	.607	.653	.697	.737	.773	.803	.826	.841
5	.551	.596	.641	.685	.727	.766	.800	.829	.852	.867
6	3.592	3.634	3.676	3.717	3.757	3.794	3.827	3.856	3.878	3.892
7	.634	.671	.710	.749	.787	.822	.854	.882	.903	.917
8	.675	.709	.745	.781	.817	.851	.882	.908	.929	.942
9	.716	.747	.779	.813	.847	.879	.909	.934	.954	.967
10	.758	.784	.814	.845	.877	.908	.936	.960	.980	.993
11	3.799	3.822	3.848	3.877	3.907	3.936	3.963	3.987	4.005	4.018
12	.840	.859	.883	.909	.937	.964	.990	4.013	.031	.043
13	.882	.897	.917	.941	.967	.993	4.017	.039	.057	.068
14	.923	.934	.952	.973	.997	4.021	.044	.065	.082	.093
15	.964	.972	.986	4.005	4.027	.050	.072	.092	.108	.119
16	4.006	4.010	4.021	4.000	4.057	4.078	4.099	4.118	4.133	4.144
17	.047	.047	.056	.070	.087	.106	.126	.144	.159	.169
18	.088	.085	.090	.102	.117	.135	.153	.170	.184	.194
19	.130	.122	.125	.134	.147	.163	.180	.196	.210	.219
20	.171	.160	.159	.166	.177	.192	.207	.223	.236	.245
21	4.212	4.198	4.194	4.198	4.207	4.220	4.235	4.249	4.261	4.270
22	.253	.235	.228	.230	.237	.248	.262	.275	.287	.295
23	.295	.273	.263	.262	.267	.277	.289	.301	.312	.320
24	.336	.310	.297	.294	.297	.305	.316	.327	.338	.345
25	.377	.348	.332	.326	.327	.334	.343	.354	.363	.370
26	4.419	4.385	4.366	4.358	4.357	4.362	4.370	4.380	4.389	4.396
27	.400	.423	.401	.390	.387	.391	.397	.406	.415	.421
28	.501	.461	.435	.422	.417	.419	.425	.432	.440	.446
29	.543	.498	.470	.454	.447	.447	.452	.459	.466	.471
30	.584	.536	.504	.486	.477	.476	.479	.485	.491	.496
31	4.625	4.573	4.539	4.518	4.507	4.504	4.506	4.511	4.517	4.522
32	.667	.611	.573	.550	.537	.533	.533	.537	.542	.547
33	.708	.649	.608	.582	.567	.561	.560	.563	.568	.572
34	.749	.686	.642	.614	.597	.589	.588	.590	.594	.597
35	.791	.724	.677	.646	.627	.618	.615	.616	.619	.622
36	4.832	4.761	4.711	4.678	4.657	4.646	4.642	4.642	4.645	4.648
37	.873	.799	.746	.710	.687	.675	.669	.668	.670	.673
38	.915	.836	.780	.742	.717	.703	.696	.695	.696	.698
39	.956	.874	.815	.774	.747	.731	.723	.721	.721	.723
40	.997	.912	.849	.806	.777	.760	.750	.747	.747	.748
41	5.039	4.949	4.884	4.838	4.807	4.788	4.778	4.773	4.773	4.774
42	.080	.987	.918	.870	.837	.817	.805	.799	.798	.799
43	.121	5.024	.953	.902	.867	.845	.832	.826	.824	.824
44	.163	.062	.988	.934	.897	.873	.859	.852	.849	.849
45	.204	.099	5.022	.966	.927	.902	.886	.878	.875	.874
46	5.245	5.137	5.057	4.998	4.957	4.930	4.913	4.904	4.900	4.900
47	.287	.175	.091	5.030	.987	.959	.941	.931	.926	.925
48	.328	.212	.126	.062	5.017	.987	.968	.957	.952	.950
49	.369	.250	.160	.094	.047	5.015	.995	.983	.977	.975
50	.411	.287	.195	.126	.078	.044	5.022	5.009	5.003	5.000

Table IV (*cont.*)

($Y = 5.0$–5.9; 0–50% response)

% response	Expected probit, Y									
	5.0	5.1	5.2	5.3	5.4	5.5	5.6	5.7	5.8	5.9
0	3.747	3.740	3.719	3.680	3.620	3.536	3.422	3.272	3.079	2.834
1	3.772	3.765	3.744	3.706	3.647	3.564	3.452	3.304	3.114	2.871
2	.797	.790	.770	.732	.675	.593	.482	.336	.148	.909
3	.822	.816	.795	.758	.702	.621	.512	.368	.183	.946
4	.847	.841	.821	.785	.729	.650	.542	.400	.217	.984
5	.872	.866	.846	.811	.756	.678	.572	.433	.252	3.021
6	3.897	3.891	3.872	3.837	3.783	3.706	3.602	3.465	3.287	3.059
7	.922	.916	.898	.863	.810	.735	.632	.497	.321	.097
8	.947	.942	.923	.890	.838	.763	.662	.529	.356	.134
9	.972	.967	.949	.916	.865	.792	.692	.561	.390	.172
10	.997	.992	.974	.942	.892	.820	.722	.593	.425	.209
11	4.022	4.017	4.000	3.968	3.919	3.848	3.752	3.625	3.459	3.247
12	.047	.042	.025	.994	.946	.877	.782	.657	.494	.284
13	.073	.068	.051	4.021	.973	.905	.812	.689	.528	.322
14	.098	.093	.077	.047	4.000	.934	.842	.721	.563	.360
15	.123	.118	.102	.073	.028	.962	.872	.753	.597	.397
16	4.148	4.143	4.128	4.099	4.055	3.990	3.902	3.785	3.632	3.435
17	.173	.168	.153	.126	.082	4.019	.932	.817	.666	.472
18	.198	.194	.179	.152	.109	.047	.962	.849	.701	.510
19	.223	.219	.204	.178	.136	.076	.992	.881	.735	.548
20	.248	.244	.230	.204	.163	.104	4.022	.913	.770	.585
21	4.273	4.269	4.256	4.230	4.191	4.132	4.052	3.945	3.804	3.623
22	.298	.294	.281	.257	.218	.161	.082	.977	.839	.660
23	.323	.320	.307	.283	.245	.189	.112	4.009	.873	.698
24	.348	.345	.332	.309	.272	.218	.142	.041	.908	.735
25	.373	.370	.358	.335	.299	.246	.172	.073	.942	.773
26	4.398	4.395	4.383	4.362	4.326	4.275	4.202	4.105	3.977	3.811
27	.423	.420	.409	.388	.353	.303	.232	.137	4.011	.848
28	.449	.445	.435	.414	.381	.331	.262	.169	.046	.886
29	.474	.471	.460	.440	.408	.360	.292	.201	.080	.923
30	.499	.496	.486	.466	.435	.388	.322	.233	.115	.961
31	4.524	4.521	4.511	4.493	4.462	4.417	4.352	4.265	4.149	3.999
32	.549	.546	.537	.519	.489	.445	.382	.297	.184	4.036
33	.574	.571	.563	.545	.516	.473	.412	.329	.219	.074
34	.599	.597	.588	.571	.544	.502	.442	.361	.253	.111
35	.624	.622	.614	.598	.571	.530	.472	.393	.288	.149
36	4.649	4.647	4.639	4.624	4.598	4.559	4.502	4.425	4.322	4.186
37	.674	.672	.665	.650	.625	.587	.532	.457	.357	.224
38	.699	.697	.690	.676	.652	.615	.562	.489	.391	.262
39	.724	.723	.716	.702	.679	.644	.592	.521	.426	.299
40	.749	.748	.742	.729	.706	.672	.622	.553	.460	.337
41	4.774	4.773	4.767	4.755	4.734	4.701	4.652	4.585	4.495	4.374
42	.799	.798	.793	.781	.761	.729	.682	.617	.529	.412
43	.825	.823	.818	.807	.788	.757	.712	.649	.564	.450
44	.850	.849	.844	.833	.815	.786	.742	.682	.598	.487
45	.875	.874	.869	.860	.842	.814	.772	.714	.633	.525
46	4.900	4.899	4.895	4.886	4.869	4.843	4.802	4.746	4.667	4.562
47	.925	.924	.921	.912	.897	.871	.832	.778	.702	.600
48	.950	.949	.946	.938	.924	.899	.862	.810	.736	.637
49	.975	.975	.972	.965	.951	.928	.892	.842	.771	.675
50	5.000	5.000	.997	.991	.978	.956	.922	.874	.805	.713

Table IV (*cont.*)

($Y = 6.0$–6.9; 0–50% response)

% response	Expected probit, Y									
	6.0	6.1	6.2	6.3	6.4	6.5	6.6	6.7	6.8	6.9
0	2.523	2.132	1.643	1.030	0.261	—	—	—	—	—
1	2.564	2.178	1.694	1.088	0.327	—	—	—	—	—
2	.606	.224	.746	.146	.394	—	—	—	—	—
3	.647	.270	.797	.205	.461	—	—	—	—	—
4	.688	.316	.849	.263	.528	—	—	—	—	—
5	.730	.362	.900	.321	.595	—	—	—	—	—
6	2.771	2.408	1.952	1.380	0.661	—	—	—	—	—
7	.812	.454	2.003	.438	.728	—	—	—	—	—
8	.854	.500	.055	.496	.795	—	—	—	—	—
9	.895	.546	.106	.555	.862	—	—	—	—	—
10	.936	.591	.158	.613	.928	0.067	—	—	—	—
11	2.978	2.637	2.209	1.671	0.995	0.144	—	—	—	—
12	3.019	.683	.261	.730	1.062	.221	—	—	—	—
13	.060	.729	.312	.788	.129	.299	—	—	—	—
14	.102	.775	.364	.846	.196	.376	—	—	—	—
15	.143	.821	.415	.905	.262	.453	—	—	—	—
16	3.184	2.867	2.467	1.963	1.329	0.530	—	—	—	—
17	.226	.913	.518	2.022	.396	.607	—	—	—	—
18	.267	.959	.570	.080	.463	.685	—	—	—	—
19	.308	3.005	.621	.138	.530	.762	—	—	—	—
20	.350	.050	.673	.197	.596	.839	—	—	—	—
21	3.391	3.096	2.724	2.255	1.663	0.916	—	—	—	—
22	.432	.142	.776	.313	.730	.993	0.062	—	—	—
23	.474	.188	.827	.372	.797	1.071	.152	—	—	—
24	.515	.234	.879	.430	.864	.148	.243	—	—	—
25	.556	.280	.930	.488	.930	.225	.333	—	—	—
26	3.598	3.326	2.982	2.547	1.997	1.302	0.423	—	—	—
27	.639	.372	3.033	.605	2.064	.379	.513	—	—	—
28	.680	.418	.085	.663	.131	.457	.603	—	—	—
29	.721	.464	.136	.722	.197	.534	.693	—	—	—
30	.763	.509	.188	.780	.264	.611	.784	—	—	—
31	3.804	3.555	3.239	2.838	2.331	1.688	0.874	—	—	—
32	.845	.601	.291	.897	.398	.766	.964	—	—	—
33	.887	.647	.342	.955	.465	.843	1.054	0.050	—	—
34	.928	.693	.394	3.014	.531	.920	.144	.156	—	—
35	.969	.739	.445	.072	.598	.997	.234	.262	—	—
36	4.011	3.785	3.497	3.130	2.665	2.074	1.324	0.369	—	—
37	.052	.831	.548	.189	.732	.152	.415	.475	—	—
38	.093	.877	.600	.247	.799	.229	.505	.581	—	—
39	.135	.923	.651	.305	.865	.306	.595	.688	—	—
40	.176	.969	.703	.364	.932	.383	.685	.794	—	—
41	4.217	4.014	3.754	3.422	2.999	2.460	1.775	0.900	—	—
42	.259	.060	.806	.480	3.066	.538	.865	1.007	—	—
43	.300	.106	.857	.539	.132	.615	.955	.113	0.035	—
44	.341	.152	.909	.597	.199	.692	2.046	.219	.162	—
45	.383	.198	.960	.655	.266	.769	.136	.326	.289	—
46	4.424	4.244	4.012	3.714	3.333	2.846	2.226	1.432	0.415	—
47	.465	.290	.063	.772	.400	.924	.316	.538	.542	—
48	.507	.336	.115	.830	.466	3.001	.406	.645	.669	—
49	.548	.382	.166	.889	.533	.078	.496	.751	.795	—
50	.589	.428	.218	.947	.600	.155	.586	.857	.922	—

Table IV (*cont.*)

($Y = 3.0$–3.9; 51–100% response)

% response	3.0	3.1	3.2	3.3	3.4	3.5	3.6	3.7	3.8	3.9
					Expected probit, Y					
51	—	—	9.205	8.249	7.504	6.922	6.467	6.111	5.834	5.618
52	—	—	.331	.355	.594	.999	.534	.170	.885	.664
53	—	—	.458	.462	.684	7.076	.600	.228	.937	.710
54	—	—	.585	.568	.774	.154	.667	.286	.988	.756
55	—	—	.711	.674	.864	.231	.734	.345	6.040	.802
56	—	—	9.838	8.781	7.954	7.308	6.801	6.403	6.091	5.848
57	—	—	.965	.887	8.045	.385	.868	.461	.143	.894
58	—	—	—	.993	.135	.462	.934	.520	.194	.940
59	—	—	—	9.100	.225	.540	7.001	.578	.246	.986
60	—	—	—	.206	.315	.617	.068	.636	.297	6.031
61	—	—	—	9.312	8.405	7.694	7.135	6.695	6.349	6.077
62	—	—	—	.419	.495	.771	.201	.753	.400	.123
63	—	—	—	.525	.585	.848	.268	.811	.452	.169
64	—	—	—	.631	.676	.926	.335	.870	.503	.215
65	—	—	—	.738	.766	8.003	.402	.928	.555	.261
66	—	—	—	9.844	8.856	8.080	7.469	6.986	6.606	6.307
67	—	—	—	.950	.946	.157	.535	7.045	.658	.353
68	—	—	—	—	9.036	.234	.602	.103	.709	.399
69	—	—	—	—	.126	.312	.669	.162	.761	.445
70	—	—	—	—	.216	.389	.736	.220	.812	.491
71	—	—	—	—	9.307	8.466	7.803	7.278	6.864	6.536
72	—	—	—	—	.397	.543	.869	.337	.915	.582
73	—	—	—	—	.487	.621	.936	.395	.967	.628
74	—	—	—	—	.577	.698	8.003	.453	7.018	.674
75	—	—	—	—	.667	.775	.070	.512	.070	.720
76	—	—	—	—	9.757	8.852	8.136	7.570	7.121	6.766
77	—	—	—	—	.848	.929	.203	.628	.173	.812
78	—	—	—	—	.938	9.007	.270	.687	.224	.858
79	—	—	—	—	—	.084	.337	.745	.276	.904
80	—	—	—	—	—	.161	.404	.803	.327	.950
81	—	—	—	—	—	9.238	8.470	7.862	7.379	6.995
82	—	—	—	—	—	.315	.537	.920	.430	7.041
83	—	—	—	—	—	.393	.604	.978	.482	.087
84	—	—	—	—	—	.470	.671	8.037	.533	.133
85	—	—	—	—	—	.547	.738	.095	.585	.179
86	—	—	—	—	—	9.624	8.804	8.154	7.636	7.225
87	—	—	—	—	—	.701	.871	.212	.688	.271
88	—	—	—	—	—	.779	.938	.270	.739	.317
89	—	—	—	—	—	.856	9.005	.329	.791	.363
90	—	—	—	—	—	.933	.072	.387	.842	.409
91	—	—	—	—	—	—	9.138	8.445	7.894	7.454
92	—	—	—	—	—	—	.205	.504	.945	.500
93	—	—	—	—	—	—	.272	.562	.997	.546
94	—	—	—	—	—	—	.339	.620	8.048	.592
95	—	—	—	—	—	—	.405	.679	.100	.638
96	—	—	—	—	—	—	9.472	8.737	8.151	7.684
97	—	—	—	—	—	—	.539	.795	.203	.730
98	—	—	—	—	—	—	.606	.854	.254	.776
99	—	—	—	—	—	—	.673	.912	.306	.822
100	—	—	—	—	—	—	.739	.970	.357	.868

Table IV (*cont.*)

($Y = 4.0$–4.9; 51–100% response)

% response	Expected probit, Y									
	4.0	4.1	4.2	4.3	4.4	4.5	4.6	4.7	4.8	4.9
51	5.452	5.325	5.229	5.158	5.108	5.072	5.049	5.035	5.028	5.025
52	.493	.363	.264	.190	.138	.101	.076	.062	.054	.051
53	.535	.400	.298	.222	.168	.129	.103	.088	.079	.076
54	.576	.438	.333	.254	.198	.157	.131	.114	.105	.101
55	.617	.475	.367	.286	.228	.186	.158	.140	.131	.126
56	5.659	5.513	5.402	5.318	5.258	5.214	5.185	5.167	5.156	5.151
57	.700	.550	.436	.351	.288	.243	.212	.193	.182	.177
58	.741	.588	.471	.383	.318	.271	.239	.219	.207	.202
59	.783	.626	.505	.415	.348	.299	.266	.245	.233	.227
60	.824	.663	.540	.447	.378	.328	.294	.271	.258	.252
61	5.865	5.701	5.574	5.479	5.408	5.356	5.321	5.298	5.284	5.277
62	.907	.738	.609	.511	.438	.385	.348	.324	.310	.303
63	.948	.776	.643	.543	.468	.413	.375	.350	.335	.328
64	.989	.814	.678	.575	.498	.441	.402	.376	.361	.353
65	6.031	.851	.712	.607	.528	.470	.429	.402	.386	.378
66	6.072	5.889	5.747	5.639	5.558	5.498	5.456	5.429	5.412	5.403
67	.113	.926	.781	.671	.588	.527	.484	.455	.437	.429
68	.155	.964	.816	.703	.618	.555	.511	.481	.463	.454
69	.196	6.001	.851	.735	.648	.583	.538	.507	.489	.479
70	.237	.039	.885	.767	.678	.612	.565	.534	.514	.504
71	6.279	6.077	5.920	5.799	5.708	5.640	5.592	5.560	5.540	5.529
72	.320	.114	.954	.831	.738	.669	.619	.586	.565	.555
73	.361	.152	.989	.863	.768	.697	.647	.612	.591	.580
74	.402	.189	6.023	.895	.798	.725	.674	.638	.617	.605
75	.444	.227	.058	.927	.828	.754	.701	.665	.642	.630
76	6.485	6.265	6.092	5.959	5.858	5.782	5.728	5.691	5.668	5.655
77	.526	.302	.127	.991	.888	.811	.755	.717	.693	.680
78	.568	.340	.161	6.023	.918	.839	.782	.743	.719	.706
79	.609	.377	.196	.055	.948	.868	.809	.770	.744	.731
80	.650	.415	.230	.087	.978	.896	.837	.796	.770	.756
81	6.692	6.452	6.265	6.119	6.008	5.924	5.864	5.822	5.796	5.781
82	.733	.490	.299	.151	.038	.953	.891	.848	.821	.806
83	.774	.528	.334	.183	.068	.981	.918	.874	.847	.832
84	.816	.565	.368	.215	.098	6.010	.945	.901	.872	.857
85	.857	.603	.403	.247	.128	.038	.972	.927	.898	.882
86	6.898	6.640	6.437	6.279	6.158	6.066	6.000	5.953	5.923	5.907
87	.940	.678	.472	.311	.188	.095	.027	.979	.949	.932
88	.981	.716	.506	.343	.218	.123	.054	6.006	.975	.958
89	7.022	.753	.541	.375	.248	.152	.081	.032	6.000	.983
90	.064	.791	.575	.407	.278	.180	.108	.058	.026	6.008
91	7.105	6.828	6.610	6.439	6.308	6.208	6.135	6.084	6.051	6.033
92	.146	.866	.644	.471	.338	.237	.162	.110	.077	.058
93	.188	.903	.679	.503	.368	.265	.190	.137	.102	.084
94	.229	.941	.713	.535	.398	.294	.217	.163	.128	.109
95	.270	.979	.748	.567	.428	.322	.244	.189	.154	.134
96	7.312	7.016	6.783	6.600	6.458	6.350	6.271	6.215	6.179	6.159
97	.353	.054	.817	.632	.488	.379	.298	.242	.205	.184
98	.394	.091	.852	.664	.518	.407	.325	.268	.230	.210
99	.436	.129	.886	.696	.548	.436	.353	.294	.256	.235
100	.477	.166	.921	.728	.578	.464	.380	.320	.281	.260

Table IV (*cont.*)

(Y = 5.0–5.9; 51–100% response)

% reponse	5.0	5.1	5.2	5.3	5.4	5.5	5.6	5.7	5.8	5.9
51	5.025	5.025	5.023	5.017	5.005	4.985	4.953	4.906	4.840	4.750
52	.050	.050	.048	.043	.032	5.013	.983	.938	.874	.788
53	.075	.075	.074	.069	.059	.041	5.013	.970	.909	.825
54	.100	.100	.100	.096	.087	.070	.043	5.002	.943	.863
55	.125	.126	.125	.122	.114	.098	.073	.034	.978	.901
56	5.150	5.151	5.151	5.148	5.141	5.127	5.103	5.066	5.012	4.938
57	.175	.176	.176	.174	.168	.155	.133	.098	.047	.976
58	.201	.201	.202	.201	.195	.183	.163	.130	.082	5.013
59	.226	.226	.227	.227	.222	.212	.193	.162	.116	.051
60	.251	.252	.253	.253	.250	.240	.223	.194	.151	.088
61	5.276	5.277	5.279	5.279	5.277	5.269	5.253	5.226	5.185	5.126
62	.301	.302	.304	.305	.304	.297	.283	.258	.220	.164
63	.326	.327	.330	.332	.331	.325	.313	.290	.254	.201
64	.351	.352	.355	.358	.358	.354	.343	.322	.289	.239
65	.376	.378	.381	.384	.385	.382	.373	.354	.323	.276
66	5.401	5.403	5.406	5.410	5.412	5.411	5.403	5.386	5.358	5.314
67	.426	.428	.432	.437	.440	.439	.433	.418	.392	.351
68	.451	.453	.458	.463	.467	.467	.463	.450	.427	.389
69	.476	.478	.483	.489	.494	.496	.493	.482	.461	.427
70	.501	.504	.509	.515	.521	.524	.523	.514	.496	.464
71	5.526	5.529	5.534	5.541	5.548	5.553	5.553	5.546	5.530	5.502
72	.551	.554	.560	.568	.575	.581	.583	.578	.565	.539
73	.577	.579	.585	.594	.603	.609	.613	.610	.599	.577
74	.602	.604	.611	.620	.630	.638	.643	.642	.634	.615
75	.627	.630	.637	.646	.657	.666	.673	.674	.668	.652
76	5.652	5.655	5.662	5.673	5.684	5.695	5.703	5.706	5.703	5.690
77	.677	.680	.688	.699	.711	.723	.733	.738	.737	.727
78	.702	.705	.713	.725	.738	.752	.763	.770	.772	.765
79	.727	.730	.739	.751	.765	.780	.793	.802	.806	.802
80	.752	.755	.764	.777	.793	.808	.823	.834	.841	.840
81	5.777	5.781	5.790	5.804	5.820	5.837	5.853	5.866	5.875	5.878
82	.802	.806	.816	.830	.847	.865	.883	.898	.910	.915
83	.827	.831	.841	.856	.874	.894	.913	.930	.944	.953
84	.852	.856	.867	.882	.901	.922	.943	.962	.979	.990
85	.877	.881	.892	.908	.928	.950	.973	.995	6.014	6.028
86	5.902	5.907	5.918	5.935	5.956	5.979	6.003	6.027	6.048	6.066
87	.927	.932	.943	.961	.983	6.007	.033	.059	.083	.103
88	.953	.957	.969	.987	6.010	.036	.063	.091	.117	.141
89	.978	.982	.995	6.013	.037	.064	.093	.123	.152	.178
90	6.003	6.007	6.020	.040	.064	.092	.123	.155	.186	.216
91	6.028	6.033	6.046	6.066	6.091	6.121	6.153	6.187	6.221	6.253
92	.053	.058	.071	.092	.118	.149	.183	.219	.255	.291
93	.078	.083	.097	.118	.146	.178	.213	.251	.290	.329
94	.103	.108	.122	.144	.173	.206	.243	.283	.324	.366
95	.128	.133	.148	.171	.200	.234	.273	.315	.359	.404
96	6.153	6.159	6.174	6.197	6.227	6.263	6.303	6.347	6.393	6.441
97	.178	.184	.199	.223	.254	.291	.333	.379	.428	.479
98	.203	.209	.225	.249	.281	.320	.363	.411	.462	.517
99	.228	.234	.250	.276	.309	.348	.393	.443	.497	.554
100	.253	.259	.276	.302	.336	.376	.423	.475	.531	.592

Table IV (*cont.*)

(Y = 6.0–6.9; 51–100% response)

% response	6.0	6.1	6.2	6.3	6.4	6.5	6.6	6.7	6.8	6.9
				Expected probit, Y						
51	4.631	4.473	4.269	4.006	3.667	3.233	2.677	1.964	1.049	—
52	.672	.519	.321	.064	.734	.310	.767	2.070	.175	0.022
53	.713	.565	.372	.122	.800	.387	.857	.176	.302	.175
54	.755	.611	.424	.181	.867	.464	.947	.283	.429	.327
55	.796	.657	.475	.239	.934	.541	3.037	.389	.555	.480
56	4.837	4.703	4.527	4.297	4.001	3.619	3.127	2.495	1.682	0.632
57	.879	.749	.578	.356	.068	.696	.218	.602	.809	.784
58	.920	.795	.630	.414	.134	.773	.308	.708	.935	.937
59	.961	.841	.681	.472	.201	.850	.398	.814	2.062	1.089
60	5.003	.887	.733	.531	.268	.927	.488	.921	.189	.242
61	5.044	4.932	4.784	4.589	4.335	4.005	3.578	3.027	2.315	1.394
62	.085	.978	.836	.647	.401	.082	.668	.133	.442	.546
63	.127	5.024	.887	.706	.468	.159	.758	.240	.569	.699
64	.168	.070	.939	.764	.535	.236	.849	.346	.695	.851
65	.209	.116	.990	.823	.602	.313	.939	.452	.822	2.004
66	5.251	5.162	5.042	4.881	4.669	4.391	4.029	3.559	2.949	2.156
67	.292	.208	.093	.939	.735	.468	.119	.665	3.075	.308
68	.333	.254	.145	.998	.802	.545	.209	.771	.202	.461
69	.375	.300	.196	5.056	.869	.622	.299	.878	.329	.613
70	.416	.346	.248	.114	.936	.700	.300	.984	.455	.766
71	5.457	5.392	5.299	5.173	5.003	4.777	4.480	4.090	3.582	2.918
72	.499	.437	.351	.231	.069	.854	.570	.197	.709	3.070
73	.540	.483	.402	.289	.136	.931	.660	.303	.835	.223
74	.581	.529	.454	.348	.203	5.008	.750	.409	.962	.375
75	.623	.575	.505	.406	.270	.086	.840	.516	4.089	.528
76	5.664	5.621	5.557	5.464	5.336	5.163	4.930	4.622	4.215	3.680
77	.705	.667	.608	.523	.403	.240	5.021	.728	.342	.832
78	.747	.713	.660	.581	.470	.317	.111	.835	.469	.985
79	.788	.759	.711	.639	.537	.394	.201	.941	.595	4.137
80	.829	.805	.763	.698	.604	.472	.291	5.047	.722	.290
81	5.870	5.851	5.814	5.756	5.670	5.549	5.381	5.154	4.849	4.442
82	.912	.896	.866	.815	.737	.626	.471	.260	.975	.594
83	.953	.942	.917	.873	.804	.703	.561	.366	5.102	.747
84	.994	.988	.969	.931	.871	.780	.652	.473	.229	.899
85	6.036	6.034	6.020	.990	.938	.858	.742	.579	.355	5.052
86	6.077	6.080	6.072	6.048	6.004	5.935	5.832	5.685	5.482	5.204
87	.118	.126	.123	.106	.071	6.012	.922	.792	.609	.356
88	.160	.172	.175	.165	.138	.089	6.012	.898	.735	.509
89	.201	.218	.226	.223	.205	.166	.102	6.004	.862	.661
90	.242	.264	.278	.281	.272	.244	.192	.111	.988	.814
91	6.284	6.310	6.329	6.340	6.338	6.321	6.283	6.217	6.115	5.966
92	.325	.355	.381	.398	.405	.398	.373	.323	.242	6.118
93	.366	.401	.432	.456	.472	.475	.463	.430	.368	.271
94	.408	.447	.484	.515	.539	.553	.553	.536	.495	.423
95	.449	.493	.535	.573	.605	.630	.643	.642	.622	.576
96	6.490	6.539	6.587	6.631	6.672	6.707	6.733	6.749	6.748	6.728
97	.532	.585	.638	.690	.739	.784	.824	.855	.875	.880
98	.573	.631	.690	.748	.806	.861	.914	.961	7.002	7.033
99	.614	.677	.741	.807	.873	.939	7.004	7.068	.128	.185
100	.656	.723	.793	.865	.939	7.016	.094	.174	.255	.338

Table IV (*cont.*)

(Y = 7.0–7.9; 51–100% response)

% response	7.0	7.1	7.2	7.3	7.4	7.5	7.6	7.7	7.8	7.9
				Expected probit, Y						
51	—	—	—	—	—	—	—	—	—	—
52	—	—	—	—	—	—	—	—	—	—
53	—	—	—	—	—	—	—	—	—	—
54	—	—	—	—	—	—	—	—	—	—
55	—	—	—	—	—	—	—	—	—	—
56	—	—	—	—	—	—	—	—	—	—
57	—	—	—	—	—	—	—	—	—	—
58	—	—	—	—	—	—	—	—	—	—
59	—	—	—	—	—	—	—	—	—	—
60	0.013	—	—	—	—	—	—	—	—	—
61	0.198	—	—	—	—	—	—	—	—	—
62	.383	—	—	—	—	—	—	—	—	—
63	.568	—	—	—	—	—	—	—	—	—
64	.753	—	—	—	—	—	—	—	—	—
65	.939	—	—	—	—	—	—	—	—	—
66	1.124	—	—	—	—	—	—	—	—	—
67	.309	0.003	—	—	—	—	—	—	—	—
68	.494	.231	—	—	—	—	—	—	—	—
69	.680	.458	—	—	—	—	—	—	—	—
70	.865	.685	—	—	—	—	—	—	—	—
71	2.050	0.913	—	—	—	—	—	—	—	—
72	.235	1.140	—	—	—	—	—	—	—	—
73	.420	.367	—	—	—	—	—	—	—	—
74	.606	.595	0.263	—	—	—	—	—	—	—
75	.791	.822	.545	—	—	—	—	—	—	—
76	2.976	2.050	0.827	—	—	—	—	—	—	—
77	3.161	.277	1.108	—	—	—	—	—	—	—
78	.347	.504	.390	—	—	—	—	—	—	—
79	.532	.732	.672	0.265	—	—	—	—	—	—
80	.717	.959	.954	.618	—	—	—	—	—	—
81	3.902	3.186	2.236	0.971	—	—	—	—	—	—
82	4.087	.414	.518	1.324	—	—	—	—	—	—
83	.273	.641	.800	.677	0.175	—	—	—	—	—
84	.458	.868	3.082	2.030	.621	—	—	—	—	—
85	.643	4.096	.364	.383	1.068	—	—	—	—	—
86	4.828	4.323	3.645	2.736	1.514	—	—	—	—	—
87	5.014	.551	.927	3.089	.961	0.438	—	—	—	—
88	.199	.778	4.209	.442	2.408	1.008	—	—	—	—
89	.384	5.005	.491	.795	.854	.579	—	—	—	—
90	.569	.233	.773	4.148	3.301	2.149	0.581	—	—	—
91	5.754	5.460	5.055	4.501	3.747	2.720	1.317	—	—	—
92	.040	.687	.337	.854	4.104	3.200	2.054	0.356	—	—
93	6.125	.915	.619	5.207	.640	.861	.790	1.316	—	—
94	.310	6.142	.901	.560	5.087	4.431	3.526	2.275	0.542	—
95	.495	.369	6.182	.914	.533	5.002	4.262	3.235	1.806	—
96	6.681	6.597	6.464	6.267	5.980	5.572	4.998	4.194	3.069	1.493
97	.866	.824	.746	.620	6.426	6.143	5.735	5.154	4.333	3.173
98	7.051	7.051	7.028	.973	.873	.713	6.471	6.114	5.596	4.853
99	.236	.279	.310	7.326	7.319	7.284	7.207	7.073	6.859	6.533
100	.421	.506	.592	.679	.766	.854	.943	8.033	8.123	8.213

Table V The weighting coefficient in Wadley's problem

Expected probit Y	Weighting coefficient w	Expected probit Y	Weighting coefficient w
1.1	0.00000004	5.1	0.34242
1.2	0.00000009	5.2	0.36344
1.3	0.0000002	5.3	0.38069
1.4	0.0000004	5.4	0.39359
1.5	0.0000008	5.5	0.40173
1.6	0.000002	5.6	0.40488
1.7	0.000003	5.7	0.40296
1.8	0.000006	5.8	0.39612
1.9	0.00001	5.9	0.38466
2.0	0.00002	6.0	0.36904
2.1	0.00004	6.1	0.34983
2.2	0.00006	6.2	0.32770
2.3	0.00011	6.3	0.30338
2.4	0.00019	6.4	0.27760
2.5	0.00031	6.5	0.25109
2.6	0.00051	6.6	0.22452
2.7	0.00081	6.7	0.19848
2.8	0.00128	6.8	0.17348
2.9	0.00197	6.9	0.14993
3.0	0.00298	7.0	0.12813
3.1	0.00443	7.1	0.10829
3.2	0.00647	7.2	0.09051
3.3	0.00926	7.3	0.07482
3.4	0.01302	7.4	0.06118
3.5	0.01798	7.5	0.04948
3.6	0.02439	7.6	0.03958
3.7	0.03251	7.7	0.03132
3.8	0.04261	7.8	0.02452
3.9	0.05491	7.9	0.01899
4.0	0.06959	8.0	0.01455
4.1	0.08677	8.1	0.01103
4.2	0.10648	8.2	0.00827
4.3	0.12863	8.3	0.00614
4.4	0.15300	8.4	0.00451
4.5	0.17926	8.5	0.00327
4.6	0.20692	8.6	0.00235
4.7	0.23540	8.7	0.00167
4.8	0.26398	8.8	0.00118
4.9	0.29189	8.9	0.00082
5.0	0.31831	9.0	0.00057

Table VI   Minimum working probit, range, and weighting
coefficient for inverse sampling

Expected probit Y	Minimum working probit $Y - Q/Z$	Range Q^2/Z	Weighting coefficient Z^2/PQ^2
5.0	3.74669	0.62666	1.27324
5.1	3.94074	0.53346	1.37842
5.2	4.12406	0.45269	1.49124
5.3	4.29816	0.38279	1.61242
5.4	4.46433	0.32241	1.74276
5.5	4.62364	0.27039	1.88305
5.6	4.77697	0.22572	2.03416
5.7	4.92511	0.18750	2.19698
5.8	5.06869	0.15493	2.37239
5.9	5.20827	0.12732	2.56132
6.0	5.34432	0.10403	2.76467
6.1	5.47726	0.084485	2.98332
6.2	5.60743	0.068187	3.21814
6.3	5.73513	0.054679	3.46995
6.4	5.86064	0.043557	3.73952
6.5	5.98418	0.034460	4.02754
6.6	6.10596	0.027073	4.33463
6.7	6.22615	0.021117	4.66135
6.8	6.34490	0.016352	5.00813
6.9	6.46235	0.012568	5.37534
7.0	6.57863	0.0095862	5.76327
7.1	6.69384	0.0072558	6.17209
7.2	6.80807	0.0054491	6.60192
7.3	6.92142	0.0040600	7.05281
7.4	7.03395	0.0030007	7.52475
7.5	7.14573	0.0021999	8.01768
7.6	7.25684	0.0015996	8.53150
7.7	7.36731	0.0011534	9.06610
7.8	7.47720	0.00082480	9.62136
7.9	7.58655	0.00058484	10.19714
8.0	7.69541	0.00041117	10.79331
8.1	7.80381	0.00028660	11.40975
8.2	7.91178	0.00019805	12.04635
8.3	8.01936	0.00013567	12.70301
8.4	8.12657	0.000092128	13.37965
8.5	8.23343	0.000062011	14.07621
8.6	8.33998	0.000041372	14.79264
8.7	8.44622	0.000027357	15.52888
8.8	8.55219	0.000017929	16.28492
8.9	8.65789	0.000011645	17.06074
9.0	8.76335	0.0000074951	17.85633

Table VII Distribution of χ^2

Degrees of freedom	Probability								
	.90	.70	.50	.30	.10	.05	.02	.01	.001
1	.016	.15	.45	1.1	2.7	3.8	5.4	6.6	10.8
2	.21	.71	1.4	2.4	4.6	6.0	7.8	9.2	13.8
3	.58	1.4	2.4	3.7	6.3	7.8	9.8	11.3	16.3
4	1.1	2.2	3.4	4.9	7.8	9.5	11.7	13.3	18.5
5	1.6	3.0	4.4	6.1	9.2	11.1	13.4	15.1	20.5
6	2.2	3.8	5.3	7.2	10.6	12.6	15.0	16.8	22.5
7	2.8	4.7	6.3	8.4	12.0	14.1	16.6	18.5	24.3
8	3.5	5.5	7.3	9.5	13.4	15.5	18.2	20.1	26.1
9	4.2	6.4	8.3	10.7	14.7	16.9	19.7	21.7	27.9
10	4.9	7.3	9.3	11.8	16.0	18.3	21.2	23.2	29.6
12	6.3	9.0	11.3	14.0	18.5	21.0	24.1	26.2	32.9
14	7.8	10.8	13.3	16.2	21.1	23.7	26.9	29.1	36.1
16	9.3	12.6	15.3	18.4	23.5	26.3	29.6	32.0	39.3
18	10.9	14.4	17.3	20.6	26.0	28.9	32.3	34.8	42.3
20	12.4	16.3	19.3	22.8	28.4	31.4	35.0	37.6	45.3
22	14.0	18.1	21.3	24.9	30.8	33.9	37.7	40.3	48.3
24	15.7	19.9	23.3	27.1	33.2	36.4	40.3	43.0	51.2
26	17.3	21.8	25.3	29.2	35.6	38.9	42.9	45.6	54.1
28	18.9	23.6	27.3	31.4	37.9	41.3	45.4	48.3	56.9
30	20.6	25.5	29.3	33.5	40.3	43.8	48.0	50.9	59.7

When χ^2 is based on more than 30 degrees of freedom, the quantity $\sqrt{(2\chi^2)} - \sqrt{(2f-1)}$ (where f is the number of degrees of freedom) has approximately the following distribution:

> 30 | -1.28 -0.52 0.00 0.52 1.28 1.64 2.05 2.33 3.09

I am indebted to Professor R. A. Fisher and Dr F. Yates, and also to Messrs Oliver and Boyd, Ltd., of Edinburgh, for permission to print Table VII as an abridgement of Table IV of their book *Statistical Tables for Biological, Agricultural and Medical Research*.

Table VIII   Distribution of t

Degrees of freedom	Probability								
	.90	.70	.50	.30	.10	.05	.02	.01	.001
1	.16	.51	1.00	1.96	6.31	12.7	31.8	63.7	637.
2	.14	.44	.82	1.39	2.92	4.30	6.96	9.92	31.6
3	.14	.42	.76	1.25	2.35	3.18	4.54	5.84	12.9
4	.13	.41	.74	1.19	2.13	2.78	3.75	4.60	8.61
5	.13	.41	.73	1.16	2.02	2.57	3.36	4.03	6.86
6	.13	.40	.72	1.13	1.94	2.45	3.14	3.71	5.96
7	.13	.40	.71	1.12	1.90	2.36	3.00	3.50	5.40
8	.13	.40	.71	1.11	1.86	2.31	2.90	3.36	5.04
9	.13	.40	.70	1.10	1.83	2.26	2.82	3.25	4.78
10	.13	.40	.70	1.09	1.81	2.23	2.76	3.17	4.59
12	.13	.40	.70	1.08	1.78	2.18	2.68	3.06	4.32
14	.13	.39	.69	1.08	1.76	2.14	2.62	2.98	4.14
16	.13	.39	.69	1.07	1.75	2.12	2.58	2.92	4.02
18	.13	.39	.69	1.07	1.73	2.10	2.55	2.88	3.92
20	.13	.39	.69	1.06	1.72	2.09	2.53	2.84	3.85
22	.13	.39	.69	1.06	1.72	2.07	2.51	2.82	3.79
24	.13	.39	.68	1.06	1.71	2.06	2.49	2.80	3.74
26	.13	.39	.68	1.06	1.71	2.06	2.48	2.78	3.71
28	.13	.39	.68	1.06	1.70	2.05	2.47	2.76	3.67
30	.13	.39	.68	1.06	1.70	2.04	2.46	2.75	3.65
40	.13	.39	.68	1.05	1.68	2.02	2.42	2.70	3.55
60	.13	.39	.68	1.05	1.67	2.00	2.39	2.66	3.46
120	.13	.39	.68	1.04	1.66	1.98	2.36	2.62	3.37
∞	.126	.385	.674	1.036	1.645	1.960	2.326	2.576	3.291

I am indebted to Professor R. A. Fisher and Dr F. Yates, and also to Messrs Oliver and Boyd, Ltd., of Edinburgh, for permission to print Table VIII as an abridgement of Table III of their book *Statistical Tables for Biological, Agricultural and Medical Research*.

Index of Authors

Index of Subjects